Lecture Notes in Computer Science 13225

More information about this series at https://link.springer.com/bookseries/558

Helle Hvid Hansen · Fabio Zanasi (Eds.)

Coalgebraic Methods in Computer Science

16th IFIP WG 1.3 International Workshop, CMCS 2022
Colocated with ETAPS 2022
Munich, Germany, April 2–3, 2022
Proceedings

Springer

Editors
Helle Hvid Hansen
University of Groningen
Groningen, The Netherlands

Fabio Zanasi
University College London
London, UK

ISSN 0302-9743 ISSN 1611-3349 (electronic)
Lecture Notes in Computer Science
ISBN 978-3-031-10735-1 ISBN 978-3-031-10736-8 (eBook)
https://doi.org/10.1007/978-3-031-10736-8

This Springer imprint is published by the registered company Springer Nature Switzerland AG
The registered company address is: Gewerbestrasse 11, 6330 Cham, Switzerland

Preface

The 16th International Workshop on Coalgebraic Methods in Computer Science (CMCS 2022) was held during April 2–3, 2022, in Munich, Germany, as a satellite event of the Joint Conference on Theory and Practice of Software, ETAPS 2022.

The aim of the workshop is to bring together researchers with a common interest in the theory of coalgebras, their logics, and their applications. Coalgebras allow for a uniform treatment of a large variety of state-based dynamical systems, such as transition systems, automata (including weighted and probabilistic variants), Markov chains, and game-based systems. Over the last two decades, coalgebra has developed into a field of its own, presenting a deep mathematical foundation, a growing field of applications, and interactions with various other fields such as reactive and interactive system theory, object-oriented and concurrent programming, formal system specification, modal and description logics, artificial intelligence, dynamical systems, control systems, category theory, algebra, analysis, etc.

Previous workshops have been organized in Lisbon (1998), Amsterdam (1999), Berlin (2000), Genoa (2001), Grenoble (2002) Warsaw (2003), Barcelona (2004), Vienna (2006), Budapest (2008), Paphos (2010), London (2012), Grenoble (2014), Eindhoven (2016), Thessaloniki (2018), and Dublin (2020, held online because of the COVID-19 pandemic). Since 2004, CMCS has been a biennial workshop, alternating with the International Conference on Algebra and Coalgebra in Computer Science (CALCO), which, in odd-numbered years, has been formed by the union of CMCS with the International Workshop on Algebraic Development Techniques (WADT).

The CMCS 2022 program featured a keynote talk by Ana Sokolova (University of Salzburg), an invited talk by Renato Neves (University of Minho), and an invited talk by Sam Staton (University of Oxford). In addition, the program included a special session on data languages featuring invited tutorials by Sławomir Lasota (University of Warsaw) and Mahsa Shirmohammadi (CNRS, University of Paris).

This volume contains the revised regular contributions (9 papers accepted out of 12 submissions) and the abstracts of the three keynote/invited talks and the two invited tutorial talks.

In addition to submissions of full-length regular papers for these post-proceedings, the workshop also solicited short 2-page submissions for presentation of work-in-progress or work published elsewhere that could be of interest to the CMCS community. Submissions and the reviewing process were handled using Easychair. The reviewing of both types of submissions was carried out single-blind, in which each regular submission received three extensive reviews, and each short submission received two short reviews that focused on relevance. PC members had 4 weeks for the reviewing and discussion of one regular and three short submissions, or two regular and zero to one short submissions. We intentionally chose a large PC to keep the review load light, in order to keep up the high quality of reviewing that has been the standard at CMCS. PC members, including PC chairs, were allowed to submit in both categories. To safeguard the integrity of the reviewing process, PC members had to declare conflict with a submission if they were a co-author, share an affiliation or recent collaboration with one of the co-authors, or if it could be otherwise perceived that the PC member would have a bias towards the decision on the submission. Using an Easychair functionality, PC members were blocked from seeing the reviews and the discussion on submissions for which they had declared a conflict. In particular, one regular submission was co-authored by one of the PC co-chairs, Helle Hvid Hansen, who therefore declared a conflict with this submission, and the discussion and decision on this submission was handled by the other PC co-chair, Fabio Zanasi.

We wish to thank all the authors who submitted to CMCS 2022, and the Program Committee members and external reviewers for their thorough reviewing and help in improving the papers that were accepted for CMCS 2022.

May 2022

Helle Hvid Hansen
Fabio Zanasi

Organization

Program Committee Chairs

Helle Hvid Hansen	University of Groningen, The Netherlands
Fabio Zanasi	University College London, UK

Steering Committee

Filippo Bonchi	University of Pisa, Italy
Marcello Bonsangue	University of Leiden, The Netherlands
Corina Cîrstea	University of Southampton, UK
Ichiro Hasuo	National Institute of Informatics, Japan
Bart Jacobs	Radboud University Nijmegen, The Netherlands
Bartek Klin	University of Oxford, UK
Alexander Kurz	Chapman University, USA
Marina Lenisa	University of Udine, Italy
Stefan Milius (Chair)	FAU Erlangen-Nürnberg, Germany
Lawrence Moss	Indiana University, USA
Dirk Pattinson	Australian National University, Australia
Daniela Petrişan	Paris Cité University, France
Jurriaan Rot	Radboud University Nijmegen, The Netherlands
Lutz Schröder	FAU Erlangen-Nürnberg, Germany
Alexandra Silva	Cornell University, USA

Program Committee

Adriana Balan	University Politehnica of Bucharest, Romania
Henning Basold	Leiden University, The Netherlands
Marta Bilkova	Czech Academy of Sciences, Czech Republic
Fredrik Dahlqvist	University College London, UK
Fredrik Nordvall Forsberg	University of Strathclyde, UK
Richard Garner	Macquarie University, Australia
Jeremy Gibbons	University of Oxford, UK
Helle Hvid Hansen	University of Groningen, The Netherlands
Ichiro Hasuo	National Institute of Informatics, Japan
Tobias Kappé	University of Amsterdam, The Netherlands
Dexter Kozen	Cornell University, USA
Clemens Kupke	University of Strathclyde, UK

Alexander Kurz	Chapman University, USA
Barbara König	Universität Duisburg-Essen, Germany
Marina Lenisa	University of Udine, Italy
Davide Sangiorgi	University of Bologna, Italy
Lutz Schröder	FAU Erlangen-Nürnberg, Germany
David Spivak	Massachusetts Institute of Technology, USA
Tarmo Uustalu	Reykyavik University, Iceland, and Talinn University of Technology, Estonia
Thorsten Wißmann	Radboud University, The Netherlands
Fabio Zanasi	University College London, UK
Maaike Zwart	IT University Copenhagen, Denmark

Additional Reviewers

Cédric Ho Thanh
Yuichi Komorida
Ken Sakayori
Ezra Schoen

Publicity Chair

Thorsten Wißmann Radboud University, The Netherlands

Sponsoring Institution

IFIP WG 1.3

Abstracts of Invited Talks

Tracing Coalgebras: A Case for Monads

Ana Sokolova

University of Salzburg, Austria
anas@cs.uni-salzburg.at

Trace semantics, also known as linear-time semantics, is the essential semantics of systems, programs, objects, seen as black boxes with some observable behaviour. Unlike its branching-time, step-wise sisters, trace semantics provides the whole observable behaviour of a system. Trace semantics is hard to compute and originally seemed difficult to study coalgebraically. This talk will provide an overview of coalgebraic approaches to trace semantics, over the last fifteen years, focusing on transition systems and automata with effects. Monads play a crucial role in the study of traces, which I will highlight in all approaches using a class of examples. I will also mention some of the context of this exciting line of work that in my view enabled and supported the coalgebraic trace theories.

Coalgebra Meets Hybrid Systems

Renato Neves

University of Minho & INESC-TEC, Portugal
nevrenato@di.uminho.pt

A main challenge of the 21st century is to engineer software systems that tightly interact with physical processes such as velocity, movement, energy, and time. Such systems are qualified as 'hybrid' to emphasise this cyber-physical interaction—which remarkably forces a shift from standard software practices to a more multifaceted view that combines computer science, control theory, and analysis.

In this talk, I will discuss how Coalgebra can help advance hybrid systems theory. In particular, I will describe previous applications of Coalgebra to tackle two central problems in the field: the lack of a uniform framework for hybrid automata (currently, the standard formalism for hybrid systems) and the lack of suitable semantics for interpreting hybrid while-loops. As alluded above, we will see that hybrid systems deviate from standard notions of computer science in many aspects. For example, whilst classical while-loops give rise to a single divergence point, hybrid while-loops give rise to a whole continuum of divergence points. Nonetheless, we will see that Coalgebra can still properly guide us in providing a semantics for the latter kind of loop.

I will conclude with a brief mention of other significant challenges in the field of hybrid systems and possible uses of Coalgebra to tackle them.

Coalgebraic Methods in Probability

Sam Staton

University of Oxford, UK
sam.staton@cs.ox.ac.uk

I will discuss some of the roles that coalgebra can play in probability theory and statistics. Infinite dimensional systems are often described as generative models, and these are often like coalgebras, as I will explain. I will look at some recent statistical models that involve symmetries, such as the "Chinese Restaurant process" and "Indian Buffet process". Since these use names implicitly, I will connect this to nominal techniques. I will also discuss our probabilistic programming library "lazyppl", which uses coinductive structures extensively. I will not assume much familiarity with probability, statistics, nominal techniques, or probabilistic programming.

The talk will draw on joint work with Ackerman, Dash, Freer, Jacobs, Kaddar, Paquet, Roy, Sabok, Stein, Wolman, Yang, and others.

Some Recent Advances in Register Automata

Sławomir Lasota

University of Warsaw, Poland
s.lasota@uw.edu.pl

I shall recall the model of register automata (RA), and relate it to the setting of orbit-finite automata. Then I will mention two recent advances concerning language expressivity of different types of RA, together with remaining open problems.

First, two complementing languages of nondeterministic RA are recognized by a deterministic RA. For two such languages which are disjoint but not necessarily complementing, one may ask if there is a deterministic (resp. unambiguous) RA whose language separates the two. It is not known if existence of a deterministic separator is decidable, and it is not known if (but conjectured that) an unambiguous separator always exists.

The second advance considers orbit-finite rational expressions (with orbit-finite unions in place of finite ones). While languages of nondeterministic RA are not definable by such rational expressions, their Parikh (commutative) images are conjectured to be so. This has been recently confirmed for automata (and grammars) with 1 register, but is still open in case of 2 registers.

Learning Weighted Automata over Fields and Principal Ideal Domains

Mahsa Shirmohammadi

CNRS, University of Paris, France
mahsa@irif.fr

In this talk, I will discuss learning algorithms for weighted automata over principal ideal domains (PIDs). An example is $\mathbb{Z}$-automata which can be seen as register automata with affine integer updates over integers. I start with discussing properties of the Hankel matrix of a weighted automaton. Then I reiterate briefly the idea behind learning weighted automata over fields. For automata over PIDs, I recall an existing algorithm (Heerdt et al., FoSSaCS 2020) for exact learning that has no complexity bounds (but only termination). I will recall a classical result of Fatou, and, inspired by its proof, draft a simpler learning algorithm for learning weighted automata over PIDs. I also briefly talk about learning algorithms for polynomial automata. I will conclude with mentioning that the automata I talk about can be seen as coalgebras. It will be interesting to see whether the learning algorithms can be simulated by general coalgebraic learning approaches.

Contents

Predicate and Relation Liftings for Coalgebras with Side Effects: An Application in Coalgebraic Modal Logic

Harsh Beohar[1(✉)], Barbara König[2], Sebastian Küpper[3], and Christina Mika-Michalski[4]

[1] The University of Sheffield, Sheffield, UK
h.beohar@sheffield.ac.uk
[2] Universität Duisburg-Essen, Duisburg, Germany
barbara_koenig@uni-due.de
[3] FernUniversität Hagen, Hagen, Germany
sebastian.kuepper@fernuni-hagen.de
[4] Hochschule Rhein-Waal, Kleve, Germany
christina.mika-michalski@hochschule-rhein-waal.de

Abstract. We study coalgebraic modal logic to characterise behavioural equivalence in the presence of side effects, i.e., when coalgebras live in a (co)Kleisli or an Eilenberg-Moore category. Our aim is to develop a general framework based on indexed categories/fibrations that is common to the aforementioned categories. In particular, we show how the coalgebraic notion of behavioural equivalence arises from a relation lifting (a special kind of indexed morphism) and we give a general recipe to construct such liftings in the above three cases. Lastly, we apply this framework to derive logical characterisations for (weighted) language equivalence and conditional bisimilarity.

Keywords: (co)Kleisli categories · Indexed morphisms · Indexed categories/fibrations

1 Introduction

Coalgebra [32] offers a categorical framework for specifying and reasoning about state-based transition systems in a generic way. In particular, new types of transition systems, behavioural equivalences (or distances), modal logics and games can be obtained by suitably instantiating the theory of coalgebras. While many types of transition systems can already be studied in the category **Set**, systems with side effects – leading to a notion of trace equivalence or conditional bisimilarity – usually require to move to a setting beyond **Set**, using Kleisli, coKleisli or Eilenberg-Moore categories, where the (co)monad specifies the side-effects.

Behavioural equivalences for such scenarios have already been studied extensively (see e.g. [1,12,16,19]). Modal logics, on the other hand, have been

Published by Springer Nature Switzerland AG 2022
H. H. Hansen and F. Zanasi (Eds.): CMCS 2022, LNCS 13225, pp. 1–22, 2022.
https://doi.org/10.1007/978-3-031-10736-8_1

considered to a lesser extent with side effects [9,21,23]; the emphasis (see the survey [27] and the recent articles [11,21,25,28]) has been on logical characterisation of various notions of bisimulation relations and metrics. The aim of the present paper is to close this gap by applying the dual adjunction setup for fibrations developed by Kupke and Rot [28] to derive logical characterisations for coalgebras with side effects. Recently, another powerful approach [9] based on graded monads has been developed to handle equivalences from the van Glabbeek spectrum and beyond. There the characterising notion is not coalgebraic behavioural equivalence, but a refinement of it, called finite depth behavioural equivalence. In addition, Kupke and Rot [28] when comparing their work with [9] noted that "trace equivalences of various kinds covered in [9] cannot be captured directly in their setup". Hence, in this paper, we will show how to capture linear notions such as trace, language, and failure equivalences in their setup.

Here, in order to treat these linear notions of equivalence uniformly, we follow the approach of Hermida and Jacobs [14] to capture bisimulation relations using the language of fibrations [15]; they have increasingly appeared in the coalgebraic literature [6,7,13,17,22,24,28,34].

$$\begin{array}{ccccc} \mathbf{Coalg}_{\mathbf{E}}(F_\lambda) & \longrightarrow & \mathbf{E} & \xrightarrow{\quad F_\lambda \quad} & \mathbf{E} \\ \big\downarrow & & \big\downarrow & & \big\downarrow \\ \mathbf{Coalg}_{\mathbf{Set}}(F) & \longrightarrow & \mathbf{Set} & \xrightarrow{\quad F \quad} & \mathbf{Set} \end{array} \tag{1}$$

The general idea is as follows and is illustrated above in (1). First, a system is modelled as a coalgebra $X \xrightarrow{\alpha} FX$ for some endofunctor $\mathbf{Set} \xrightarrow{F} \mathbf{Set}$. Second a fibration $\mathbf{E}$ of binary relations on the working category of sets is realised, whose fibres are all the relations on the underlying state space. Third, a mechanism $\mathcal{P}(X \times X) \xrightarrow{\lambda} \mathcal{P}(FX \times FX)$ (aka relation lifting) is defined, which amounts to the lifting of F to an endofunctor F_λ on $\mathbf{E}$. Now one can study the coalgebras induced by F_λ and, more importantly, this category $\mathbf{Coalg}_{\mathbf{E}}(F_\lambda)$ can be again arranged (see (1)) as a fibration on $\mathbf{Coalg}_{\mathbf{Set}}(F)$. Lastly, the applicability is shown by characterising bisimulation relations on X as coalgebras of a certain endofunctor living in the fibre above (X, α).

One of the objectives of this paper is to extend this 'categorical' picture (1) w.r.t. coalgebraic notion of behavioural equivalence for dynamical systems having side effects, i.e., those systems that can be modelled as coalgebras living in a (co)Kleisli or Eilenberg-Moore category for some (co)monad on $\mathbf{Set}$. Typical examples, in the context of this paper, are the following: nondeterministic (linear weighted) automata modelled as coalgebras in the Kleisli category for the powerset (multiset) monad (see Sect. 6); conditional transition systems (which facilitate formal modelling of software product lines) modelled as coalgebas in the coKleisli category for the writer comonad $\mathbb{K} \times \_$ (see Sect. 7). We left out case studies in Eilenberg-Moore categories in this paper; however, they are shown to satisfy the assumptions of this paper [3].

We explore the general conditions on relation liftings (technically they are called indexed morphisms in the paper like the ones indicated by λ in (1)) to ensure that behavioural equivalences can be viewed as coalgebras living in the fibre above a given coalgebra (X, α) with side effects. Another contribution is a recipe for obtaining relation and predicate liftings (a special type of indexed morphisms) whose definition and correctness proof are otherwise (at least in the Kleisli case) quite cumbersome to establish. Predicate liftings are instrumental in providing interpretation to various modalities for coalgebras in (co)Kleisli or Eilenberg-Moore categories, just like in the case of **Set** (cf. [17,30,33]). Technically, our study focuses on lifting an indexed morphism for a given endofunctor F on **Set** to an indexed morphism for a (co)Kleisli extension/Eilenberg-Moore lifting $\bar{F}$ of F. And to the best of our knowledge, this question is open at least for coalgebras with side effects.

Once we have captured behavioural equivalence in a fibration, we can then apply the Kupke-Rot setup [28] based on dual adjunctions (see the survey [27] on coalgebraic modal logic) to establish the logical characterisation of behavioural equivalence. In particular, we first construct the Kupke-Rot setup for behavioural equivalences and show that the sufficient conditions for adequacy (i.e., behavioural equivalence is contained in logical equivalence) and expressivity (i.e., converse of adequacy) given in [28] are satisfied. This setup is later used to derive the logical characterisation for (weighted) language equivalence and conditional bisimilarity; note that these notions were not studied in [28].

While several ingredients (especially encompassing fibrations) used in this paper are already known, our paper contains the following original contributions:

- We capture behavioural equivalences on coalgebras beyond **Set** as a fibred notion by characterising them as special types of coalgebras.
- We give concrete recipes for defining predicate and relation liftings (which is both tedious and error-prone) in (co)Kleisli and Eilenberg-Moore categories.
- We extend the dual adjunction framework for fibrations by Kupke and Rot to side effects, in particular to Kleisli categories. Here we need a mechanism to factor the state space of a coalgebra by behavioural equivalence, which is difficult if the category has no coequalisers. We provide a technique based on reflective subcategories to circumvent this issue.

This paper is organised as follows. Section 2 sets the relevant categorical preliminaries required for this paper. It is assumed that the reader is already familiar with basic category theory, particularly, how a Kleisli or an Eilenberg-Moore category is induced by a monad. Section 3 introduces the assumptions that ensure behavioural equivalence is a fibred notion (in the sense of (1)). Section 4 is devoted to coalgebraic modal logic where general adequacy and expressivity results for behavioural equivalence are derived from [28]. Section 5 gives the recipe to construct relation/predicate liftings for coalgebras with side effects. In the next sections, the results of this paper are applied in the context of non-deterministic automata and conditional transition systems. Section 8 concludes this paper with some discussions on future research. Note that proofs as well as additional material on linear weighted and generalised Moore automata in Kleisli and Eilenberg-Moore categories can be found in the technical report [3].

2 Preliminaries

Coalgebraic Preliminaries [18,32]

Let $\mathbf{C}$ be a category and let $\mathbf{C} \xrightarrow{F} \mathbf{C}$ be an endofunctor modelling the branching type of the system of interest. Then the behaviour of a state-based system will be modelled as an F-coalgebra (or, simply coalgebra), i.e., as a morphism $X \xrightarrow{\alpha} FX$ in the category $\mathbf{C}$.

Definition 1. *A coalgebra homomorphism f between (X, α) and (Y, β) is a morphism $X \xrightarrow{f} Y \in \mathbf{C}$ satisfying $Ff \circ \alpha = \beta \circ f$. The collection of coalgebras and their homomorphisms forms a category denoted* $\mathbf{Coalg}_{\mathbf{C}}(F)$.

Moreover, one can define behavioural equivalence on the (concrete) states of a coalgebra under the assumption that there is a functor $\mathbf{C} \xrightarrow{|\_|} \mathbf{Set}$. By the concrete state-space of a coalgebra (X, α), we mean the set $|X|$. Typically, in our case studies, the functor $|\_|$ will be a forgetful functor and will have a left/right adjoint ι. For instance, ι is left adjoint to $|\_|$ when $\mathbf{C} = \mathbf{Kl}(T)$ or $\mathbf{C} = \mathbf{EM}(T)$ (for some monad $\mathbf{Set} \xrightarrow{T} \mathbf{Set}$), while it is right adjoint when $\mathbf{C} = \mathbf{coKl}(G)$ for some comonad $\mathbf{Set} \xrightarrow{G} \mathbf{Set}$.

Definition 2. *Two states $x, x' \in |X|$ of a coalgebra (X, α) are* behaviourally equivalent *iff there is a coalgebra homomorphism f such that $|f|x = |f|x'$.*

Example 3. An interesting example of coalgebras living in Kleisli categories is nondeterministic automata (NDA). Following [12] an NDA is a coalgebra living in $\mathbf{C} = \mathbf{Kl}(\mathcal{P})$, which is isomorphic to the category $\mathbf{Rel}$ of sets as objects and relations as maps. Recall that a Kleisli extension $\mathbf{Rel} \xrightarrow{\bar{F}} \mathbf{Rel}$ of $\mathbf{Set} \xrightarrow{F} \mathbf{Set}$ (i.e. $\bar{F} \circ \iota = \iota \circ F$) is in correspondence [29, Theorem 2.2] with a distributive law $FT \xrightarrow{\vartheta} TF$ such that the following diagrams commute in $\mathbf{Set}$.

$$
\begin{array}{ccc}
FX & = & FX \\
{\scriptstyle F\eta_X}\downarrow & & \downarrow{\scriptstyle \eta_{FX}} \\
FTX & \xrightarrow{\vartheta_X} & TFX
\end{array}
\qquad
\begin{array}{ccccc}
FTTX & \xrightarrow{\vartheta_{TX}} & TFTX & \xrightarrow{T\vartheta_X} & TTFX \\
{\scriptstyle F\mu_X}\downarrow & & & & \downarrow{\scriptstyle \mu_{FX}} \\
FTX & & \xrightarrow{\vartheta_X} & & TFX
\end{array}
\tag{2}
$$

Consider $T = \mathcal{P}, F\_ = A \times \_ + 1$ (where $1 = \{\bullet\}$) with the distributive law [16]:

$$
Act \times \mathcal{P}X + 1 \xrightarrow{\vartheta_X} \mathcal{P}(Act \times X + 1) \qquad (a, U) \mapsto \{a\} \times U, \ \bullet \mapsto \{\bullet\}. \tag{3}
$$

This induces a functor $\mathbf{Rel} \xrightarrow{\bar{F}} \mathbf{Rel}$ which acts on a relation $X \xrightarrow{f} Y$, seen as a Kleisli arrow $X \xrightarrow{f'} \mathcal{P}Y$, as follows: $\bar{F}f = \vartheta_Y \circ Ff'$. Notice that $\bar{F}$-coalgebras model implicit nondeterminism (i.e. this side-effect is hidden to an outside observer) [12], thus behavioural equivalence typically coincides with *language equivalence* (instead of bisimilarity) in this case.

Predicate Liftings as Indexed Morphisms

Predicates and their liftings are quite common within the literature on (coalgebraic) modal logic. In particular, a predicate is used as the semantics of a logical formula [30], or as a relation on the state space of a coalgebra [14]. In the basic setting, when $\mathbf{C} = \mathbf{Set}$, the predicates on a set X are given by the subsets of X. Now, given a function $X \xrightarrow{f} Y$, a predicate V on Y (i.e. $V \subseteq Y$) can be transformed into a predicate on X by the pullback operation $f^{-1}V \subseteq X$ in **Set**. Note that this operation is functorial in nature; thus this 'logical' structure can be organised as a functor $\mathbf{Set}^{\mathrm{op}} \xrightarrow{\hat{\mathcal{P}}} \mathbf{Cat}$ [17], where $\hat{\mathcal{P}}X$ is the poset $(\mathcal{P}X, \subseteq)$ viewed as a category. As noted by Jacobs in [17], *predicate logic on a category is given by an indexed category* and *predicate liftings are (endo)morphisms of indexed categories.*

Definition 4. *An indexed category is a* **Cat***-valued presheaf, i.e., a contravariant functor Φ from* **C** *to* **Cat**. *In addition, a morphism between two indexed categories $\mathbf{C}^{op} \xrightarrow{\Phi} \mathbf{Cat}$ and $\mathbf{D}^{op} \xrightarrow{\Psi} \mathbf{Cat}$ is a pair of a functor $\mathbf{C} \xrightarrow{G} \mathbf{D}$ and a natural transformation $\Phi \xrightarrow{\lambda} \Psi \circ G^{op}$.*

Note 5. Often the application of Φ on $f \in \mathbf{C}$ is denoted as f^*. We also omit the use of superscript 'op' on functors and use the phrases 'indexed morphism' and 'predicate lifting' interchangeably.

Remark 6. Another, equivalent, way to organise logic is by specifying the fibration of predicates over a category [15]. The transformation of a fibration over **C** into a contravariant pseudofunctor $\mathbf{C} \longrightarrow \mathbf{Cat}$ is given by taking the fibres at each object in **C**. Conversely one has to invoke the so-called *Grothendieck construction* to get a fibration, which glues all the fibres $(\Phi X)_{X \in \mathbf{C}}$ to form a total category of predicates $\mathbb{E}(\Phi)$ defined as follows.

$$\frac{X \in \mathbf{C} \wedge U \in \Phi X}{(X, U) \in \mathbb{E}(\Phi)} \qquad \frac{X \xrightarrow{f} Y \in \mathbf{C} \wedge U \xrightarrow{\bar{f}} f^*V \in \Phi X}{(X, U) \xrightarrow{f, \bar{f}} (Y, V) \in \mathbb{E}(\Phi)}$$

Moreover, there is an obvious 'forgetful' functor $\mathbb{E}(\Phi) \xrightarrow{p} \mathbf{C}$ given by $(X, U) \mapsto X$ that induces a (split) fibration on **C** [15,17]. In the parlance of concrete categories, the functor p is topological [2, Definition 21.1] when Φ has fibred limits. Often, in applications, the fibres $(\Phi X, \preceq)$ (at each $X \in \mathbf{C}$) form a poset (rather than a full-fledged category); we label such an indexed category/fibration as *thin.* We restrict ourselves to thin fibrations in this paper. Note that, under this restriction, a map $(X, U) \xrightarrow{f} (Y, V) \in \mathbb{E}(\Phi)$ is *Cartesian* iff $U = f^*V$.

Example 7. The contravariant powerset functor $\mathbf{Set} \xrightarrow{\hat{\mathcal{P}}} \mathbf{Cat}$ is an example of an indexed category such that $\mathbb{E}(\hat{\mathcal{P}}) \longrightarrow \mathbf{Set}$ is a bifibration [15]. This is because the reindexing functor f^{-1} (for any function f) has a left adjoint given by the direct image functor $f_!$. Moreover, as an example of a predicate lifting, consider $F = \mathcal{P}$ over $\mathbf{C} = \mathbf{Set}$ (which describes the branching type of unlabelled transition systems) with $\hat{\mathcal{P}}X \xrightarrow{\lambda_X} \hat{\mathcal{P}}\mathcal{P}X$ given by $U \mapsto \mathcal{P}U$. It is well known that the above predicate lifting encodes the box modality $\Box$ from logic [17].

3 Behavioural Equivalence Through Indexed Morphisms

Indexed morphisms not only induce modalities of interest in Computer Science; but they can also be used to characterise behavioural equivalence. The original idea [14] is to work with an indexed category $\mathbf{Set}^{\mathrm{op}} \xrightarrow{\Psi} \mathbf{Cat}$ of binary relations, i.e., Ψ is the composition $\mathbf{Set}^{\mathrm{op}} \xrightarrow{\text{-}\times\text{-}} \mathbf{Set}^{\mathrm{op}} \xrightarrow{\hat{\mathcal{P}}} \mathbf{Cat}$. In particular, ΨX is the set of all relations on X. Then, for a relation lifting $\Psi X \xrightarrow{\lambda_X} \Psi F X$ and a coalgebra $X \xrightarrow{\alpha} FX \in \mathbf{Set}$, bisimilarity is the largest fixpoint of the functional:

$$\Psi X \xrightarrow{\lambda_X} \Psi F X \xrightarrow{\alpha^*} \Psi X. \tag{4}$$

Unfortunately, this idea of working with relations on the concrete state space immediately does not generalise to coalgebras with side effects; e.g., in the case of conditional transition systems (CTSs) viewed as coalgebras living in the coKleisli category of the writer comonad $\mathbb{K} \times \_$ (see Sect. 7). The problem essentially lies in associating a fibre to be the set of all binary relations on the state space. There are situations (as in CTSs) where the fibres will only be some subset of all the relations on the state space. As a result, we impose the following restriction:

A1 our working category $\mathbf{C}$ has binary products $\otimes$.

Thus we can define an indexed category Ψ of relations as the composition:

$$\mathbf{C}^{\mathrm{op}} \xrightarrow{\text{-}\otimes\text{-}} \mathbf{C}^{\mathrm{op}} \xrightarrow{|\text{-}|} \mathbf{Set}^{\mathrm{op}} \xrightarrow{\hat{\mathcal{P}}} \mathbf{Cat}. \tag{5}$$

We view the elements of ΨX (for some object $X \in C$) as 'abstract' relations on X. Furthermore, **A1** also ensures that for any object $X \in \mathbf{C}$ there is a function

$$|X \otimes X| \xrightarrow{\langle |\pi_1^X|, |\pi_2^X| \rangle} |X| \times |X|,$$

where $X \otimes X \xrightarrow{\pi_1^X, \pi_2^X} X$ are the two projection arrows in $\mathbf{C}$. And thanks to these functions, we can define abstract equality in the fibre ΨX. In particular,

$$\equiv_X \;=\; \langle |\pi_1^X|, |\pi_2^X| \rangle^{-1} \; =_{|X|} .$$

Notice that in some cases (like when $\mathbf{C}$ is a Kleisli or Eilenberg-Moore category) the abstract equality $\equiv_X$ coincides with the equality on the concrete state space $=_{|X|}$ because the forgetful functor is product preserving. However, in the context of CTSs, we will see that the two notions of equality differ.

Proposition 8. *Under Assumption **A1** the square drawn in* (6) *commutes for any arrow* $X \xrightarrow{f} Y$ *in* $\mathbf{C}$. *As a result, there is a functor* $\mathbf{C} \xrightarrow{\text{Eq}} \mathbb{E}(\Psi)$ *(henceforth called* equality *functor) that maps an object* X *to the abstract equality* $\equiv_X$.

$$\begin{array}{ccc} |X \otimes X| & \xrightarrow{\langle|\pi_1^X|, |\pi_2^X|\rangle} & |X| \times |X| \\ {\scriptstyle |f \otimes f|}\Big\downarrow & (6) & \Big\downarrow{\scriptstyle |f| \times |f|} \\ |Y \otimes Y| & \xrightarrow{\langle|\pi_1^Y|, |\pi_2^Y|\rangle} & |Y| \times |Y| \end{array}$$

The next proposition (originally from [17]) is a general result on indexed categories useful in lifting an endofunctor on $\mathbf{C}$ to an endofunctor on the given fibration $\mathbb{E}(\Phi)$. Moreover the category of coalgebras of the lifted endofunctor can be structured again as a fibration on the given category of coalgebras in which our original system of interest is modelled.

Proposition 9. *Consider the diagram in* (7), *then the following statements hold for a given functor* $\mathbf{C} \xrightarrow{F} \mathbf{C}$ *and an indexed morphism* $\Phi \xrightarrow{\lambda} \Phi F$.

- *The map* λ *induces a map* $\mathbb{E}(\Phi) \xrightarrow{F_\lambda} \mathbb{E}(\Phi)$ *of fibrations given by* $(X, U) \mapsto (FX, \lambda_X U)$.
- *The category of coalgebras induced by* F_λ *forms a fibration on* $\mathbf{Coalg}_{\mathbf{C}}(F)$. *The corresponding indexed category* $\mathbf{Coalg}_{\mathbf{C}}(F)^{op} \xrightarrow{\Phi_\lambda^F} \mathbf{Cat}$ *is given by the mapping:* $(X, \alpha) \mapsto \mathbf{Coalg}_{\Phi X}(\alpha^* \circ \lambda_X)$.

$$\begin{array}{ccc} \mathbf{Coalg}_{\mathbb{E}(\Phi)}(F_\lambda) = \mathbb{E}(\Phi_\lambda^F) & \longrightarrow & \mathbb{E}(\Phi) \\ {\scriptstyle p_\lambda^F}\Big\downarrow & (7) & \Big\downarrow{\scriptstyle p} \\ \mathbf{Coalg}_{\mathbf{C}}(F) & \longrightarrow & \mathbf{C} \end{array}$$

Now recall (4) and Ψ as indexed category of relations on **Set** (i.e. substitute $\otimes$ by $\times$ and $|\_|$ by the identity functor in (5)), an arbitrary bisimulation relation R on a coalgebra $X \xrightarrow{\alpha} FX \in \mathbf{Set}$ is the relation $R \in \Psi X$ satisfying $R \subseteq \alpha^* \lambda_X R$. In other words, bisimulation relations on the state space X are again coalgebras of the functor $\alpha^* \circ \lambda_X$ living in the fibre ΨX. Next we show that the same holds for behavioural equivalence in general, however, under the following assumptions:

A2 the given morphism $\Psi \xrightarrow{\lambda} \Psi F$ preserves Eq, i.e., $F_\lambda \circ \text{Eq} = \text{Eq} \circ F$. Equivalently, this means that $\lambda_X(\equiv_X) = \equiv_{FX}$ for every $X \in \mathbf{C}$.
A3 the functor Eq has a left adjoint $\mathbb{Q}$.

Remark 10. Assumption **A3** already appeared in [14] to model quotient types in the context of type theory. However, our usage is in the unit κ of $\mathbb{Q} \dashv \text{Eq}$ to construct a witnessing coalgebra homomorphism in Theorem 12. This idea is already known in type theory; for instance, see [10, Theorem 3.7] where a similar result was proven albeit under the stronger assumption that the final coalgebra for F exists. So when $\mathbf{C} = \mathbf{Set}$, $\mathbb{Q}$ maps an relation R on X to the quotient generated by the smallest equivalence containing R; the unit κ_X (for any set X) is the usual quotient function mapping an element to its equivalence class.

Theorem 11. *Given an indexed morphism* $\Psi \xrightarrow{\lambda} \Psi \circ F$, *then under Assumptions* ***A1*** *and* ***A2****, the behavioural equivalence induced by a coalgebra homomorphism* $f \in \mathbf{Coalg}_{\mathbf{C}}(F)$ *on a coalgebra* $(X, \alpha) \in \mathbf{Coalg}_{\mathbf{C}}(F)$ *is a* $\alpha^* \circ \lambda$*-coalgebra living in the fibre* ΨX, *i.e.,* $f^*(\equiv_Y) \subseteq \alpha^* \lambda_X(f^* \equiv_Y)$.

Theorem 12. *Under Assumptions* ***A1****,* ***A2****, and* ***A3*** *for every* $\alpha^* \circ \lambda$*-coalgebra* R *there is a coalgebra homomorphism* $f \in \mathbf{Coalg}_{\mathbf{C}}(F)$ *such that* $R \subseteq f^*(\equiv_{\mathsf{cod}(f)})$, *where* $\mathsf{cod}(f)$ *denotes the codomain of* f. *Moreover,* $R = f^*(\equiv_{\mathsf{cod}(f)})$ *when the unit of* $\mathbb{Q} \dashv \mathrm{Eq}$ *is Cartesian.*

Remark 13. An application of Theorem 12 could be in establishing the completeness of coalgebraic games (as in the spirit of [26]). For instance, if the winning positions of Duplicator viewed as a relation R is a coalgebra in the fibre of Ψ, then Theorem 12 can be used to show that winning positions of Duplicator are behaviourally equivalent. In the future, we would like to test this application by working out a notion of 2-player games for coalgebra with side effects.

4 Coalgebraic Modal Logic

The 'partial' characterisation of behavioural equivalence as a fibred notion (cf. Theorems 11 and 12) enables us to use the dual adjunction framework of Kupke and Rot [28] in (8) to develop a logical characterisation of behavioural equivalence. It should be noted that, although this framework can handle behavioural preorders and distances, we prove our results only for behavioural equivalence, i.e. in the context of Assumptions **A1** and **A2**.

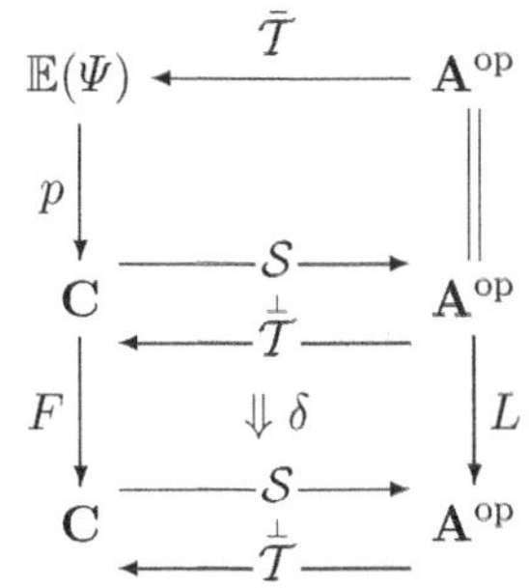

(8)

Below we explain the role of various functors drawn in (8) in an incremental manner; subsequently, we will establish our general adequacy and expressivity results (Theorems 18 and 21) for behavioural equivalences.

The fibration $\mathbb{E}(\Psi) \xrightarrow{p} \mathbf{C}$ will be used to define (internally) a behavioural equivalence of interest. Often it is defined as a colimit of a diagram resembling the final sequence in a fibre (cf. [13]). More abstractly, we assume that the indexed category Ψ^F_λ (recall this notation from Proposition 9) has indexed final objects, for some indexed morphism $\Psi \xrightarrow{\lambda} \Psi F$.

Lemma 14. *Suppose* $\mathbf{C}^{op} \xrightarrow{\Phi} \mathbf{Cat}$ *has indexed final objects (i.e., the final object exists in each fibre* ΦX*) and the reindexing functor* f^* *preserves these final objects. Then there is a functor* $\mathbf{C} \xrightarrow{\mathbb{1}} \mathbb{E}(\Phi)$ *that is right adjoint to* p*.*

Usually, $\mathbb{1}$ is used (called the *truth functor* [15] in the context of logic) when the underlying fibration $\mathbb{E}(\Phi) \xrightarrow{p} \mathbf{C}$ has indexed final objects. However when Ψ^F_λ satisfies the conditions of the previous lemma, it results in a functor $\mathbf{Coalg}_{\mathbf{C}}(F) \xrightarrow{\mathbb{1}^\lambda} \mathbf{Coalg}_{\mathbb{E}(\Psi)}(F_\lambda)$ (which we call *the behavioural conformance* functor) that maps a coalgebra $X \xrightarrow{\alpha} FX$ to the terminal element in $\mathbf{Coalg}_{\Psi X}(\alpha^* \circ \lambda_X)$ denoted as 1^λ_X. Note that 1^λ_X in our applications will correspond to the largest behavioural equivalence on a given system. Moreover, it is not hard to arrive at the adjoint situation as indicated in (9).

$$\begin{array}{ccc}
\mathbf{Coalg}_{\mathbb{E}(\Psi)}(F_\lambda) = \mathbb{E}(\Psi^F_\lambda) & \longrightarrow & \mathbb{E}(\Psi) \\
p^F_\lambda \Big\downarrow \dashv \Big\uparrow \mathbb{1}^\lambda & (9) & \Big\downarrow p \\
\mathbf{Coalg}_{\mathbf{C}}(F) & \longrightarrow & \mathbf{C}
\end{array}$$

So the behavioural conformance functor is right adjoint to the forgetful functor that witnesses the fibration of behavioural conformance on coalgebras.

Example 15. Consider the indexed category Ψ induced by binary relations on sets and a labelled transition system modelled as a coalgebra $X \xrightarrow{\alpha} (\mathcal{P}X)^{Act}$, i.e., our $\mathbf{C} = \mathbf{Set}$, $F = (\mathcal{P}\_)^{Act}$. Consider the function $\Psi X \xrightarrow{\lambda_X} \Psi FX$ that maps a relation $R \in \Psi X$ to a relation $\lambda_X R \in \Psi FX$ s.t. $q\, \lambda_X R\, q'$, $q, q' \in (\mathcal{P}X)^{Act}$ iff:

$$\forall_{a,x} \exists_{x'}\ (x \in qa \implies x' \in q'a \wedge x\, R\, x') \wedge \forall_{a,x'} \exists_x\ (x' \in q'a \implies x \in qa \wedge x\, R\, x').$$

It is well known (as first noted in [14]) that a bisimulation relation is a $\alpha^* \circ \lambda$-coalgebra. Moreover, bisimilarity $\leftrightarroweq_X$ (the largest bisimulation relation on X) corresponds to the final object in $\Psi_\lambda(X, \alpha)$, i.e., $\mathbb{1}^\lambda(X, \alpha) = (X, \alpha, \leftrightarroweq_X)$.

As for the dual adjunction $\mathcal{S} \dashv \mathcal{T}$ in (8), it provides a connection (cf. [31]) between states and theories (the formulae satisfied by a state). The syntax of the logic is given by a functor $\mathbf{A} \xrightarrow{L} \mathbf{A}$ and it is assumed that the initial algebra $L\mathcal{A} \xrightarrow{h} \mathcal{A} \in \mathbf{A}$ exists for L, which models the typical Lindenbaum algebra induced by the term algebra. Lastly, the natural transformation δ gives the one-step interpretation to the formulae which can be given its mate θ as described below (cf. [20, Proposition 2]).

Proposition 16. *Given* $\mathbf{C} \xrightarrow{F} \mathbf{C}, \mathbf{A} \xrightarrow{L} \mathbf{A}, \mathbf{C} \underset{\mathcal{T}}{\overset{\mathcal{S}}{\rightleftarrows}} \mathbf{A}^{op}$ *with* $\mathcal{S} \dashv \mathcal{T}$*, there is a correspondence between* $F\mathcal{T} \xrightarrow{\delta} \mathcal{T}L$ *and* $\mathcal{S}F \xrightarrow{\theta} L\mathcal{S}$*.*

Now given a coalgebra $X \xrightarrow{\alpha} FX \in \mathbf{C}$, the semantics $\mathcal{A} \xrightarrow{[\![\text{-}]\!]_X} \mathcal{S}X$ of the logic (L, δ) is given by the universal property of the initial algebra $L\mathcal{A} \xrightarrow{h} \mathcal{A}$. In particular, it is the unique arrow in $\mathbf{A}$ that makes the following diagram (drawn on the left) commutative. And the transpose of the semantics map $[\![\text{-}]\!]_X$ under $\mathcal{S} \dashv \mathcal{T}$ gives a 'theory' map $X \xrightarrow{(\!|\text{-}|\!)_X} Q\mathcal{A}$; it is the unique arrow in $\mathbf{C}$ that makes the following diagram on the right commutative.

$$
\begin{array}{ccccc}
L\mathcal{A} & \xrightarrow{L[\![\text{-}]\!]_X} & L\mathcal{S}X & \xrightarrow{\theta_X} & \mathcal{S}FX \\
{\scriptstyle h}\downarrow & & (10) & & \downarrow{\scriptstyle \mathcal{S}\alpha} \\
\mathcal{A} & & \xrightarrow{\quad [\![\text{-}]\!]_X \quad} & & \mathcal{S}X
\end{array}
\qquad
\begin{array}{ccccc}
X & & \xrightarrow{\quad (\!|\text{-}|\!)_X \quad} & & \mathcal{T}\mathcal{A} \\
{\scriptstyle \alpha}\downarrow & & & & \downarrow{\scriptstyle \mathcal{T}h} \\
FX & \xrightarrow{F(\!|\text{-}|\!)_X} & F\mathcal{T}\mathcal{A} & \xrightarrow{\delta_{\mathcal{A}}} & \mathcal{T}L\mathcal{A}
\end{array}
$$

Once these niceties are set up, one can argue when a logic (L, δ) is adequate and expressive. Intuitively, a logic (L, δ) is adequate if behaviourally equivalent states satisfy the same logical formulae; while an adequate logic is expressive if logically equivalent states are also behaviourally equivalent. The formulation below is a straightforward formulation of adequacy and expressivity given in [28] using the language of indexed categories.

Definition 17. *Suppose the behavioural conformance functor* $\mathbb{1}^{\lambda}$ *exists (for some* λ*) with* $\mathbf{A}^{op} \xrightarrow{\bar{\mathcal{T}}} \mathbb{E}(\Psi)$ *such that* $p \circ \bar{\mathcal{T}} = \mathcal{T}$*. Then a logic* (L, δ) *is* adequate *(resp.* expressive*) w.r.t.* $\bar{\mathcal{T}}$ *if* $(X, 1^{\lambda}_X) \xrightarrow{(\!|\text{-}|\!)_X} \bar{\mathcal{T}}\mathcal{A}$ *is a (resp. Cartesian) map in* $\mathbb{E}(\Psi)$*, for every coalgebra* $X \xrightarrow{\alpha} FX \in \mathbf{C}$*.*

The role of $\bar{\mathcal{T}}$ is to encode a relationship between the theories of any two states (cf. [28]); so we let $\bar{\mathcal{T}} = \mathrm{Eq} \circ \mathcal{T}$ in the context of behavioural equivalence. Next we state the main result of this section, which is a refinement of adequacy and expressivity results given in [28].

Theorem 18. *Under the assumptions of Theorem 11, if* $\bar{\mathcal{T}}$ *has a left adjoint* $\bar{\mathcal{S}}$*, the logic* (L, δ) *is adequate. Moreover it is expressive if* $|\delta_{\mathcal{A}}|$ *is injective.*

In short, the Kupke-Rot logical setup for behavioural equivalence can be summarised as drawn left in (11). Now if our indexed category Ψ satisfies **A3** (like in the case of coKleisli and Eilenberg-Moore categories), then $\bar{\mathcal{S}} = \mathcal{S} \circ \mathbb{Q}$ as indicated in (11). However, in the case of Kleisli categories we will construct $\bar{\mathcal{S}}$ under some restrictions (cf. Theorem 21).

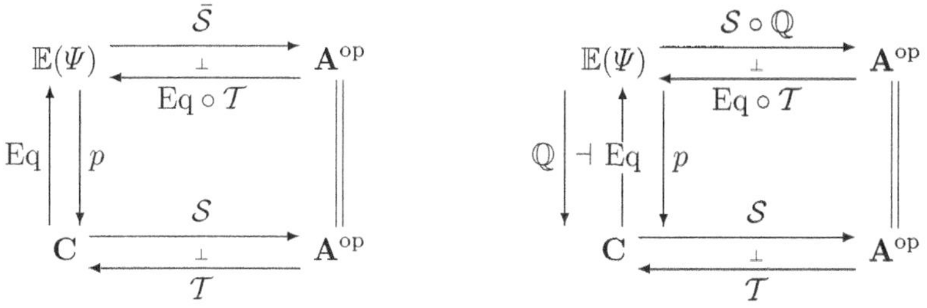

(11)

Construction of $\bar{\mathcal{S}}$ for Kleisli Categories

Unfortunately, arbitrary (co)limits in general do not exists in a Kleisli category. For instance, one of our working categories $\mathbf{Kl}(\mathcal{P}) \cong \mathbf{Rel}$ (the category of sets and relations) does not have all coequalisers, but **Rel** has a reflective subcategory $\mathbf{Set}^{\mathrm{op}}$ that does. The presence of these coequalisers in the reflective subcategory will then be used to construct $\bar{\mathcal{S}}$.

Definition 19. *A subcategory* $\mathbf{B} \xrightarrow{\jmath} \mathbf{C}$ *is* reflective *when the inclusion functor* $\jmath$ *has a left adjoint* $\mathfrak{r}$ *(often called as reflector).*

Theorem 20 ([1]). *If* $\mathbf{B} \xrightarrow{\jmath} \mathbf{C}$ *is a reflective subcategory of* $\mathbf{C}$ *and* $\mathbf{C} \xrightarrow{F} \mathbf{C}$ *preserves* $\mathbf{B}$*, i.e.,* $\forall_{B,f \in \mathbf{B}}\ (FB \in \mathbf{B} \wedge Ff \in \mathbf{B})$ *and* $F \circ \jmath = \jmath \circ F$*, then* $\mathbf{Coalg}_B(F) \underset{\bar{\jmath}}{\overset{\bar{\mathfrak{r}}}{\leftrightarrows}} \mathbf{Coalg}_C(F)$ *with* $\bar{\mathfrak{r}} \dashv \bar{\jmath}$*. Here,* $\bar{\jmath}$ *is the obvious inclusion.*

The reflector $\bar{\mathfrak{r}}$ typically results in a form of (on-the-fly) determinisation (cf. Example 27). Moreover, in our case studies, these reflective subcategories will also take the place of algebras in (11), and if these reflective subcategories have coequalisers, then we can construct $\bar{\mathcal{S}}$ in general.

So let $\mathbf{B} = \mathbf{A}^{\mathrm{op}}, \mathcal{S} = \mathfrak{r}, \mathcal{T} = \jmath$, and $(X, R) \in \mathbb{E}(\Psi)$. Then the idea is to use the following series of transformations (depicted below on the left) to construct $\bar{\mathcal{S}}$ as the equaliser of the parallel arrows $\mathcal{S}p'_1, \mathcal{S}p'_2 \in \mathbf{A}$. Below p_i (for $i \in \{1,2\}$) are the obvious projection functions and each p'_i is the transpose of p_i under the free-forgetful adjunction $\iota \dashv |\_|$.

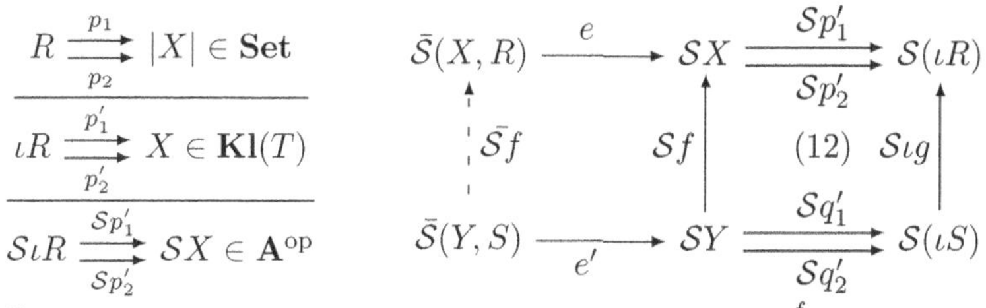

Let $(\bar{\mathcal{S}}(X,R), e)$ be the equaliser of $\mathfrak{r}p'_i$ in $\mathbf{A}$. Now $(X,R) \xrightarrow{f} (Y,S) \in \mathbb{E}(\Psi)$ means that $X \xrightarrow{f} Y \in \mathbf{Kl}(T)$ and $R \subseteq (|f| \times |f|)^{-1}S$. So there is a function $R \xrightarrow{g} S$ such that $|f| \circ p_i = q_i \circ g$ with p_i, q_i being the obvious projections when the relations R, S are viewed as spans in **Set**. Moreover $f \circ p'_i = q'_i \circ \iota g$ due to the naturality of the counit of $\iota \dashv |\_|$. So the two squares in (12) commute and the universal property of equalisers gives the unique $\bar{S}f$.

Theorem 21. *Under* $\boldsymbol{A1}$ *and* $\mathbf{A}^{\mathrm{op}}$ *being a reflective subcategory of* $\mathbf{Kl}(T)$ *having coequalisers, the above defined* $\bar{S}$ *is a functor and left adjoint to* $\bar{T} = \mathrm{Eq} \circ T$.

5 Lifting of Predicate and Relation Liftings

In this paper, the indexed categories corresponding to predicates (or relations) on our working category (like (co)Kleisli or Eilenberg-Moore categories) are always induced by lifting an indexed category on the underlying base category **Set** (for instance, recall Ψ from (5)). In a similar spirit, our aim is to construct indexed morphisms on our working category by lifting an indexed morphism on **Set**. So consider an indexed category Φ of predicates given by $\Phi = \hat{\mathcal{P}} \circ |\_|$ and an endofunctor $\mathbf{C} \xrightarrow{\bar{F}} \mathbf{C}$ modelling the branching type of behaviour of interest.

Lifting of Predicate Liftings. Next we give a recipe to construct a predicate lifting, i.e., an indexed morphism of type $\Phi \xrightarrow{\lambda} \Phi\bar{F}$. In particular, we need an endofunctor $\mathbf{Set} \xrightarrow{G} \mathbf{Set}$, a predicate lifting $\hat{\mathcal{P}} \xrightarrow{\sigma} \hat{\mathcal{P}}G$, and a natural transformation γ as indicated below.

$$\begin{array}{ccc} \mathbf{C} & \xrightarrow{\quad \bar{F} \quad} & \mathbf{C} \\ {\scriptstyle |\_|}\downarrow & \Downarrow \gamma & \downarrow{\scriptstyle |\_|} \\ \mathbf{Set} & \xrightarrow[\quad G \quad]{} & \mathbf{Set} \end{array}$$

As a result, we can define λ by the composition:

$$\hat{\mathcal{P}}|X| \xrightarrow{\sigma_{|X|}} \hat{\mathcal{P}}G|X| \xrightarrow{\gamma_X^*} \hat{\mathcal{P}}|FX|. \tag{13}$$

Theorem 22. *The above mapping* λ *is an indexed morphism.*

Note that in the case of coKleisli and Eilenberg-Moore categories, we simply let $G = F$ and $\bar{F}$ be a coKleisli extension/Eilenberg-Moore lifting of F, which results in a distributive law of type $|\bar{F}\_| \xrightarrow{\gamma} G|\_|$; in the case of Eilenberg-Moore categories, such natural transformations are also known as EM-laws [19].

In the case of Kleisli categories, the situation is slightly complicated. This stems from the fact that the distributive law $FT \xrightarrow{\vartheta} TF$ (which induces a Kleisli extension $\bar{F}$ of F) results in a natural transformation in the 'wrong' direction $F|\_| \xrightarrow{\vartheta} |\bar{F}\_|$. However, in various applications, G is typically associated with the branching type of a deterministic version of the corresponding system of interest (such as $G = \_^{Act} \times 2$ in the case of NDA), if it exists. The next result helps in finding such a distributive law γ for a given G in a more elementary way.

Lemma 23. *Let $\bar{F}$ be a Kleisli extension of F induced by a distributive law $FT \xrightarrow{\vartheta} TF$. Then a natural transformation $TF \xrightarrow{\gamma} GT$ compatible with ϑ and μ (i.e., Square 14 commutes) induces a distributive law $|\_| \circ \bar{F} \longrightarrow G \circ |\_|$. Moreover, the converse also holds.*

$$\begin{array}{ccc} TFTX & \xrightarrow{\gamma_{TX}} & GTTX \\ {\scriptstyle T\vartheta_X}\downarrow & & \\ TTFX & & \downarrow{\scriptstyle G\mu_X} \\ {\scriptstyle \mu_{FX}}\downarrow & & \\ TFX & \xrightarrow{\gamma_X} & GTX \end{array} \qquad (14)$$

Remark 24. Note that the compatibility property in the above lemma appeared in [19] as part of an 'extension' natural transformation. In short, the properties of an extension natural transformation are more stringent than of Lemma 23.

Lifting of Relation Liftings. The above idea can also be used to construct a relation lifting, i.e., an indexed morphism of type $\Psi \longrightarrow \Psi\bar{F}$, where Ψ is the indexed category of abstract relations given in (5). So now given a relation lifting $\hat{\mathcal{P}}(X \times X) \xrightarrow{\sigma_X} \hat{\mathcal{P}}(GX \times GX)$ of G, then we can define $\Psi \xrightarrow{\bar{\lambda}} \Psi\bar{F}$:

$$\begin{array}{llll} \Psi X \xrightarrow{\langle \pi_1^X, \pi_2^X \rangle_!} \hat{\mathcal{P}}(|X| \times |X|) \xrightarrow{\sigma_{|X|}} & \hat{\mathcal{P}}(G|X| \times G|X|) & & \\ & \downarrow{\scriptstyle (\gamma_X \times \gamma_X)^{-1}} & & (15) \\ & \hat{\mathcal{P}}(|\bar{F}X| \times |\bar{F}X|) & \xrightarrow{\langle \pi_1^{FX}, \pi_2^{FX} \rangle^{-1}} & \Psi\bar{F}X. \end{array}$$

Here, we use the fact that $\mathbb{E}(\hat{\mathcal{P}})$ is a bifibration [15], i.e., for any function $X \xrightarrow{f} Y$ the reindexing functor f^{-1} has the direct image functor $f_!$ as its left adjoint. Now under the following assumption we can show that the $\bar{\lambda}$ is indeed an indexed morphism.

A4 The square drawn in (6) is a weak pullback in **Set** for every $f \in \mathbf{C}$.

The above assumption ensures that, in the context of $\mathbb{E}(\hat{\mathcal{P}})$, the square in (6) satisfies the Beck-Chevalley condition, i.e., the following equation holds

$$\langle |\pi_1^X|, |\pi_2^X| \rangle_! \circ (f \otimes f)^{-1} = (|f| \times |f|)^{-1} \circ \langle |\pi_1^Y|, |\pi_2^Y| \rangle_!.$$

In turn this equation is used in diagram chasing to show that $\bar{\lambda}$ is a natural transformation. Furthermore, **A4** trivially holds when **C** is a Kleisli or Eilenberg-Moore category (cf. Corollary 26) because the canonical function $\langle |\pi_1^X|, |\pi_2^X| \rangle$ is a bijection for each $X \in \mathbf{C}$. In other words, when **C** is a Kleisli or Eilenberg-Moore category, Ψ can be alternatively defined as the composition $\hat{\mathcal{P}} \circ (|\_| \times |\_|)$.

Theorem 25. *Under **A4** $\bar{\lambda}$ as defined in* (15) *is an indexed morphism.*

Corollary 26. *If the forgetful functor* $\mathbf{C} \xrightarrow{|\_|} \mathbf{Set}$ *is product preserving, then **A4** is always satisfied. As a result,* $\bar{\lambda}$ *defined in* (15) *is an indexed morphism.*

6 Nondeterministic Automata (NDA)

Recall the necessary parameters from Example 3 for coalgebraic modelling of an NDA, i.e., $T = \mathcal{P}$, $\mathbf{C} = \mathbf{Kl}(\mathcal{P}) \cong \mathbf{Rel}$, $F = Act \times \_ + 1$, the distributive law ϑ given in (3), and the free-forgetful adjunction $\iota \dashv |\_|$ associated with any Kleisli category. To apply Theorem 21, we also recall the reflective subcategory $\mathbf{Set}^{\mathrm{op}}$ of $\mathbf{Rel}$ from [1]. Below $X \xrightarrow{f} Y \in \mathbf{Set}$ and $X \xrightarrow{g} Y \in \mathbf{Rel}$:

$$\jmath X = X \qquad Y \xrightarrow{\jmath f} X \in \mathbf{Rel} \qquad y\ \jmath f\ x \iff fx = y$$

$$\mathfrak{r} X = \mathcal{P}X \qquad \mathfrak{r}X \xrightarrow{\mathfrak{r}g} \mathfrak{r}Y \in \mathbf{Set}^{\mathrm{op}} \qquad \mathfrak{r}g(V) = \{x \mid \exists_{y \in V}\ x\ g\ y\}.$$

Next we illustrate the definition of $\bar{S}$ and how the unit of $\bar{S} \dashv \bar{T}$ maps an NDA to the largest subautomaton (respecting language equivalence) obtained after backward determinisation of the given NDA. It is worthwhile to note that the abstract and concrete state space coincide (up to isomorphism) in the case of NDA because forgetful functor preserves products, i.e., $\mathcal{P}(X + X) = |X + X| \cong |X| \times |X| = \mathcal{P}X \times \mathcal{P}X$. Therefore, as mentioned earlier, we will simplify our presentation by working with the indexed category $\hat{\mathcal{P}} \circ (|\_| \times |\_|)$.

Example 27. Consider the NDA drawn on the next page with the accepting state z as a coalgebra $X \xrightarrow{\alpha} Act \times X + 1 \in \mathbf{Kl}(\mathcal{P})$.

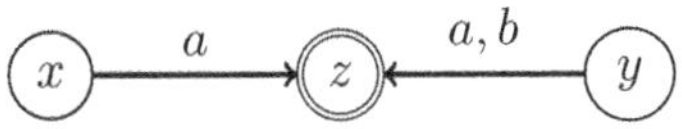

Logical equivalence $\simeq$ is the least equivalence R that equates $\{x, y\}$ with $\{y\}$ (both accept the language $\{a, b\}$) and $\{x, y, z\}$ with $\{y, z\}$ (both accept the language $\{a, b, \epsilon\}$). Also, for any $U, U' \subseteq X$ such that $U\ R\ U'$ we have $(U, U')\ p'_1\ x \iff x \in U$ and $(U, U')\ p'_2\ x \iff x \in U'$. Note that p'_i are transpose of p_i for $i \in \{1, 2\}$ (see (12)). So the equaliser $\bar{S}(X, \simeq)$ of $\mathfrak{r}p'_i$ is the set:

$$\bar{S}(X, \simeq) = \{W \in \mathcal{P}X \mid \forall_{U,V}\ (U \cap W \neq \emptyset \wedge U \simeq V) \implies V \cap W \neq \emptyset\}.$$

The arrow $\bar{S}(X, \simeq) \xrightarrow{\beta} F\bar{S}(X, \simeq) \in \mathbf{Set}^{\mathrm{op}}$ is defined by the following (depicted below) universal property of equaliser in $\mathbf{Set}$.

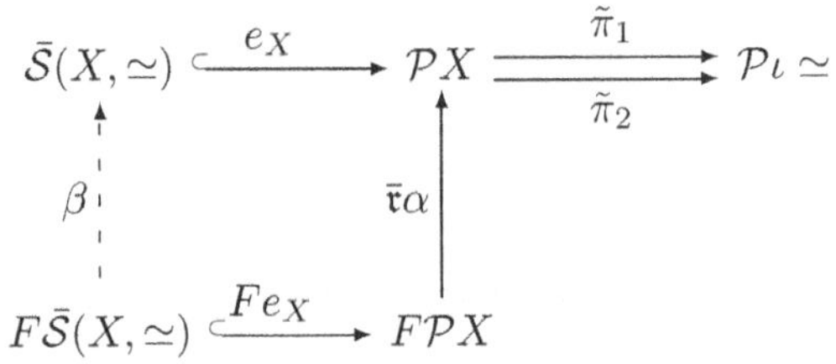

Here, $\bar{\mathfrak{r}}\alpha$ is the backward determinisation of the given coalgebra (as described, e.g. in [1] as a deterministic automaton accepting the reverse language), i.e. it maps $(a, U) \mapsto \{x \mid \exists_{x' \in U}(a, x') \in \alpha(x)\}$ and $\bullet \mapsto \{x \mid \bullet \in \alpha(x)\}$. Thus, in essence, β acts like $\bar{\mathfrak{r}}\alpha$ on the elements of $\bar{\mathcal{S}}(X, \simeq)$.

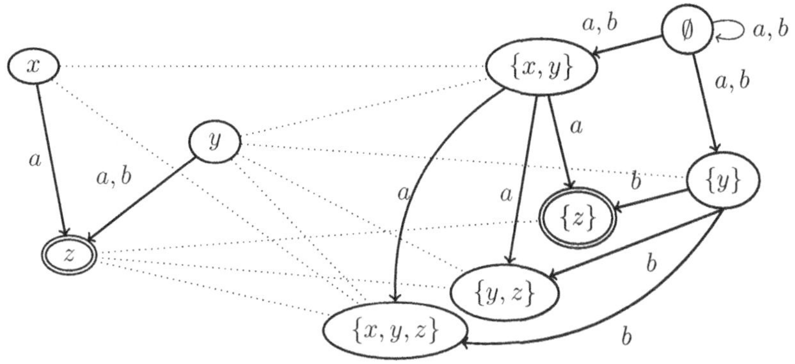

In this example, we obtain as β the automaton drawn above with six states. The relation κ_X indicated by dotted line is the transpose of e_X w.r.t. $\mathfrak{r} \dashv \jmath$; concretely, $x \ \kappa_X \ U \iff x \in U$. Furthermore $\kappa_X \in \mathbf{Rel}$ is a witnessing coalgebra homomorphism because $\bar{F}(\kappa_X) \circ \alpha = \beta \circ \kappa_X$. Note that $|\kappa_X|$ maps both $\{x, y\}, \{y\}$ to $\{\{x, y\}, \{y\}, \{y, z\}, \{x, y, z\}\}$, witnessing the fact that they are language equivalent. Hence, the coequaliser gives us the largest sub-automaton of the backwards determinisation that respects logical equivalence (removing $\emptyset$, $\{y, z\}$, and $\{x, y, z\}$ will result in the smallest such sub-automaton).

Predicate Liftings for NDAs. To apply techniques from Sect. 5, we set $G = \_^{Act} \times 2$ and define γ as follows:

$$\mathcal{P}(Act \times X + 1) \xrightarrow{\gamma_X} (\mathcal{P}X)^{Act} \times 2 \qquad\qquad \bar{U} \mapsto (\gamma_X^{Act}\bar{U}, \gamma_X^2\bar{U}),$$
$$\text{where } \gamma_X^{Act}\bar{U}(a) = \{x \mid (a, x) \in \bar{U}\} \qquad \text{and} \qquad \gamma_X^2\bar{U} = 1 \iff \bullet \in \bar{U}.$$

Moreover, from [19] we know that γ is compatible with θ and $\bigcup$ in the sense of Lemma 23. Next consider the family of liftings $\hat{\mathcal{P}}X \xrightarrow{\sigma_X^a} \hat{\mathcal{P}}(X^{Act} \times 2)$ (for each $a \in Act$) and $\hat{\mathcal{P}}X \xrightarrow{\sigma_X^{\downarrow}} \hat{\mathcal{P}}(X^{Act} \times 2)$: $U \mapsto \{(p, b) \in X^{Act} \times 2 \mid p(a) \in U\}$ and $U \mapsto \{(p, 1) \mid p \in X^{Act}\}$, respectively.

Lemma 28. *The above mappings σ_X^a and $\sigma_X^{\downarrow}$ are indexed morphisms.*

Thanks to Theorem 22, $\lambda^a = \gamma^{-1} \circ \sigma^a, \lambda^{\downarrow} = \gamma^{-1} \circ \sigma^{\downarrow}$ are valid predicate liftings for the functor $\mathbf{Rel} \xrightarrow{\overline{Act \times \_ + 1}} \mathbf{Rel}$. Moreover, for any $\mathbb{U} \subseteq \mathcal{P}X$ we find:

$$\begin{aligned} \lambda_X^a(\mathbb{U}) &= \gamma_X{}^{-1}\sigma_{\mathcal{P}X}^a(\mathbb{U}) \\ &= \gamma_X{}^{-1}\{(p, b) \in (\mathcal{P}X)^{Act} \times 2 \mid p(a) \in \mathbb{U}\} \\ &= \{\bar{U} \in Act \times X + 1 \mid \{x \mid (a, x) \in \bar{U}\} \in \mathbb{U}\} \end{aligned} \qquad \begin{aligned} \lambda_X^{\downarrow}(\mathbb{U}) &= \gamma_X{}^{-1}\sigma_{\mathcal{P}X}^{\downarrow}(\mathbb{U}) \\ &= \gamma_X{}^{-1}\{(p, 1) \mid p \in (\mathcal{P}X)^{Act}\} \\ &= \{\bar{U} \in Act \times X + 1 \mid \bullet \in \bar{U}\}. \end{aligned}$$

The above indexed morphisms $\lambda^a, \lambda^{\downarrow}$ induce modalities that can be interpreted on determinised automata. Given an NDA $(X, Act, \rightarrow, \downarrow)$ with $\downarrow \subseteq X$ being the termination predicate (or $X \xrightarrow{\alpha} \mathcal{P}(Act \times X + 1) \in \mathbf{Set}$), then the determinised automaton has state space $\mathcal{P}X$ with dynamics given by the SOS rules or abstractly by the composition $\gamma_X \circ \bigcup \circ \mathcal{P}\alpha$:

$$\frac{U \subseteq X \quad U_\alpha = \{x' \mid \exists_{x \in U}\; x \xrightarrow{a} x'\}}{U \xrightarrow{a} U_\alpha} \qquad \frac{U \subseteq X \qquad \exists_{x \in U}\; x \downarrow}{U \downarrow}.$$

In turn, we can now rewrite the two modalities to a simpler form:

$$\begin{aligned} |\alpha|^{-1}\lambda^a_X \mathbb{U} &= |\alpha|^{-1}\{\bar{U} \mid \{x \mid (a,x) \in \bar{U}\} \in \mathbb{U}\} & |\alpha|^{-1}\lambda^{\downarrow}_X \mathbb{U} &= |\alpha|^{-1}\{\bar{U} \mid \bullet \in \bar{U}\} \\ &= \{U \mid U \xrightarrow{a} U_\alpha \implies U_\alpha \in \mathbb{U}\} & &= \{U \mid U \downarrow\}. \end{aligned}$$

Language Equivalence Through Relation Lifting. First note that products exists in $\mathbf{Kl}(\mathcal{P})$ and are given by disjoint union. Moreover, Assumption **A4** is trivially satisfied since the forgetful functor preserves all limits (cf. Corollary 26). So we can create a relation lifting of $\bar{F}$ from the following relation lifting $\hat{\mathcal{P}}(X \times X) \xrightarrow{\bar{\sigma}_X} \hat{\mathcal{P}}(GX \times GX)$ of G (below $R \subseteq X \times X$):

$$(p,b)\; \bar{\sigma}_X R\; (p',b') \iff b = b' \wedge \forall_{a \in Act}\; pa\; R\; p'a.$$

Lemma 29. *The mapping $\bar{\sigma}$ defined above is an indexed morphism.*

So $\Psi \xrightarrow{\bar{\lambda}} \Psi\bar{F}$ given in (15) is an indexed morphism. Concretely, it maps a relation $R \subseteq \mathcal{P}X \times \mathcal{P}X$ to a relation $\bar{\lambda}_X R \subseteq \mathcal{P}FX \times \mathcal{P}FX$: $\bar{U}\; \bar{\lambda}_X R\; \bar{U}'$ iff

$$\left(\bullet \in \bar{U} \iff \bullet \in \bar{U}'\right) \wedge \left(\forall_{a \in Act}\; \{x \mid (a,x) \in \bar{U}\}\; R\; \{x \mid (a,x) \in \bar{U}'\}\right).$$

Lemma 30. *The indexed morphism $\bar{\sigma}$ preserves arbitrary intersections at each component; therefore, so does the predicate lifting $\bar{\lambda}$. Moreover, $\bar{\lambda}$ satisfies **A2**.*

Theorem 31. *Let $X \xrightarrow{\alpha} FX \in \mathbf{Rel}$ be an NDA. Then language equivalence $\simeq_X \subseteq \mathcal{P}X \times \mathcal{P}X$ on the determinised system is a $\Psi(\alpha) \circ \bar{\lambda}_X$-coalgebra, i.e., $\simeq_X \subseteq (|\alpha| \times |\alpha|)^{-1}(\bar{\lambda}_X \simeq_X)$. Moreover, $\simeq_X = (|f| \times |f|)^{-1} \simeq_Y$ for any coalgebra homomorphism f between (X, α) and (Y, β); thus, there is a behavioural conformance functor $\mathbf{Coalg}_{\mathbf{Rel}}(\bar{F}) \xrightarrow{\mathbb{1}^{\bar{\lambda}}} \mathbf{Coalg}_{\mathbb{E}(\Psi)}(\bar{F}_{\bar{\lambda}})$.*

Logical Characterisation of Language Equivalence. Recall the adjoint situation $\mathfrak{r} \dashv \jmath$ that witnesses $\mathbf{Set}^{\mathrm{op}}$ is a reflective subcategory of $\mathbf{Rel}$. We use this dual adjunction to model our logic because (intuitively) conjunction is not needed to characterise language equivalence. Thus we fix $\mathbf{A} = \mathbf{Set}$, $S = \mathfrak{r}$, and $T = \jmath$. Moreover, a left adjoint $\bar{S}$ of $\bar{T}$ exists due to Theorem 21.

Since to establish language equivalence one needs to ascertain whether a word in $Act^\star$ is accepting or not, so we take our syntax functor $L = Act \times \_ + 1$. Note that the initial algebra of L exists and is given by $\mathcal{A} = Act^\star$. As for the one-step semantics given by a natural transformation δ, we are going to define it (indirectly) by defining its mate $LSX = Act \times \mathcal{P}X + 1 \xrightarrow{\theta_X} \mathcal{P}(Act \times X + 1) = \mathcal{S}\bar{F}X \in \mathbf{Set}$ that acts on objects like the distributive law ϑ_X (see (3)). Note that, however, they differ in their naturality conditions.

Proposition 32. *The above defined mapping θ is a natural transformation.*

The algebra $Act \times Act^\star + 1 \xrightarrow{h} Act^\star$ is given by the unary concatenation of words and the constant ε (i.e., $h(a, w) = aw$ and $h\bullet = \varepsilon$). Consider the map $X \xrightarrow{(\!|\_|\!)} \mathcal{A} \in \mathbf{Rel}$ that maps a state to the language accepted by it.

Corollary 33. *The above map $(\!|\_|\!)$ is indeed the theory map for a given NDA. So the logic (L, δ) is both adequate and expressive for language equivalence on determinised systems.*

7 Conditional Transition Systems: An Application in coKleisli Categories

We next consider conditional transition systems (CTSs) [1], strongly related to the featured transition systems used for modelling software product lines [8]. A conditional transition system is a compact representation of several transition systems – one for each condition or product – where transitions are labelled by products. Here we consider the simpler case of conditional transition systems without upgrades and action labels; their full treatment [4,5] is left for the future. In our earlier work, CTSs were coalgebras living in the Kleisli category induced by the reader monad $\_^{\mathbb{K}}$ (for a fixed set of conditions $\mathbb{K}$). It is however more convenient to treat CTSs as coalgebras in coKleisli categories, hence we start with a relevant comonad $G = \mathbb{K} \times \_$ whose counit is given by the projection of the second component and comultiplication Δ is given by the diagonal map, i.e. the map $\mathbb{K} \times X \xrightarrow{\Delta_X} \mathbb{K} \times \mathbb{K} \times X$ is given by $(k, x) \mapsto (k, k, x)$ (for $k \in \mathbb{K}, x \in X$).

Consider the coKleisli category $\mathbf{coKl}(G)$ whose objects, just like in any Kleisli category, are sets; an arrow $X \xrightarrow{f} Y$ corresponds to a function $GX \xrightarrow{f} Y$. Now there is a forgetful functor $\mathbf{coKl}(G) \xrightarrow{|\text{-}|} \mathbf{Set}$; however, in contrast to the Kleisli setting, it is now left adjoint to the inclusion $\mathbf{Set} \xrightarrow{\iota} \mathbf{coKl}(G)$. Concretely, this forgetful functor maps an object $X \mapsto GX$ and an arrow $X \xrightarrow{f} Y \in \mathbf{coKl}(G)$ to a function $|f|$ mapping $(k, x) \mapsto (k, f(k, x))$.

Next to model the branching type of CTSs, take $\bar{F}$ to be the coKleisli extension [5] of $F = \mathcal{P}$ given by the following distributive law $\mathbb{K} \times \mathcal{P}X \xrightarrow{\gamma_X} \mathcal{P}(\mathbb{K} \times X)$: $\gamma_X(k, U) = \{k\} \times U$; concretely, $\bar{F}X = \mathcal{P}X$ and $\bar{F}f(k, U) = \{f(k, x) \mid x \in U\}$ for $X \xrightarrow{f} Y \in \mathbf{coKl}(G)$. A CTS modelled as a coalgebra $X \xrightarrow{\alpha} \bar{F}X \in \mathbf{coKl}(G)$ is a function $\mathbb{K} \times X \xrightarrow{\alpha} \mathcal{P}X$ that assigns to each state $x \in X$ and each condition $k \in \mathbb{K}$, the successors of x under condition k.

As mentioned earlier in Sect. 5, it is easier (than in the Kleisli case) to lift a predicate lifting $\hat{\mathcal{P}} \xrightarrow{\sigma} \hat{\mathcal{P}}F$ of F to define a predicate lifting $\hat{\mathcal{P}}|\_| \xrightarrow{\lambda} \hat{\mathcal{P}}|\bar{F}\_|$ of

$\bar{F}$. In particular, λ is given by the composition in (13) and, moreover, Theorem 22 ensures that λ is indeed an indexed morphism once we have fixed the predicate lifting σ of F. To this end, we simply take σ that corresponds to the box modality (cf. Example 7). To answer whether these definitions give the right kind of 'box' modality for CTSs, let us first instantiate λ_X for any $U \subseteq \mathbb{K} \times X$:

$$\lambda_X U = \gamma_X^{-1} \sigma_X U = \gamma_X^{-1} \{U' \mid U' \subseteq U\} = \{(k, U') \mid \{k\} \times U' \subseteq U\}. \tag{16}$$

Now given a coalgebra $\mathbb{K} \times X \xrightarrow{\alpha} \mathcal{P}X \in \mathbf{Set}$, we derive the interpretation of box modality for CTSs in the following way (below $x \xrightarrow{k} x' \iff x' \in \alpha x k$):

$$|\alpha|^{-1} \lambda_X U = \{(k, x) \mid \forall_{x'}\ x \xrightarrow{k} x' \implies (k, x') \in U\}. \tag{17}$$

Conditional Bisimilarity Through Relation Lifting
Next we introduce conditional bisimilarity in which two states might be bisimilar for all conditions or only under certain conditions.

Definition 34. *Given a CTS* $X \xrightarrow{\alpha} \bar{\mathcal{P}}X \in \mathbf{coKl}(G)$*, a* conditional bisimulation *is a relation* $R \subseteq (\mathbb{K} \times X) \times (\mathbb{K} \times X)$ *satisfying:*

1 $\forall_{k,k' \in \mathbb{K}, x, x' \in X}\ (k, x)\ R\ (k', x') \implies k = k'$.
2 $\forall_{x_1, x_2, x_3, k} \big(x_1 \xrightarrow{k} x_3 \wedge (k, x_1)\ R\ (k, x_2)\big) \Rightarrow \exists_{x_4 \in X} \big(x_2 \xrightarrow{k} x_4 \wedge (k, x_3)\ R\ (k, x_4)\big)$.

Two states $x, x' \in X$ *are* conditional bisimilar under k *iff there is a conditional bismulation relation* R *such that* $(k, x)\ R\ (k, x')$*. Moreover, two states* x, x' *are* conditional bisimilar*, denoted* $x \leftrightarroweq_X x'$*, iff* x *and* x' *are conditional bisimilar under every condition* $k \in \mathbb{K}$*.*

In order to capture conditional bisimilarity, we first need a fibration Ψ of binary relations on the state space. The first choice for Ψ is to consider the set of all binary relations on the underlying state space, i.e., $\Psi = \hat{\mathcal{P}}(|\_| \times |\_|)$. Unfortunately, Assumption **A3** fails to hold which we explain next.

Remark 35. We argue that Eq cannot have a left adjoint since it does not preserve finite limits (in particular, terminal objects). Clearly, $\iota 1 = 1 = \{\bullet\}$ is the terminal object in $\mathbf{coKl}(G)$ because ι is the right adjoint of $|\_|$. Now suppose $\mathrm{Eq}1 = (1, =_{\mathbb{K} \times 1})$ is the terminal object in $\mathbb{E}(\Psi)$. Then, for any (X, R), there is a unique arrow $(X, R) \xrightarrow{!_X} \mathrm{Eq}1$, i.e., $X \xrightarrow{!_X} 1 \in \mathbf{coKl}(G)$ and $R \subseteq (|!_X| \times |!_X|)^{-1} =_{\mathbb{K} \times 1}$. But we argue that $!_X$ is *not* a map in $\mathbb{E}(\Psi)$ for $|\mathbb{K}| \geq 2$. To see this, let $k, k' \in \mathbb{K}$ with $k \neq k'$ and let $R \subseteq (\mathbb{K} \times X) \times (\mathbb{K} \times X)$ be an equivalence relation such that $(k, x)\ R\ (k', x)$. Then we find a contradiction

$$R \subseteq (|!_X| \times |!_X|)^{-1} =_{\mathbb{K} \times 1} \implies (k, !_X(k, x)) = (k', !_X(k', x)) \implies k = k'.$$

So we need to restrict ourselves to relations satisfying the first property of conditional bisimulation (see Definition 34). An elegant way to address this is by working with the indexed category of abstract relations given in (5); thus, enabling the applicability of constructions given in Sect. 5. Note that binary products $\otimes$ exist in $\mathbf{coKl}(G)$ and is given by the Cartesian product of two sets. Thus Ψ in (5) is well defined.

Lemma 36. *Assumptions* ***A3*** *and* ***A4*** *are valid.*

Now to invoke the definition of $\bar{\lambda}$ in (15), it suffices to define a relation lifting σ of the set endofunctor $F = \mathcal{P}$. We take σ to be the relation lifting associated with bisimulation relations as defined in Example 15.

Theorem 37. *Alternatively, the indexed morphism* $\Psi \xrightarrow{\bar{\lambda}} \Psi\bar{\mathcal{P}}$ *given in* (15) *can be defined as follows:* $\bar{\lambda}_X R = \{(k, U, U') \mid \forall_{x \in U} \exists_{x' \in U'} (k, x, x') \in R \wedge \forall_{x' \in U'} \exists_{x \in U} (k, x, x') \in R\}$. *Moreover,* $\bar{\lambda}$ *satisfies Assumption* ***A2****.*

Corollary 38. *Let* $X \xrightarrow{\alpha} \mathcal{P}X \in \mathbf{coKl}(G)$ *be a CTS. A relation* R *on* $\mathbb{K} \times X$ *is a conditional bisimulation iff* $\langle \pi_1^X, \pi_2^X \rangle^{-1} R$ *is an* $\alpha^* \circ \bar{\lambda}$*-coalgebra in* ΨX*. Moreover, for any* $(X, \alpha) \xrightarrow{f} (Y, \beta) \in \mathbf{coKl}(G)$ *we have* $\underline{\leftrightarrow}_X = (|f| \times |f|)^{-1} \underline{\leftrightarrow}_Y$*; thus, there is a functor* $\mathbf{Coalg}_{\mathbf{coKl}(G)}(\bar{\mathcal{P}}) \xrightarrow{\mathbb{1}^{\bar{\lambda}}} \mathbf{Coalg}_{\mathbb{E}(\Psi)}(\bar{\mathcal{P}}_{\bar{\lambda}})$.

Modal Characterisation of Conditional Bisimilarity

$$\mathbf{coKl}(\mathbb{K} \times \_) \underset{\iota}{\overset{|\_|}{\rightleftarrows}} \mathbf{Set} \underset{\mathcal{T}}{\overset{\mathcal{S}}{\rightleftarrows}} \mathbf{BA}^{\mathrm{op}} \qquad (\text{both } \perp)$$

Consider the above adjoint situations where the adjoint situation on the right is the well known duality (see, for instance [27]) between $\mathbf{Set}$ and the opposite category of Boolean algebras $\mathbf{BA}$; $\mathcal{S}$ is the contravariant powerset functor $\hat{\mathcal{P}}$ and $\mathcal{T}$ maps a Boolean algebra to its set of ultrafilters. We follow [20] and use the proposed syntax functor $\mathbf{BA} \xrightarrow{L} \mathbf{BA}$ and the interpretation $\mathcal{P}_\omega \mathcal{T} \xrightarrow{\delta} \mathcal{T}L$ induced by the box modality on (unlabelled) transition systems. I.e., the initial algebra $\mathcal{A}$ of L can be viewed as the Lindenbaum-Tarski algebra generated by the following grammar:

$$\top \mid \perp \mid \neg\varphi \mid \varphi \wedge \varphi' \mid \Box\varphi$$

together with the axioms of 'propositional' logic and the following ones: $\Box\top = \top$ and $\Box(\varphi \wedge \varphi') = \Box\varphi \wedge \Box\varphi'$.

Note that the above logic with finite conjunctions is expressive (i.e., δ is injective [20]) only for image-finite transition systems, so we restrict F to be finite powerset functor $\mathcal{P}_\omega$. Moreover, since $\bar{\mathcal{P}}_\omega$ is a coKleisli extension of $\mathcal{P}_\omega$, i.e., $\bar{\mathcal{P}}_\omega \circ \iota = \iota \circ \mathcal{P}_\omega$, we consider the following logical interpretation

$$\bar{\mathcal{P}} \circ \iota \circ \mathcal{T} = \iota \circ \mathcal{P} \circ \mathcal{T} \xrightarrow{\iota\delta} \iota \circ \mathcal{T} \circ L$$

for image-finite CTSs (i.e., coalgebras of type $X \longrightarrow \bar{\mathcal{P}}_\omega \in \mathbf{coKl}(G)$).

Corollary 39. *Since the lifting $\bar{\lambda}$ preserves equalities, the logic $(L, \iota\delta)$ defined above is adequate for conditional bisimilarity. Moreover, since δ is injective in each component [20] when $F = \mathcal{P}_\omega$, so the function $|\iota\delta_{\mathcal{A}}|$ is injective for $\bar{F} = \bar{\mathcal{P}}_\omega$. Thus the logic $(L, \iota\delta)$ is expressive for image-finite CTSs.*

8 Conclusions

To recapitulate, we gave a systematic way to construct both predicate and relation liftings in (co)Kleisli categories and Eilenberg-Moore categories. Relation liftings form the basis to define behavioural equivalence as a coalgebra of certain lifted endofunctor in the fibre of relations, although in some cases (such as CTSs) such fibres can be subtle to define.

Once behavioural equivalence is captured as a fibred notion, the Kupke and Rot setup becomes applicable to obtain its corresponding logical characterisation. In particular, we gave a recipe to find the left adjoint $\bar{S}$ of $\bar{T}$ which is a sufficient condition for both adequacy and expressivity. For coKleisli and Eilenberg-Moore categories, the construction (11) of $\bar{S}$ is based on the existence of coequalisers in the underlying categories, while in the Kleisli case one has to resort to a reflective subcategory having coequalisers (cf. Theorem 21).

In the future, we plan to develop extend the 2-player game [26] to coalgebras with side effects. Lastly, we also plan to investigate whether the given recipe of constructing predicate/relation liftings can be extended to more general monads (like the ones on pseudometric spaces). This should help in developing quantitative modal logics for coalgebras with side effects; thus, providing a pertinent litmus test for the categorical unification of quantitiative expressivity as claimed in the recent work [25].

References

1. Adámek, J., Bonchi, F., Hülsbusch, M., König, B., Milius, S., Silva, A.: A coalgebraic perspective on minimization and determinization. In: Birkedal, L. (ed.) FoSSaCS 2012. LNCS, vol. 7213, pp. 58–73. Springer, Heidelberg (2012). https://doi.org/10.1007/978-3-642-28729-9_4
2. Adámek, J., Herrlich, H., Strecker, G.E.: Abstract and Concrete Categories: The Joy of Cats. No. 17, Reprints in Theory and Applications of Categories (2006). Originally published by: Wiley, New York (1990)
3. Beohar, H., König, B., Küpper, S., Mika-Michalski, C.: Predicate, relation liftings and modal logics for coalgebras with side effects. CoRR abs/2110.09911 (2021). https://arxiv.org/abs/2110.09911
4. Beohar, H., König, B., Küpper, S., Silva, A.: Conditional transition systems with upgrades. Sci. Comput. Program. **186**, 102320 (2020). https://doi.org/10.1016/j.scico.2019.102320
5. Beohar, H., König, B., Küpper, S., Silva, A., Wißmann, T.: A coalgebraic treatment of conditional transition systems with upgrades. Log. Methods Comput. Sci. **14**(1) (2018). https://doi.org/10.23638/LMCS-14(1:19)2018. https://lmcs.episciences.org/4330

6. Bonchi, F., König, B., Petrisan, D.: Up-to techniques for behavioural metrics via fibrations. In: Schewe, S., Zhang, L. (eds.) 29th International Conference on Concurrency Theory (CONCUR 2018). Leibniz International Proceedings in Informatics (LIPIcs), vol. 118, pp. 17:1–17:17. Schloss Dagstuhl-Leibniz-Zentrum fuer Informatik, Dagstuhl (2018). https://doi.org/10.4230/LIPIcs.CONCUR.2018.17. http://drops.dagstuhl.de/opus/volltexte/2018/9555
7. Bonchi, F., Petrişan, D., Pous, D., Rot, J.: Coinduction up-to in a fibrational setting. In: Proceedings of the Joint Meeting of the Twenty-Third EACSL Annual Conference on Computer Science Logic (CSL) and the Twenty-Ninth Annual ACM/IEEE Symposium on Logic in Computer Science (LICS), CSL-LICS 2014. Association for Computing Machinery, New York (2014). https://doi.org/10.1145/2603088.2603149
8. Cordy, M., Classen, A., Perrouin, G., Schobbens, P.Y., Heymans, P., Legay, A.: Simulation-based abstractions for software product-line model checking. In: Proceedings of ICSE 2012. International Conference on Software Engineering, pp. 672–682. IEEE (2012)
9. Dorsch, U., Milius, S., Schröder, L.: Graded monads and graded logics for the linear time - branching time spectrum. In: Fokkink, W., van Glabbeek, R.J. (eds.) 30th International Conference on Concurrency Theory (CONCUR 2019). Leibniz International Proceedings in Informatics (LIPIcs), vol. 140, pp. 36:1–36:16. Schloss Dagstuhl-Leibniz-Zentrum fuer Informatik, Dagstuhl (2019). https://doi.org/10.4230/LIPIcs.CONCUR.2019.36. http://drops.dagstuhl.de/opus/volltexte/2019/10938
10. Ghani, N., Johann, P., Fumex, C.: Indexed induction and coinduction, fibrationally. Log. Methods Comput. Sci. **9**(3) (2013). https://doi.org/10.2168/LMCS-9(3:6)2013. https://lmcs.episciences.org/738
11. Gorín, D., Schröder, L.: Simulations and bisimulations for coalgebraic modal logics. In: Heckel, R., Milius, S. (eds.) CALCO 2013. LNCS, vol. 8089, pp. 253–266. Springer, Heidelberg (2013). https://doi.org/10.1007/978-3-642-40206-7_19
12. Hasuo, I., Jacobs, B.P.F., Sokolova, A.: Generic trace semantics via coinduction. Log. Methods Comput. Sci. **3**(4) (2007). https://doi.org/10.2168/LMCS-3(4:11)2007. https://lmcs.episciences.org/864
13. Hasuo, I., Kataoka, T., Cho, K.: Coinductive predicates and final sequences in a fibration. Math. Struct. Comput. Sci. **28**(4), 562–611 (2018). https://doi.org/10.1017/S0960129517000056
14. Hermida, C.A., Jacobs, B.P.F.: Structural induction and coinduction in a fibrational setting. Inf. Comput. **145**(2), 107–152 (1998)
15. Jacobs, B.P.F.: Categorical Logic and Type Theory, Studies in Logic and the Foundations of Mathematics, vol. 141, 1st edn. Elsevier, Amsterdam (1999)
16. Jacobs, B.P.F.: Trace semantics for coalgebras. Electron. Notes Theor. Comput. Sci. **106**, 167–184 (2004). Proceedings of the Workshop on Coalgebraic Methods in Computer Science (CMCS)
17. Jacobs, B.P.F.: Predicate logic for functors and monads (2010). http://www.cs.ru.nl/~bart/PAPERS/predlift-indcat.pdf
18. Jacobs, B.P.F.: Introduction to Coalgebra: Towards Mathematics of States and Observation. Cambridge Tracts in Theoretical Computer Science, Cambridge University Press, Cambridge (2016). https://doi.org/10.1017/CBO9781316823187
19. Jacobs, B.P.F., Silva, A., Sokolova, A.: Trace semantics via determinization. J. Comput. Syst. Sci. **81**(5), 859–879 (2015). 11th International Workshop on Coalgebraic Methods in Computer Science, CMCS 2012 (Selected Papers)

20. Jacobs, B.P.F., Sokolova, A.: Exemplaric expressivity of modal logics. J. Log. Comput. **20**(5), 1041–1068 (2010)
21. Kissig, C., Kurz, A.: Generic trace logics (2011). arXiv:1103.3239
22. Klin, B.: The least fibred lifting and the expressivity of coalgebraic modal logic. In: Fiadeiro, J.L., Harman, N., Roggenbach, M., Rutten, J. (eds.) CALCO 2005. LNCS, vol. 3629, pp. 247–262. Springer, Heidelberg (2005). https://doi.org/10.1007/11548133_16
23. Klin, B., Rot, J.: Coalgebraic trace semantics via forgetful logics. Log. Methods Comput. Sci. **12**(4) (2017). https://doi.org/10.2168/LMCS-12(4:10)2016. https://lmcs.episciences.org/2622
24. Komorida, Y., Katsumata, S., Hu, N., Klin, B., Hasuo, I.: Codensity games for bisimilarity. In: 2019 34th Annual ACM/IEEE Symposium on Logic in Computer Science (LICS), pp. 1–13 (2019). https://doi.org/10.1109/LICS.2019.8785691
25. Komorida, Y., Katsumata, S.Y., Kupke, C., Rot, J., Hasuo, I.: Expressivity of quantitative modal logics: categorical foundations via codensity and approximation. Accepted for publication in LICS 2021, June 2021
26. König, B., Mika-Michalski, C.: (Metric) bisimulation games and real-valued modal logics for coalgebras. In: Proceedings of CONCUR 2018. LIPIcs, vol. 118, pp. 37:1–37:17. Schloss Dagstuhl - Leibniz Center for Informatics (2018)
27. Kupke, C., Pattinson, D.: Coalgebraic semantics of modal logics: an overview. Theor. Comput. Sci. **412**(38), 5070–5094 (2011). https://doi.org/10.1016/j.tcs.2011.04.023. CMCS Tenth Anniversary Meeting
28. Kupke, C., Rot, J.: Expressive logics for coinductive predicates. In: Fernández, M., Muscholl, A. (eds.) 28th EACSL Annual Conference on Computer Science Logic (CSL 2020). Leibniz International Proceedings in Informatics (LIPIcs), vol. 152, pp. 26:1–26:18. Schloss Dagstuhl-Leibniz-Zentrum fuer Informatik, Dagstuhl (2020). https://doi.org/10.4230/LIPIcs.CSL.2020.26
29. Mulry, P.S.: Lifting theorems for Kleisli categories. In: Brookes, S., Main, M., Melton, A., Mislove, M., Schmidt, D. (eds.) MFPS 1993. LNCS, vol. 802, pp. 304–319. Springer, Heidelberg (1994). https://doi.org/10.1007/3-540-58027-1_15
30. Pattinson, D.: Coalgebraic modal logic: soundness, completeness and decidability of local consequence. Theor. Comput. Sci. **309**(1), 177–193 (2003)
31. Pavlovic, D., Mislove, M., Worrell, J.B.: Testing semantics: connecting processes and process logics. In: Johnson, M., Vene, V. (eds.) AMAST 2006. LNCS, vol. 4019, pp. 308–322. Springer, Heidelberg (2006). https://doi.org/10.1007/11784180_24
32. Rutten, J.: Universal coalgebra: a theory of systems. Theor. Comput. Sci. **249**(1), 3–80 (2000)
33. Schröder, L.: Expressivity of coalgebraic modal logic: the limits and beyond. Theor. Comput. Sci. **390**(2), 230–247 (2008)
34. Sprunger, D., Katsumata, S., Dubut, J., Hasuo, I.: Fibrational bisimulations and quantitative reasoning. In: Cîrstea, C. (ed.) CMCS 2018. LNCS, vol. 11202, pp. 190–213. Springer, Cham (2018). https://doi.org/10.1007/978-3-030-00389-0_11

Discrete Density Comonads and Graph Parameters

Samson Abramsky[1], Tomáš Jakl[2(✉)], and Thomas Paine[3]

[1] University College London, London, UK
s.abramsky@ucl.ac.uk
[2] University of Cambridge, Cambridge, UK
tomas.jakl@cl.cam.ac.uk
[3] University of Oxford, Oxford, UK
thomas.paine@stcatz.ox.ac.uk

Abstract. Game comonads have brought forth a new approach to studying finite model theory categorically. By representing model comparison games semantically as comonads, they allow important logical and combinatorial properties to be exressed in terms of their Eilenberg-Moore coalgebras. As a result, a number of results from finite model theory, such as preservation theorems and homomorphism counting theorems, have been formalised and parameterised by comonads, giving rise to new results simply by varying the comonad.

In this paper we study the limits of the comonadic approach in the combinatorial and homomorphism-counting aspect of the theory, regardless of whether any model comparison games are involved. We show that any standard graph parameter has a corresponding comonad, classifying the same class. This comonad is constructed via a simple Kan extension formula, making it the initial solution to this problem and, furthermore, automatically admitting a homomorphism-counting theorem.

Keywords: density comonads · graph parameters · Lovász' theorem

1 Introduction

An important feature of the emerging theory of game comonads [1,2,4,12] is that game comonads classify a number of important classes of finite relational structures. We say that a comonad $\mathbb{C}$ *classifies a class* Δ if a finite structure A is in the class Δ precisely when A admits a $\mathbb{C}$-coalgebra. For example, the Ehrenfeucht-Fraïssé comonad $\mathbb{E}_k$ classifies the structures of tree-depth $\leq k$ and, similarly, the pebbling comonad $\mathbb{P}_k$ classifies tree-width $<k$.

In this paper we study the theoretical limits of the comonadic approach. In particular, we aim to identify classes of structures which can be classified by comonads. We can readily predict two necessary requirements. Since the problem

Supported by EPSRC grant EP/T00696X/1: Resources and Coresources: a junction between categorical semantics and descriptive complexity.

Published by Springer Nature Switzerland AG 2022
H. H. Hansen and F. Zanasi (Eds.): CMCS 2022, LNCS 13225, pp. 23–44, 2022.
https://doi.org/10.1007/978-3-031-10736-8_2

is stated in the language of category theory, we know that the classes of structures classified by comonads need to be closed under isomorphisms and, moreover, since finite coalgebras are closed under binary coproducts + (i.e. disjoint unions), this must also be the case for the classes classifiable by comonads.

In fact, we show that one further, very natural requirement suffices in order to be able to classify a class of structures by a comonad. It suffices to assume that the class is closed under connected substructures.

We define a class Δ of finite relational structures or graphs to be *component-based* if it is closed under

- isomorphisms,
- finite coproducts, and
- summands (i.e. if $A + B$ is in Δ then so are A and B).

We can now state our first main result.

Theorem 1. *Any* component-based *class Δ can be classified by a comonad.*

The theorem applies to a wide variety of classes of structures studied in the literature. In particular, these assumptions hold for all classes of finite structures classified by our game comonads and, moreover, for a number of typical examples of classes of structures for which a given graph parameter is bounded by a constant. For example, we obtain comonads for planar graphs, bipartite graphs, or graphs of degree or clique-width bounded by a constant. Moreover, we show that the constructed comonad $\mathbb{C}$ is weakly initial among the comonads classifying Δ, meaning that for any comonad $\mathbb{D}$ classifying Δ, there is a comonad morphism $\mathbb{C} \Rightarrow \mathbb{D}$. This initiality allows us to obtain a characterisation of comonads that classify monotone *nowhere dense classes* [27].

Another important aspect of game comonads is that they classify various well-known binary relations between relational structures. We say that a comonad $\mathbb{C}$ *classifies relation* $\asymp$ whenever $A \asymp B$ holds precisely when the cofree $\mathbb{C}$-coalgebras on A and B are isomorphic. For example, the comonad $\mathbb{E}_k$ classifies the relation that expresses that Duplicator has a winning strategy in the bijective k-round variant of the Ehrenfeucht-Fraïssé game [4] and, similarly, $\mathbb{P}_k$ classifies the existence of a winning strategy in the bijective k-pebble game [1]. Furthermore, it was recently shown that the relation classified by $\mathbb{E}_k$ admits a Lovász-type theorem. In particular, finite structures A, B have isomorphic cofree $\mathbb{E}_k$-coalgebras if, and only if, they admit the same homomorphism counts from finite structures of tree-depth $\leq k$, i.e. when there is a bijection between $\hom(C, A)$ and $\hom(C, B)$ for every finite C of tree-depth $\leq k$. Similar Lovász-type theorems have also been shown for $\mathbb{P}_k$ and the pebble-relation $\mathbb{PR}_k$ comonads [13,26].

We show that the comonad constructed in the proof of Theorem 1 automatically admits a Lovász-type theorem for the class of structures it classifies. In fact, we show that such a comonad always has *finite rank*[1], which ensures that

[1] Comonads of finite rank should not be confused with finitary comonads, which is a weaker notion.

the category of coalgebras for the comonad is locally finitely presentable (cf. [14, Proposition 1.12.1], see also [29, Appendix B]) and therefore, by a recent result of Luca Reggio [29, Corollary 5.15], admits the following Lovász-type result.

Corollary 2. *Let Δ be a component-based class of finite structures and let $\asymp$ be the binary relation on finite structures such that, for any two finite structures A, B,*

$$A \asymp B \iff \hom(C, A) \cong \hom(C, B) \quad \textit{for all } C \in \Delta$$

The comonad classifying Δ by Theorem 1 also classifies $\asymp$.

As an example, we obtain that the comonad obtained by Theorem 1 which classifies planar graphs also classifies quantum isomorphism (cf. [25]), and similarly, the comonad for coproducts of cycles classifies co-spectrality, and the comonad for bipartite graphs classifies isomorphic bipartite double covers.

The density comonad construction is the main technical tool of this paper. In fact, we develop most of our theory by means of discrete density comonads, that is, density comonads of functors with discrete domain. A general overview of the necessary categorical terminology and results is given in Sect. 2. Discrete density comonads are introduced in Sect. 3 and Theorem 1 is proved in Sect. 4. In Sect. 5 we take a look at how graph parameters correspond to coalgebra numbers of graded comonads, compare discrete density comonads with game comonads, and characterise comonads classifying monotone nowhere dense classes of graphs. Lastly, in Sect. 6 we prove Corollary 2 by showing that, under mild conditions, discrete density comonads have finite rank.

2 Preliminaries

In this section we fix notation and recall some basic facts about comonads and the density construction. We assume the reader is familiar with elementary category theory notions such as functors, natural transformations, adjunctions, limits and colimits (see e.g. [5] or [9]).

Throughout the paper we use the following notation. Given a natural transformation $\lambda\colon E \Rightarrow F$ between functors $E, F\colon \mathcal{A} \to \mathcal{B}$ and functors $G\colon \mathcal{B} \to \mathcal{B}'$ and $H\colon \mathcal{A}' \to \mathcal{A}$, we denote by $G\lambda$ and λH the obvious natural transformations of type $GE \Rightarrow GF$ and $EH \Rightarrow FH$, respectively.

2.1 Comonads and Coalgebras

A *comonad* (on category $\mathcal{A}$) is a triple $(\mathbb{C}, \varepsilon, \delta)$ where $\mathbb{C}\colon \mathcal{A} \to \mathcal{A}$ is an endofunctor, and $\varepsilon\colon \mathbb{C} \Rightarrow \mathrm{Id}$ and $\delta\colon \mathbb{C} \Rightarrow \mathbb{C}^2$ are natural transformations such that the following diagrams commute.

$$\begin{array}{ccc} \mathbb{C} & \xrightarrow{\delta} & \mathbb{C}^2 \\ {\scriptstyle\delta}\Downarrow & & \Downarrow{\scriptstyle\delta\mathbb{C}} \\ \mathbb{C}^2 & \xrightarrow[\mathbb{C}\delta]{} & \mathbb{C}^3 \end{array} \qquad \begin{array}{ccc} \mathbb{C} & \xrightarrow{\delta} & \mathbb{C}^2 \\ & {\scriptstyle\mathrm{id}}\searrow & \Downarrow{\scriptstyle\varepsilon\mathbb{C}} \\ & & \mathbb{C} \end{array} \qquad \begin{array}{ccc} \mathbb{C} & \xrightarrow{\delta} & \mathbb{C}^2 \\ & {\scriptstyle\mathrm{id}}\searrow & \Downarrow{\scriptstyle\mathbb{C}\varepsilon} \\ & & \mathbb{C} \end{array}$$

A morphism $\alpha\colon A \to \mathbb{C}(A)$ is an *(Eilenberg–Moore)* $\mathbb{C}$ *-coalgebra*[2] if the following diagrams commute.

$$\begin{array}{ccc} A & & \\ {\scriptstyle\alpha}\big\downarrow & \searrow^{\mathrm{id}} & \\ \mathbb{C}(A) & \xrightarrow[\varepsilon_A]{} & A \end{array} \qquad\qquad \begin{array}{ccc} A & \xrightarrow{\alpha} & \mathbb{C}^2(A) \\ {\scriptstyle\alpha}\big\downarrow & & \big\downarrow{\scriptstyle\delta_A} \\ \mathbb{C}(A) & \xrightarrow[\mathbb{C}\alpha]{} & \mathbb{C}^2(A) \end{array} \tag{1}$$

We say that A *admits a coalgebra* if there exists a morphism $A \to \mathbb{C}(A)$ which is a $\mathbb{C}$-coalgebra.

Coalgebras form a category $\mathtt{EM}(\mathbb{C})$ where morphisms between coalgebras $(A, \alpha) \to (B, \beta)$ are morphisms $h\colon A \to B$ such that $\beta \circ h = \mathbb{C}(h) \circ \alpha$. Moreover, there is a pair of adjoint functors

$$U^{\mathbb{C}}\colon \mathtt{EM}(\mathbb{C}) \to \mathcal{A} \quad \text{and} \quad F^{\mathbb{C}}\colon \mathcal{A} \to \mathtt{EM}(\mathbb{C})$$

between $\mathtt{EM}(\mathbb{C})$ and the underlying category $\mathcal{A}$. The left adjoint is just a forgetful functor, it sends a coalgebra $\alpha\colon A \to \mathbb{C}(A)$ to its underlining object $U^{\mathbb{C}}(A, \alpha) = A$. The right adjoint returns the *cofree coalgebra* $F^{\mathbb{C}}(A)$ *on* A, represented by the morphism $\delta_A\colon \mathbb{C}(A) \to \mathbb{C}^2(A)$.

2.2 Comonad Morphisms

Given two comonads $(\mathbb{C}, \varepsilon^{\mathbb{C}}, \delta^{\mathbb{C}})$ and $(\mathbb{D}, \varepsilon^{\mathbb{D}}, \delta^{\mathbb{D}})$ on $\mathcal{A}$, a natural transformation $\lambda\colon \mathbb{C} \Rightarrow \mathbb{D}$ is a *comonad morphism* if the following two diagrams of natural transformations commute.

$$\begin{array}{ccc} \mathbb{C} & \xRightarrow{\lambda} & \mathbb{D} \\ & {\scriptstyle\varepsilon^{\mathbb{C}}}\searrow \quad \swarrow{\scriptstyle\varepsilon^{\mathbb{C}}} & \\ & \mathrm{Id} & \end{array} \qquad\qquad \begin{array}{ccccc} \mathbb{C} & & \xRightarrow{\lambda} & & \mathbb{D} \\ {\scriptstyle\delta^{\mathbb{C}}}\Big\Downarrow & & & & \Big\Downarrow{\scriptstyle\delta^{\mathbb{D}}} \\ \mathbb{C}^2 & \xRightarrow{\mathbb{C}\lambda} & \mathbb{C}\mathbb{D} & \xRightarrow{\lambda\mathbb{D}} & \mathbb{D}^2 \end{array} \tag{2}$$

Note that comonad morphisms can be equivalently presented as functors $L\colon \mathtt{EM}(\mathbb{C}) \to \mathtt{EM}(\mathbb{D})$ such that the following diagram of functors commutes.

$$\begin{array}{ccc} \mathtt{EM}(\mathbb{C}) & \xrightarrow{L} & \mathtt{EM}(\mathbb{D}) \\ & {\scriptstyle U^{\mathbb{C}}}\searrow \quad \swarrow{\scriptstyle U^{\mathbb{D}}} & \\ & \mathcal{A} & \end{array}$$

The functor L is constructed from the comonad morphism given as a natural transformation λ by sending $A \xrightarrow{\alpha} \mathbb{C}(A)$ to $A \xrightarrow{\alpha} \mathbb{C}(A) \xrightarrow{\lambda_A} \mathbb{D}(A)$. For details, see e.g. [30].

[2] All coalgebras in this text are Eilenberg–Moore coalgebras. We do not work with functor coalgebras at any point.

2.3 Density Comonads

Next, we review basics of the theory of density comonads, introduced independently by Appelgate and Tierney [8] and Kock [21] (who studied the dual notion of codensity monads). The *density comonad* of a functor $M\colon \mathcal{A} \to \mathcal{B}$ is a functor $\mathbb{D}_M\colon \mathcal{B} \to \mathcal{B}$ with a natural transformation $\eta\colon M \Rightarrow \mathbb{D}_M M$.

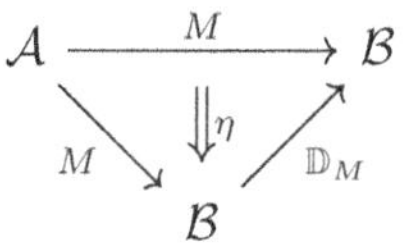

Moreover, η is required to be the initial natural transformation with this property. In other words, for any functor $K\colon \mathcal{B} \to \mathcal{B}$ and a natural transformation $\varphi\colon M \Rightarrow KM$ there is a *unique* $\varphi^*\colon \mathbb{D}_M \Rightarrow K$ such that $\varphi = \varphi^* M \circ \eta$, i.e. diagramatically

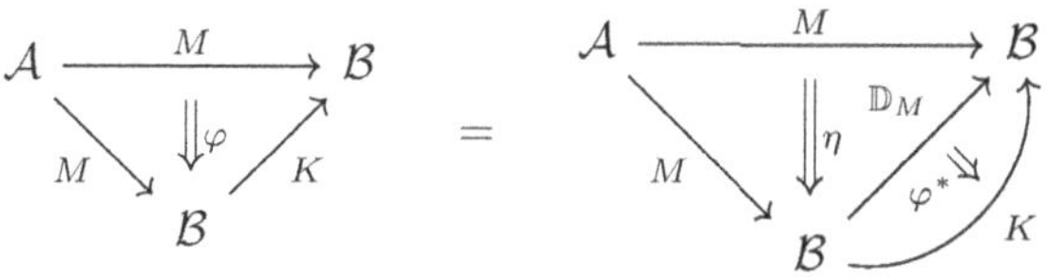

Density comonads are special types of left Kan extensions. They do not exist for all functors. However, when $\mathcal{A}$ is a small category and $\mathcal{B}$ is cocomplete then $\mathbb{D}_M$ exists for every functor $M\colon \mathcal{A} \to \mathcal{B}$. In such case, $\mathbb{D}_M(B)$ is computed as the colimit of the diagram:

$$M \downarrow B \xrightarrow{V} \mathcal{A} \xrightarrow{M} \mathcal{B}$$

where V is the forgetful functor from the comma category $M \downarrow B$, which consists of pairs (A, f) where $f\colon M(A) \to B$ is a morphism in $\mathcal{B}$, and morphisms $(A, f) \to (A', f')$ between such pairs are morphisms $g\colon A \to A'$ in $\mathcal{A}$ making the following triangle commute.

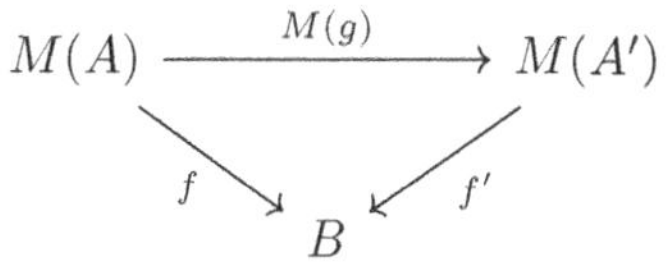

We may express the same fact by the formula:

$$\mathbb{D}_M(B) \;=\; \operatorname*{colim}_{A\in\mathcal{A},\ M(A)\to B} M(A). \tag{3}$$

Note that $\mathcal{A}$ does not have to be small nor $\mathcal{B}$ cocomplete in general. It is enough that the colimit above exists. In such case we speak of *pointwise* density comonads. Denote by

$$\iota_f\colon M(A) \to \mathbb{D}_M(B)$$

the inclusion morphism of the copy of $M(A)$ corresponding to the morphism $f\colon M(A) \to B$ into the colimit. Then, the component $\eta_A\colon M(A) \to \mathbb{D}_M(M(A))$ of the natural transformation $\eta\colon M \Rightarrow \mathbb{D}_M M$ is given as ι_f for f equal to the identity morphism $\mathrm{id}\colon M(A) \to M(A)$.

2.4 The Comonad Structure

The initiality of $\eta\colon M \Rightarrow \mathbb{D}_M M$ ensures that we can equip $\mathbb{D}_M$ with a comonad structure. In particular, the identity natural transformation $M \Rightarrow \mathrm{Id} \circ M$ uniquely factors as the composition of η with the counit $\varepsilon\colon \mathbb{D}_M \Rightarrow \mathrm{Id}$ and, similarly, $\mathbb{D}_M(\eta) \circ \eta\colon M \Rightarrow \mathbb{D}_M \circ \mathbb{D}_M \circ M$ factors through the comultiplication $\delta\colon \mathbb{D}_M \Rightarrow \mathbb{D}_M^2$. In other words, the counit and the comultiplication are uniquely determined by the equations

$$\varepsilon M \circ \eta = \mathrm{id} \quad \text{and} \quad \delta M \circ \eta = \mathbb{D}_M(\eta) \circ \eta. \tag{4}$$

Moreover, these two equations guarantee that the functor

$$M^{\dagger}\colon \mathcal{A} \to \mathtt{EM}(\mathbb{D}_M) \tag{5}$$

which sends $A \in \mathcal{A}$ to the coalgebra $\eta_A\colon M(A) \to \mathbb{D}_M(M(A))$ is well-defined.

Lastly, we recall three equations of density comonads, following from (4), which we use extensively throughout the paper. For morphisms $f\colon M(A) \to B$ and $h\colon B \to C$, the following three triangles commute.

$$\begin{array}{ccc} M(A) & & \\ {\scriptstyle \iota_f}\downarrow & \searrow^{\iota_{h\circ f}} & \\ \mathbb{D}_M(B) & \xrightarrow[\mathbb{D}_M(h)]{} & \mathbb{D}_M(C) \end{array} \qquad \begin{array}{ccc} M(A) & & \\ {\scriptstyle \iota_f}\downarrow & \searrow^{f} & \\ \mathbb{D}_M(B) & \xrightarrow[\varepsilon_B]{} & B \end{array} \qquad \begin{array}{ccc} M(A) & & \\ {\scriptstyle \iota_f}\downarrow & \searrow^{\iota_{\iota_f}} & \\ \mathbb{D}_M(B) & \xrightarrow[\delta_B]{} & \mathbb{D}_M(\mathbb{D}_M(B)) \end{array}$$

(DC1) (DC2) (DC3)

2.5 Comonad Morphisms from Composites

Let M be the composite of functors

$$\mathcal{A}_0 \xrightarrow{M_0} \mathcal{A} \xrightarrow{M_1} \mathcal{B}$$

such that the density comonads $\mathbb{D}_M$ and $\mathbb{D}_{M_1}$ exist. Let

$$\eta\colon M \Rightarrow \mathbb{D}_M M \qquad \eta^1\colon M_1 \Rightarrow \mathbb{D}_{M_1} M_1$$

be the corresponding initial natural transformations. By initiality of η, there is a unique natural transformation

$$\lambda\colon \mathbb{D}_M \Rightarrow \mathbb{D}_{M_1}$$

such that:

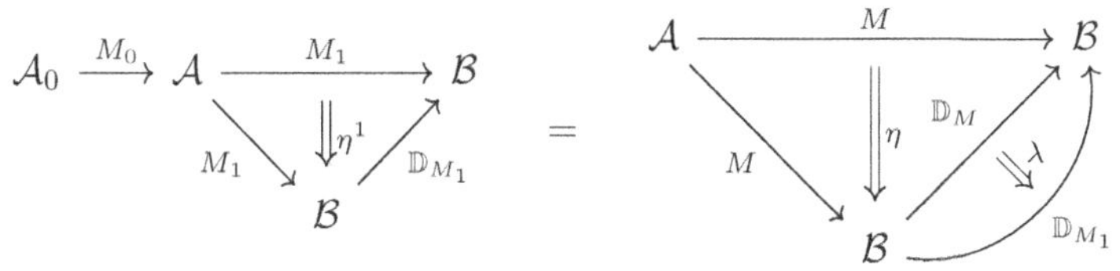

Lemma 3. $\lambda\colon \mathbb{D}_M \Rightarrow \mathbb{D}_{M_1}$ *is a comonad morphism.*

In fact, $\mathbb{D}_{(-)}$ is a functor from the category of functors $X \to \mathcal{B}$ which admit density comonads into the category of comonads and comonad morphisms (cf. page 73 in [15], see also [22]).

3 Discrete Density Comonads

Recall that a category is *discrete* whenever it only has identity morphisms. By *discrete density comonads* we mean density comonads for functors whose domain is discrete. An important feature of discrete density comonads is that the formula in (3) simplifies dramatically. Indeed, assume that

$$M\colon \mathcal{A} \to \mathcal{B}$$

is a fixed functor from a small and discrete category $\mathcal{A}$. Then, since $\mathcal{A}$ is discrete, there are no morphisms between objects $M(A)$, given by morphisms $M(A) \to B$, in the colimit formula (3). Therefore, the density comonad $\mathbb{D}_M : \mathcal{B} \to \mathcal{B}$ is computed as a coproduct, that is, the colimit of a discrete diagram. Concretely, for an object B in $\mathcal{B}$,

$$\mathbb{D}_M(B) \;=\; \coprod_{A\in\mathcal{A}} \; \coprod_{f\colon M(A)\to B} M(A). \tag{6}$$

Note that $\mathbb{D}_M$ exists whenever the above coproduct exists in $\mathcal{B}$, for every object $B \in \mathcal{B}$. In particular, $\mathbb{D}_M$ exists whenever $\mathcal{B}$ has all coproducts.

As with general density comonads, we have inclusion morphisms

$$\iota_f\colon M(A) \to \mathbb{D}_M(B)$$

for every $f\colon M(A) \to B$, which satisfy axioms (DC1)–(DC3) from Sect. 2.4.

For the proof of Theorem 1, the category $\mathcal{B}$ is either the category $\mathcal{R}(\sigma)$ of σ-structures (i.e. relational structures in a fixed relational signature σ) or the category **Graph** of graphs (where by graphs we mean undirected loopless graphs). The morphisms in these categories are the structure-preserving functions: σ-structure homomorphisms $f\colon A \to B$ satisfy that $R^A(x_1,\ldots,x_n)$ implies $R^B(f(x_1),\ldots,f(x_n))$ and, likewise, graph homomorphisms preserve the edge relation. For example, in the former case, we may describe the comonad $\mathbb{D}_M$ explicitly as follows. For a σ-structure B, the universe of $\mathbb{D}_M(B)$ consists of tuples

$$(A, f, x)$$

where $f\colon M(A) \to B$ is a homomorphism of relational structures and x is an element of $M(A)$. Further, an n-ary relation R in σ is interpreted as the set of all tuples

$$(A, f, x_1), \quad \dots, \quad (A, f, x_n)$$

such that $R(x_1, \dots, x_n)$ in $M(A)$.

Example 4. Let $\mathcal{A}$ be the discrete subcategory of graphs, consisting of only the triangle graph, and let $M\colon \mathcal{A} \to$ **Graph** be the inclusion of $\mathcal{A}$ into the category of graphs. Then, given an arbitrary graph G, the graph computed as $\mathbb{D}_M(G)$ is the disjoint union of $k \times l$ triangles, where k is the number of triangles in G and l is the number of automorphisms of the triangle graph, i.e. $l = 6$.

4 The Abstract Classification Theorem

In this section we prove Theorem 1. Since the entire argument can be carried out at the abstract categorical level, we actually prove a general categorical statement that can be applied in other scenarios too. In the following we fix a functor

$$M\colon \mathcal{A} \to \mathcal{B}$$

from a discrete category $\mathcal{A}$ into $\mathcal{B}$. We further assume that the pointwise density comonad $\mathbb{D}_M$ exists on $\mathcal{B}$ (i.e. it is given by the formula (6)).

We start with a useful observation. Recall that an object C is *connected* iff, for every morphism $f\colon C \to \coprod_i A_i$ into a coproduct, the morphism f factors uniquely through one of the inclusion morphisms $\iota_i\colon A_i \to \coprod_i A_i$.[3] Whenever a connected C is such that $A \cong C + X$ for some A, X, we say that C is a *component* of A.

Lemma 5. *Let $\xi\colon X \to \mathbb{D}_M(X)$ be a $\mathbb{D}_M$-coalgebra and let $\iota_C\colon C \to X$ be a component inclusion. Furthermore, let $f\colon M(A) \to X$ be the morphism for which $\xi \circ \iota_C$ decomposes through ι_f, as shown below.*

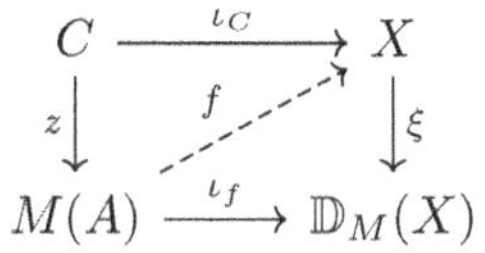

(7)

Then, also the two triangles in the diagram above commute, that is, $\iota_C = f \circ z$ and $\iota_f = \xi \circ f$.

[3] Equivalently, C is connected iff $\hom(C, -)$ preserves coproducts.

Proof. The first equality is obtained immediately from the triangle law of coalgebras (cf. (1)) together with (DC2) from Sect. 2.4 as

$$\iota_C = \varepsilon_C \circ \xi \circ \iota_C = \varepsilon_C \circ \iota_f \circ z = f \circ z.$$

To show that also $\iota_f = \xi \circ f$ we apply the square law of coalgebras (cf. (1)). Observe that

- $\delta_X \circ \xi \circ \iota_C = \delta_X \circ \iota_f \circ z = \iota_{\iota_f} \circ z$ by (DC3), and
- $\mathbb{D}_M(\xi) \circ \xi \circ \iota_C = \mathbb{D}_M(\xi) \circ \iota_f \circ z = \iota_{\xi \circ f} \circ z$ by (DC1).

Because C is connected, the factorisation $C \to M(A) \to \mathbb{D}_M(\mathbb{D}_M(C))$ into the coproduct must be unique, hence $\iota_f = \xi \circ f$. □

In the following we need to assume that $\mathcal{B}$ is a *componental category*, i.e. that

- every object in $\mathcal{B}$ is (isomorphic to) a coproduct of connected objects, and
- inclusion morphisms into coproducts $\iota_i \colon a_i \to \coprod_i a_i$ are monomorphisms.

We say that an object of $\mathcal{B}$ is *essentially in* $\mathcal{A}$ if it is isomorphic to $M(A)$, for some A in $\mathcal{A}$.

Lemma 5 directly implies a version of Theorem 1 for connected objects.

Lemma 6. *If $\mathcal{B}$ is a componental category, then a connected object of $\mathcal{B}$ is essentially in $\mathcal{A}$ iff it admits a $\mathbb{D}_M$-coalgebra.*

Since the non-trivial direction is proved similarly to Lemma 8 below, we omit its proof. Next, we show a useful feature of componental categories.

Lemma 7. *In a componental category, a component inclusion $C \to Y$ which factors through a monomorphism $X \to Y$ is a component of X as well.*

Proof. Assume Y is equal to the coproduct $\coprod_i C_i$ and X is equal to $\coprod_j D_j$ for some collections of connected objects $\{C_i\}_i$ and $\{D_j\}_j$. By assumption, the component inclusion $\iota \colon C \to Y$ factors as $m \circ h$ for some monomorphism $m \colon X \to Y$ and a morphism $h \colon C \to X$.

Since C is connected, h factors through a component inclusion $\iota_j \colon D_j \to X$ as shown in the left diagram below:

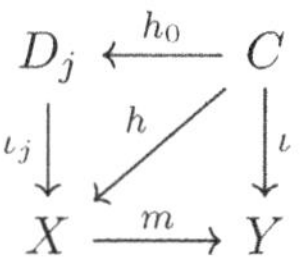

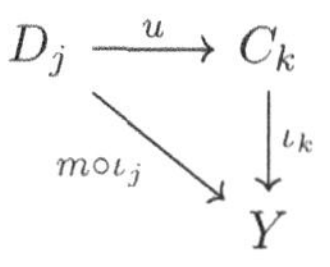

Since D_j is connected, $m \circ \iota_j$ factors through some inclusion $\iota_k \colon C_k \to Y$, as shown in the right diagram above. But then $i = k$ and $u \circ h_0 = \mathrm{id}$ since $\iota_i = m \circ h = m \circ \iota_j \circ h_0 = \iota_k \circ u \circ h_0$ and component inclusions are unique. Furthermore, since both m and ι_j are monomorphisms by our assumptions, so must be u because $\iota_\ell \circ u = m \circ \iota_j$. Consequently, u is an isomorphism because it is both a monomorphism and a split epimorphism. □

To make progress, we need to assume that $\mathcal{A}$ is *component-based*. This means that, whenever $B \in \mathcal{B}$ is essentially in $\mathcal{A}$ then so is every component of B. Note that this condition mirrors the third item in the definition of component-based classes. With this we show the main technical lemma of this section.

Lemma 8. *If $\mathcal{A}$ is component-based and $\mathcal{B}$ is a componental category, then any component C of an object X of $\mathcal{B}$ which admits a $\mathbb{D}_M$-coalgebra $\xi \colon X \to \mathbb{D}_M(X)$ is essentially in $\mathcal{A}$.*

Proof. Let $\iota_C \colon C \to X$ be the inclusion morphism of C into X. Furthermore, let $f \colon M(A) \to X$ be the morphism such that $\xi \circ \iota_C$ decomposes through $\iota_f \colon M(A) \to \mathbb{D}_M(X)$ and recall that, by Lemma 5, the following diagram commutes.

$$\begin{array}{ccc} C & \xrightarrow{\iota_C} & X \\ {\scriptstyle z}\downarrow & \nearrow^{f} & \downarrow{\scriptstyle \xi} \\ M(A) & \xrightarrow{\iota_f} & \mathbb{D}_M(X) \end{array} \tag{8}$$

Observe that f is a monomorphism since ι_f is. Therefore, by Lemma 7, C is a component of $M(A)$. Consequently, C is essentially in $\mathcal{A}$ because $\mathcal{A}$ is component-based. □

The main classification theorem, which we state in full, is obtained as a consequence of the previous lemma.

Theorem 9. *Let $M \colon \mathcal{A} \to \mathcal{B}$ be a functor from a discrete component-based category $\mathcal{A}$ into a componental category $\mathcal{B}$ such that the pointwise density comonad $\mathbb{D}_M$ exists.*

Then an object $b \in \mathcal{B}$ is isomorphic to a coproduct of objects essentially in $\mathcal{A}$ if and only if b admits a $\mathbb{D}_M$-coalgebra.

Proof. The left-to-right implication follows the fact that $M^\dagger(A)$ is a coalgebra on $M(A)$, for every $A \in \mathcal{A}$ (cf. (5)), and that coalgebras are closed under coproducts that exist in $\mathcal{B}$. Conversely, if $\xi \colon X \to \mathbb{D}_M(X)$ is a coalgebra then, by our assumptions, X is isomorphic to a coproduct $\coprod_i C_i$ of connected objects by Lemma 8 and all those components are essentially in $\mathcal{A}$. □

Observe that both the category of relational structures $\mathcal{R}(\sigma)$ and the category of graphs **Graph** are componental categories.[4] Therefore, the previous theorem immediately yields Theorem 1. Indeed, given a component-based class Δ of relational structures or graphs, let Δ_C be the subclass of Δ consisting of connected structures only. We then set $\mathcal{A}$ to be a discrete subcategory of $\mathcal{R}(\sigma)$ or **Graph** consisting of one representative from every isomorphism class in Δ_C. Since we picked only one representative from every equivalence class, the category $\mathcal{A}$ is small. Therefore, the density comonad $\mathbb{D}_M$, for the inclusion functor

[4] Note that for $\mathcal{R}(\sigma)$, the connected objects are those structures whose Gaifman graphs are connected.

$M: \mathcal{A} \to \mathcal{R}(\sigma)$, exists because both $\mathcal{R}(\sigma)$ and **Graph** have all (small) coproducts. Observe that the comonad $\mathbb{D}_M$ classifies Δ. Indeed, by Theorem 9, a finite relational structure B has a $\mathbb{D}_M$-coalgebra if and only if there exist $C_1, \ldots, C_n$ in Δ_C such that $B \cong C_1 + \cdots + C_n$. In turn, this is equivalent to B being in Δ, which follows from being component-based as then $C_1 + \cdots + C_n$ is in Δ iff all the individual structures $C_1, \ldots, C_n$ are.

Remark 10. In the proof of Theorem 1, in the previous paragraph, we made sure that all objects in the image of M are connected. This is stronger than assuming that $\mathcal{A}$ is component-based. By carefully inspecting the proof of Theorem 9 and the preceding lemmas one can check that this extra assumption allows us to drop the requirement that $\mathcal{B}$ is componental. See Lemma 21 below for details.

Remark 11. Theorem 1 says that, for a class Δ of structures closed under isomorphisms and finite coproducts, if Δ is also closed under summands then it can be classified by a comonad. However, the converse does not hold. Let C and D be two graphs with no homomorphism $C \to D$ nor any homomorphisms $D \to C$. This happens, for example, if C is the triangle and D is the cycle on five vertices. Take $\mathcal{A}$ to be the discrete subcategory of **Graph** consisting of $C + D$ only and let $M: \mathcal{A} \to$ **Graph** be the subcategory inclusion. Then, despite $C + D$ admitting a $\mathbb{D}_M$-coalgebra, no connected graph admits a $\mathbb{D}_M$-coalgebra (by Lemma 6). It is easy to see that the class of finite structures classified by $\mathbb{D}_M$ is the class consisting of graphs isomorphic to

$$C + \cdots + C + D + \cdots + D$$

where both C and D appear in at least one copy in the coproduct. Consequently, the class of structures classified by $\mathbb{D}_M$ is not closed under summands.

4.1 Examples

The category $\mathcal{R}(\sigma)$ of σ-structures is a componental category. Therefore, in our applications we only need to check that a class Δ of σ-structures is closed under finite coproducts and summands. These are fairly weak conditions, satisfied by many well-known examples of classes from the literature. In particular, this includes classes of finite structures closed under finite coproducts which are

1. monotone, i.e. class closed under taking substructures,
2. hereditary, i.e. class closed under taking induced substructures, or
3. closed under taking graph minors.

Further examples include

4. Fraïssé classes closed under free amalgamations, or
5. classes of coproducts of connected cores.

Recall that a core is a structure with the property that all of its endomorphisms are automorphisms. An example of a class from (5) is the class of coproducts of cycles. Note that the discrete density comonad for this class captures co-spectrality, see Sect. 6.

As an example of a non-example, take the class of graphs that can be drawn on a surface of genus n, for $n > 1$. This class is characterised by a finite set of forbidden minors. However, it is not closed under taking coproducts and hence is not a component-based class. On the other hand, any minor-closed class can be completed under finite coproducts. The resulting class is then still minor closed [11, Lemma 5] and hence is classified by a comonad.[5]

Remark 12. The proof of Theorem 1 is carried out abstractly, in the language of category theory, and thus can be dualised. In the dual statement we have monads instead of comonads and instead of component-based classes we have classes closed under isomorphisms, products, and *factors*, i.e. with the property that if $A \times B$ is in the class then so are A and B. For such classes there is a monad which classifies the class, i.e. a finite structure is in the class iff it admits an algebra for the monad. An example of a class of graphs which can be classified in this way is the class of connected non-bipartite graphs, cf. Chapter 8 in [19].

5 Graph Parameters

A *graph parameter* is a mapping $\mu\colon \mathbf{Graph}_{\mathsf{fin}} \to \overline{\mathbb{R}}$, from the class of finite graphs $\mathbf{Graph}_{\mathsf{fin}}$ to the class of extended real numbers $\overline{\mathbb{R}} = [-\infty, +\infty]$, which gives the same value to any two isomorphic graphs. Moreover, we say that it is *standard*[6] if $\mu(G_1 + G_2) = \max\{\mu(G_1), \mu(G_2)\}$.

Standard graph parameters cover many well-known examples of graph parameters from the literature, such as

- clique number, chromatic number, max-degree,
- tree-depth, tree-width, path-width, clique-width, etc.

In this section we show that every standard graph parameter μ gives rise to a *graded comonad* $(\mathbb{C}_k)_k$, that is, a sequence of comonads $(\mathbb{C}_k)_k$ indexed by extended real numbers and comonad morphisms $g_{k,l}\colon \mathbb{C}_k \Rightarrow \mathbb{C}_l$, for every $k \leq l$ in $\overline{\mathbb{R}}$, such that $g_{k,l} = g_{j,l} \circ g_{k,j}$ for any $k \leq j \leq l$ in $\overline{\mathbb{R}}$.[7] Given a graded comonad $(\mathbb{C}_k)_k$ we define the *coalgebra number* $\kappa^{\mathbb{C}}(G)$, of a graph G, to be the infimum of $k \in \overline{\mathbb{R}}$ such that G admits a $\mathbb{C}_k$-coalgebra [1,4]. We show that the coalgebra number for the constructed graded comonad agrees with the standard graph parameter μ we started with. In other words, we have that $\mu(G) \leq k$ iff G admits a $\mathbb{C}_k$-coalgebra.

Note that every graded comonad $(\mathbb{C}_k)_k$ trivially determines a graph parameter, by setting $\mu(G) := \kappa^{\mathbb{C}}(G)$. We have already mentioned graded comonads characterising graph parameters this way. For example, the (graded)

5 We are grateful to Anuj Dawar for pointing out these facts.

6 Also known as *maxing*, e.g. in [24].

7 In fact, this is a special type of graded comonad, with the grading being over the fixed monoid $(\overline{\mathbb{R}}, \min, +\infty)$. For details see [4].

Ehrenfeucht-Fraïssé comonad $(\mathbb{E}_k)_k$ characterises tree-depth [4], the pebbling comonad $(\mathbb{P}_k)_k$ characterises tree-width [1] and the pebble-relation comonad $(\mathbb{PR}_k)_k$ classifies path-width [26].

To start with, observe that there is a one-to-one correspondence between *graph properties*, i.e. graph parameters valued in $\{0,1\}$, and classes of finite graphs which are closed under isomorphisms. Furthermore, it is easy to see that the correspondence restricts to that of standard graph properties and component-based classes of graphs:

Lemma 13. *Given a standard graph property* $\mu\colon \mathbf{Graph}_{\mathsf{fin}} \to \{0,1\}$, *the class of graphs* G *such that* $\mu(G)=0$ *is closed under isomorphisms, finite coproducts, and summands. In fact, every such class is obtained from a standard graph property this way.* □

Therefore, by Theorem 1, there is a comonad $\mathbb{C}^{\mu}$ which classifies the class Δ of finite graphs G such that $\mu(G)=0$, for every standard graph property μ. We construct $\mathbb{C}^{\mu}$ explicitly, as the pointwise density comonad for the inclusion functor

$$\mathcal{A} \to \mathbf{Graph}, \tag{9}$$

where $\mathcal{A}$ is a discrete subcategory of finite connected graphs consisting precisely of one graph from every isomorphism class in Δ. Then, by Theorem 9, $\mathbb{C}^{\mu}$ classifies Δ.

5.1 Grading Graph Parameters

We use this to construct a sequence of comonads for a given standard graph parameter μ. For every extended real number k, we turn μ into a graph property

$$\mu_{\leq k}\colon \mathbf{Graph} \to \{0,1\}$$

by setting $\mu_{\leq k}(G)=0$ iff $\mu(G)\leq k$. Then, the density comonad $\mathbb{C}^{\mu}_{k}$, defined as $\mathbb{C}^{\mu}$ for $\mu := \mu_{\leq k}$, classifies finite graphs G such that $\mu(G)\leq k$.

Moreover, we can make sure that there is a linearly ordered chain of embeddings of discrete categories

$$\mathcal{A}_{-\infty} \hookrightarrow \ldots \hookrightarrow \mathcal{A}_k \hookrightarrow \mathcal{A}_l \hookrightarrow \ldots \hookrightarrow \mathcal{A}_{+\infty} \qquad \text{(with } k \leq l\text{)}$$

where each $\mathcal{A}_k$ is a category as in (9), for the class of graphs G such that $\mu(G)\leq k$. Then, by Lemma 3 in Sect. 2.5, the composite

$$\mathcal{A}_k \hookrightarrow \mathcal{A}_l \to \mathbf{Graph},$$

for $k\leq l$, gives rise to a comonad morphism $g_{k,l}\colon \mathbb{C}^{\mu}_{k} \Rightarrow \mathbb{C}^{\mu}_{l}$. In fact, we have $g_{k,l} = g_{j,l}\circ g_{k,j}$ for every $k\leq j\leq l$, by functoriality of $\mathbb{D}_{(-)}$. Hence, $(\mathbb{C}^{\mu}_{k})_k$ is a graded comonad with the property that $\kappa^{\mathbb{C}^{\mu}}(G)=\mu(G)$ for every finite graph G.

Remark 14. The procedure to produce sequences of comonads for standard graph parameters can be defined dually for graph parameters μ with the property that $-\mu$ is standard, i.e. graph parameters such that $\mu(G_1 + G_2) = \min\{\mu(G_1), \mu(G_2)\}$. This is done by constructing a sequence of standard graph properties

$$\mu_{\geq k} \colon \textbf{Graph} \to \{0, 1\}$$

and inducing comonads classifying the classes of graphs such that $\mu(G) \geq k$ in a similar spirit as before. This then covers examples of graph parameters such as min-degree and girth.

5.2 Comparison with Game Comonads

For some graph parameters and classes of structures we already knew how to construct comonads that classify them. In particular, this holds for the classes of structures classified by the comonads $\mathbb{P}_k$, $\mathbb{E}_k$, and $\mathbb{PR}_k$. In this section, we explain the relationship between those comonads and discrete density comonads constructed directly for given classes.

In fact, we show that discrete density comonads are weakly initial in the category of comonads that classify the same class. To this end, denote by $\mathbb{D}_\Delta$ the discrete density comonad constructed as in (9) above, for a component-based class Δ.

Proposition 15. *Let Δ be a component-based class of relational structures or graphs and let $\mathbb{C}$ be a comonad that classifies a class Γ. Then, $\Delta \subseteq \Gamma$ if, and only if, there exists a comonad morphism $\mathbb{D}_\Delta \Rightarrow \mathbb{C}$.*

Observe that the right-to-left direction is immediate as a comonad morphism $\mathbb{D}_\Delta \Rightarrow \mathbb{C}$ lifts to a functor $L \colon \mathtt{EM}(\mathbb{D}_\Delta) \to \mathtt{EM}(\mathbb{C})$ making the following diagram commute (cf. Sect. 2.2).

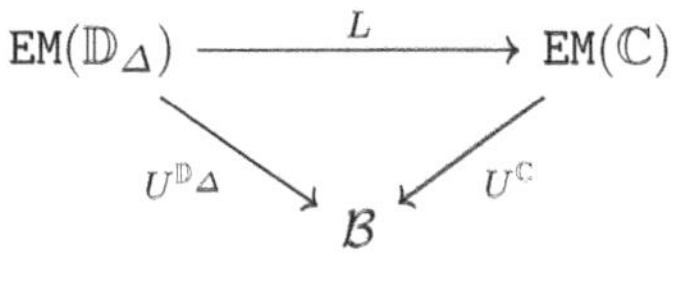

(Here $\mathcal{B}$ is either the category of relational structures or graphs.) For a structure B in Δ, let $\beta \colon B \to \mathbb{D}_\Delta(B)$ be a $\mathbb{D}_\Delta$-coalgebra, which exists because $\mathbb{D}_\Delta$ classifies Δ. Then, by the commutativity of the above triangle $L(\beta)$ is a $\mathbb{C}$-coalgebra $B \to \mathbb{C}(B)$ making $B \in \Gamma$ because $\mathbb{C}$ classifies Γ.

We carry out the left-to-right direction of the proof abstractly, for arbitrary categories rather than just relational structures or graphs. Let $\mathbb{D} := \mathbb{D}_M$ be the density comonad of a functor $M \colon \mathcal{A} \to \mathcal{B}$ from a discrete category $\mathcal{A}$. Further, assume that $\mathbb{C}$ is a comonad on $\mathcal{B}$ such that for every $A \in \mathcal{A}$, there exists a coalgebra

$$\varphi_A \colon M(A) \to \mathbb{C}(M(A)).$$

Observe that, since $\mathcal{A}$ is a discrete category, the collection of morphisms $\{\varphi_A \mid A \in \mathcal{A}\}$ trivially forms a natural transformation $\varphi\colon M \Rightarrow \mathbb{C}M$. Since $\mathbb{D}$ is a density comonad of M, there is a natural transformation $\varphi^*\colon \mathbb{D} \Rightarrow \mathbb{C}$ such that

$$\varphi = \varphi^* M \circ \eta \tag{10}$$

where $\eta\colon M \Rightarrow \mathbb{D}M$ is the initial natural transformation determining $\mathbb{D}$ (cf. Sect. 2.3). Then, Proposition 15 follows from the following lemma, which is a direct consequence of Theorem II.1.1 in [15].

Lemma 16. *$\varphi^*\colon \mathbb{D} \Rightarrow \mathbb{C}$ is a morphism of comonads.*

Example 17. Proposition 15 gives us that that for our running examples of comonads $\mathbb{E}_k$, $\mathbb{P}_k$, $\mathbb{PR}_k$, there exist comonad morphisms

$$\mathbb{D}_{\mathcal{TD}_k} \Rightarrow \mathbb{E}_k, \quad \mathbb{D}_{\mathcal{TW}_k} \Rightarrow \mathbb{P}_k, \quad \text{and} \quad \mathbb{D}_{\mathcal{PW}_k} \Rightarrow \mathbb{PR}_k,$$

where $\mathcal{TD}_k$, $\mathcal{TW}_k$, and $\mathcal{PW}_k$ are the classes of finite structures of tree-depth, tree-width, and path-width $\leq k$, respective.

Note that unlike the game comonads $\mathbb{E}_k$, $\mathbb{P}_k$, and $\mathbb{PR}_k$, the discrete density comonads $\mathbb{D}_{\mathcal{TD}_k}$, $\mathbb{D}_{\mathcal{TW}_k}$, and $\mathbb{D}_{\mathcal{PW}_k}$ do not classify infinite structures with the corresponding properties.

5.3 Nowhere Dense Comonads

A direct consequence of Proposition 15 is that we can characterise comonads that classify monotone nowhere dense classes of graphs in terms of non-existence of certain comonad morphisms. Recall that a class Δ is *somewhere dense* if there exists a natural number p such that, for every n, the p-th subdivision K_n^p of all edges in the clique graph K_n on n vertices is a subgraph of some graph in Δ.[8] Then, a class is *nowhere dense* [27] if it is not somewhere dense.

It is immediate that a monotone class of graphs Δ (i.e. a class closed under substructures) is somewhere dense if and only if $\mathbf{Cli}_p \subseteq \Delta$, for some p, where

$$\mathbf{Cli}_p = \{K_n^p \mid n \in \mathbb{N}\}$$

is the class of p-th subdivisions of all cliques. We can now state the characterisation.

Proposition 18. *Assume $\mathbb{C}$ classifies a monotone class of graphs Δ. Then, Δ is nowhere dense if, and only if, there is no comonad morphism $\mathbb{D}_{\mathbf{Cli}_p} \Rightarrow \mathbb{C}$ for any $p \in \mathbb{N}$.*

Proof. Define $\overline{\mathbf{Cli}_p}$ to be the closure of $\mathbf{Cli}_p$ under finite coproducts. Observe that $\overline{\mathbf{Cli}_p}$ is component-based and, since the connected objects in $\overline{\mathbf{Cli}_p}$ are precisely the objects in $\mathbf{Cli}_p$, the comonad $\mathbb{D}_{\mathbf{Cli}_p}$ classifies $\overline{\mathbf{Cli}_p}$. Moreover, since any class classified by a comonad needs to be closed under finite coproducts, $\mathbf{Cli}_p \subseteq \Delta$ iff $\overline{\mathbf{Cli}_p} \subseteq \Delta$. The result follows by monotonicity of Δ and by Proposition 15. □

[8] A subdivision of a set of edges in a graph replaces each edge in the set by a path of length 2 through a new vertex.

6 Lovász-Type Theorems for Free

A classic result of Lovász [23] says that two finite structures are isomorphic if and only if they admit the same number of homomorphisms from all finite structures. This result has been extended in many different ways. In one type of generalisation, isomorphisms are replaced by a selected equivalence relation $\asymp$ on finite structures, and the class of all finite structures by a class of selected finite structures Δ. Then a typical Lovász-type theorem expresses that, for finite structures A, B,

$$A \asymp B \iff \hom(C, A) \cong \hom(C, B) \quad \text{for every } C \in \Delta.$$

A number of well-known equivalence relations on finite structures have been characterised in this way. See Fig. 1 for an overview of some Lovász-type results[9].

Δ	$\asymp$	reference
cycles	co-spectrality	(folklore)
trees	fractional isomorphism	[28]
bipartite graphs	isomorphic bipartite double covers	[10,24]
planar graphs	quantum isomorphism	[25]
tree-depth $\leq k$	Duplicator wins the bijective k-round Ehrenfeucht–Fraïssé game	[17]
tree-width $< k$	Duplicator wins the bijective k-pebble game	[16]
path-width $< k$	Duplicator wins the bijective k-pebble relation game	[26]
admitting a k-pebble tree cover of height $\leq n$	Duplicator wins the bijective n-round k-pebble E.F. game	[13]
synchronization trees of height $\leq k$	equivalence in graded modal logic of modal depth $\leq k$	[13]
an inner-product compatible class	existence of a certain unitary map between homomorphism tensor spaces	[18]

Fig. 1. Examples of Lovász-type theorems

Note that the equivalence relations $\asymp$ in Fig. 1 corresponding to winning strategies of Duplicator can be equivalently described as logical equivalences with

[9] The *bipartite double cover* of a graph G is the product graph $G \times K_2$ where K_2 is the clique on two vertices. The fact that isomorphic bipartite double covers correspond to counting homomorphisms from bipartite graphs was worked out by Böker in his master thesis [10]. He later observed (in private communication) that the same result already follows from Section 5.4.2 in [24].

respect to a fragment of first-order logic with counting quantifiers. A comonadic proof of the first two Lovász-type theorems that identify a logic fragment was established in [13] and later adapted in [26] to obtain the new result for path-width. For the comonadic proof to work it is necessary that the comonad $\mathbb{C}$ classifies the relation $\asymp$, i.e. that $A \asymp B$ holds precisely whenever the cofree coalgebras $F^{\mathbb{C}}(A)$ and $F^{\mathbb{C}}(B)$ are isomorphic.

Anuj Dawar has asked (in private communication) whether there are comonads covering the other listed cases as well. In our terminology, this means finding comonads that classify both the class Δ as well as the corresponding $\asymp$ relation, in the same row of the table. We answer this question in the positive for any component-based class Δ. We show that the discrete density comonad $\mathbb{C}$ that classifies such class also classifies the relation $\asymp$, with $A \asymp B$ whenever $\hom(C, A) \cong \hom(C, B)$ for every $C \in \Delta$, thereby proving Corollary 2.

The main ingredient of our proof is the following recent result due to Luca Reggio, proved abstractly for locally finitely presentable categories.

Theorem 19. (Corollary 5.15 in [29]). *Let $\mathbb{C}$ be a comonad of finite rank on* **Graph** *or $\mathcal{R}(\sigma)$. Then, for two finite structures A, B,*

$$F^{\mathbb{C}}(A) \cong F^{\mathbb{C}}(B) \quad \textit{iff} \quad \hom(C, A) \cong \hom(C, B),$$

for every finite C which admits a $\mathbb{C}$-coalgebra.

We see that in order to prove Corollary 2, it is enough to show that the comonad $\mathbb{C}$ constructed in the proof of Theorem 1 has finite rank. In the next section we define finite rank comonads and show that the discrete density comonad constructed in the proof of Theorem 1 does have finite rank.

6.1 Finite Rank Comonads

We work in the general setting of categories, rather than the more concrete setting of relational structures or graphs. Recall that an object C of a category $\mathcal{B}$ is *finitely presentable* [7] if the functor $\hom(C, -)$ preserves filtered colimits.

Let $\mathbb{C}$ be a comonad over a category $\mathcal{B}$ and let $U \colon \mathtt{EM}(\mathbb{C}) \to \mathcal{B}$ be the usual forgetful functor. We say that a comonad $\mathbb{C}$ has *finite rank* if

1. $\mathbb{C}$ is *finitary*, i.e. its underlying functor preserves filtered colimits,
2. if every morphism of the form $f \colon C \to U(\xi)$, from a finitely presentable C, admits a factorisation

$$f = C \xrightarrow{f_0} U(\xi_0) \xrightarrow{U(\gamma)} U(\xi)$$

for some $\gamma \colon Y \to X$ such that $U(\xi_0)$ is finitely presentable, and
3. this factorisation is essentially unique, i.e. if $g \colon C \to U(\xi_0)$ satisfies $f = U(\gamma) \circ g$ then for some factorisation of γ into $\lambda \colon \xi_0 \to \xi_0'$ and $\gamma' \colon \xi_0' \to \xi$ such that $U(\xi_0')$ is finitely presentable, $U(\lambda) \circ f_0 = U(\lambda) \circ g$.

Observe that if $U(\gamma)$ in the second item is a monomorphism, then essential uniqueness is automatic. In fact, this is the case in our construction.

In the following we fix a functor

$$M\colon \mathcal{A} \to \mathcal{B}$$

from a discrete category $\mathcal{A}$ and assume that the pointwise density comonad $\mathbb{D}_M$ for M exists.

Proposition 20. *If all objects in the image of M are finitely presentable, then $\mathbb{D}_M$ is finitary.*

Proof (by courtesy of an anonymous referee). It is enough to check the proof for $M\colon \mathbf{1} \to \mathcal{B}$ where $\mathbf{1} = \{\star\}$ is the discrete category with one object. Indeed, the density comonad for any $\mathcal{A} \to \mathcal{B}$ with $\mathcal{A}$ discrete is computed as a coproduct of density comonads for individual restrictions $\mathbf{1} \to \mathcal{B}$ and coproducts commute with colimits. Then, for an $M\colon \mathbf{1} \to \mathcal{B}$, the density comonad is

$$\mathbb{D}_M(X) = \mathcal{B}(M(\star), X) \cdot M(\star)$$

where $\cdot$ denotes copower. Consequently, since $M(\star)$ is finitary in $\mathcal{B}$ copowers commute with colimits, for a (small) directed diagram $D\colon I \to \mathcal{B}$ with a colimit $\operatorname{colim}_i D(i)$,

$$\begin{aligned}\mathbb{D}_M(\operatorname*{colim}_i D(i)) &= \mathcal{B}(M(\star), \operatorname*{colim}_i D(i)) \cdot M(\star)\\ &\cong (\operatorname*{colim}_i \mathcal{B}(M(\star), D(i))) \cdot M(\star)\\ &\cong \operatorname*{colim}_i (\mathcal{B}(M(\star), D(i)) \cdot M(\star)) \cong \operatorname*{colim}_i \mathbb{D}_M(D(i)).\end{aligned}$$ □

Next, we prove a technical lemma, which is (in some sense) a strengthening of Lemma 8, and which is needed in the proof of Proposition 22 below.

Lemma 21. *Assume that all objects in the image of M are connected. Let C be a component of X and let $\xi\colon X \to \mathbb{D}_M(X)$ be a $\mathbb{D}_M$-coalgebra. Then, C can be equipped with a $\mathbb{D}_M$-coalgebra $\gamma\colon C \to \mathbb{D}_M(C)$ such that the inclusion $\iota_C\colon C \to X$ is a coalgebra homomorphism $(C, \gamma) \to (X, \xi)$.*

Proof. As in the proof of Lemma 8, we arrive at the diagram (8). This time $M(A)$ is connected. Therefore, since f is a monomorphism (because so is ι_f), by Lemma 7, z is an isomorphism. Next we show that f is a coalgebra morphism from $\eta_A\colon M(A) \to \mathbb{D}_M(M(A))$ to $\xi\colon X \to \mathbb{D}_M(X)$. To this end, recall that $\eta_A = \iota_g$ for $g = \mathrm{id}\colon M(A) \to M(A)$. Therefore, we obtain the desired $\mathbb{D}_M(f) \circ \eta_A = \iota_{f \circ \mathrm{id}} = \iota_f = \xi \circ f$ by (DC1). Consequently, ι_C is also a coalgebra morphism, for C equipped with the coalgebra structure of η_A transported along the isomorphism z. □

Lastly, we show that also conditions (2) and (3) are satisfied for discrete density comonads.

Proposition 22. *Assume that all objects in the image of M are connected and finitely presentable, and that $\mathcal{B}$ is a componental category with finite coproducts. Then, every morphism $f\colon A \to U(\xi)$ in $\mathcal{B}$, for finitely presentable A in $\mathcal{B}$, admits a unique factorisation (up to isomorphism)*

$$f \;=\; A \xrightarrow{\;g\;} U(\xi_0) \xrightarrow{\;U(\gamma)\;} U(\xi)$$

where $U(\xi_0)$ is finitely presentable.

Proof. In the following, we denote by U the forgetful functor $U^{\mathbb{C}}\colon \mathtt{EM}(\mathbb{D}_M) \to \mathcal{B}$. By Theorem 9, we may assume that the underlying object X of $\xi\colon X \to \mathbb{D}_M(X)$ is a coproduct $\coprod_{i\in I} C_i$ of a collection of connected, finitely presentable objects C_i essentially in $\mathcal{A}$.

Recall that $\coprod_{i\in I} C_i$ is isomorphic to the directed colimit of the following directed diagram

$$\{\coprod_{i\in F} C_i \mid F \text{ is a finite subset of } I\}$$

with the obvious morphisms between these finite coproducts. Since A is finitely presentable, $f\colon A \to X$ decomposes as

$$f = A \xrightarrow{g} \coprod_{i\in F} C_i \xrightarrow{\iota_F} X$$

for some finite $F \subseteq I$. By Lemma 21, for each $i \in F$, the inclusion morphism $\iota_i\colon C_i \to X$ is a coalgebra morphism $(C_i, \xi_i) \to (X, \xi)$, for some comonad coalgebra $\xi_i\colon C_i \to \mathbb{D}_M(C_i)$.

Lastly, because the forgetful functor $U\colon \mathtt{EM}(\mathbb{D}_M) \to \mathcal{B}$ creates colimits (see e.g. Proposition 20.12 in [6]), $\coprod_{i\in F} C_i$ can be equipped with the coalgebra structure of the coproduct of the coalgebras $\xi_i\colon C_i \to \mathbb{D}_M(C_j)$. Moreover, the morphism $\iota_F\colon \coprod_{i\in F} C_i \to X$ is a coalgebra morphism because each of its components is. Also, $\coprod_{i\in F} C_j$ is finitely presentable because it is a finite coproduct of finitely presentable objects (see e.g. Proposition 1.3 in [7]). Finally, g is unique because $\mathcal{B}$ is a componental category and ι_F is the inclusion morphism of $\coprod_F C_i$ into the coproduct $\coprod_I C_i$ and hence a monomorphism. □

As a corollary of Propositions 20 and 22 we obtain the main theorem of this section.

Theorem 23. *Let $M\colon \mathcal{A} \to \mathcal{B}$ be a functor from a discrete category $\mathcal{A}$ to a componental category $\mathcal{B}$ with finite coproducts and assume that the pointwise density comonad $\mathbb{D}_M$ for M exists. If all objects in the image of M are connected and finitely presentable then $\mathbb{D}_M$ has finite rank.*

Observe that the constructed functor $M\colon \mathcal{A} \to \mathcal{B}$ in the proof of Theorem 1 (cf. the paragraph following Theorem 9) automatically satisfies the assumptions of Theorem 23. Indeed, M is an inclusion of a class of finite connected structures and finite structures are precisely the finitely presentable objects in the category of relational structures or the category of graphs. Therefore, Theorem 23 together with Theorem 19 concludes the proof of Corollary 2.

7 Conclusion

In this paper we have shown that classes of structures closed under isomorphism, disjoint unions, and summands can always be classified by a comonad, and moreover this comonad admits a Lovász-type theorem. We have also shown that standard graph parameters give rise to graded comonads, i.e. sequences of comonads indexed by real numbers, such that the graph parameter is captured by the coalgebra number of a given structure. Both results cover a huge range of examples of structure classes and graph parameters from the literature.

The comonads we construct are, in some sense, the minimal solutions to this problem in that they are weakly initial among the comonads that classify the same class of structures. Our proofs show is that the classifying comonads (or graded comonads) can be very simple and do not need to be specifically tailored for the concept at hand.

Conversely, however, the power of game comonads is that they shed light on previously known constructions and reveal new connections between them. This can lead to new results. For example, the links between game comonads, logic fragments and combinatorial parameters established in [1,4] are leveraged in [20] and [26] to obtain new results for other combinatorial properties simply by changing the comonad at hand.

In [3] the common structure exhibited by game comonads is axiomatised in terms of *arboreal categories* and *arboreal covers*. This suggests one line of further development, by relating the general results using discrete density comonads of the present paper to the axiomatic setting of [3]. This can provide a basis for general transfer results of this kind between comonads arising from arboreal covers.

Another potential source of useful comonads is when a particular class of structures is given by a *construction*, similar to the inductive definition of clique-width or the algebraic definition of planar graphs found e.g. in [25]. We hope to explore comonads arising from inductive constructions in future work.

Acknowledgements. We thank the referees for careful reading and many useful suggestions for improvements. SA thanks Mikołaj Bojańczyk and Bartek Klin for very stimulating and enjoyable discussions in Warsaw in 2018 on finding a comonad to classify clique-width, which led to some initial results. TJ thanks Anuj Dawar who asked for a comonad for co-spectrality, which was another stimulus for this research. We thank all the members of the EPSRC project on Resources and Coresources for their encouragement and feedback.

References

1. Abramsky, S., Dawar, A., Wang, P.: The pebbling comonad in finite model theory. In: Proceedings of the 32nd Annual ACM/IEEE Symposium on Logic in Computer Science (LICS), pp. 1–12. IEEE (2017)
2. Abramsky, S., Marsden, D.: Comonadic semantics for guarded fragments. In: Proceedings of the 36th Annual ACM/IEEE Symposium on Logic in Computer Science (LICS), pp. 1–13. IEEE (2021)

3. Abramsky, S., Reggio, L.: Arboreal categories and resources. In: 48th International Colloquium on Automata, Languages, and Programming (ICALP 2021), pp. 115:1–115:20. Schloss Dagstuhl-Leibniz-Zentrum für Informatik (2021)
4. Abramsky, S., Shah, N.: Relating structure and power: Comonadic semantics for computational resources. J. Log. Comput. **31**(6), 1390–1428 (2021)
5. Abramsky, S., Tzevelekos, N.: Introduction to categories and categorical logic. In: New Structures for Physics, pp. 3–94. Springer Berlin, Heidelberg (2010). https://doi.org/10.1007/978-3-642-12821-9
6. Adámek, J., Herrlich, H., Strecker, G.E.: Abstract and Concrete Categories. John Wiley & Sons Inc., New York (1990)
7. Adámek, J., Rosický, J.: Locally Presentable and Accessible Categories, vol. 189. Cambridge University Press, Cambridge (1994)
8. Appelgate, H., Tierney, M.: Categories with models. In: Eckmann, B. (ed.) Seminar on Triples and Categorical Homology Theory. LNM, vol. 80, pp. 156–244. Springer, Heidelberg (1969). https://doi.org/10.1007/BFb0083086
9. Awodey, S.: Category Theory. Oxford University Press, New York (2010)
10. Böker, J.: Structural similarity and homomorphism counts. Master's thesis, RWTH Aachen (2018)
11. Bulian, J., Dawar, A.: Fixed-parameter tractable distances to sparse graph classes. Algorithmica **79**(1), 139–158 (2017)
12. Conghaile, A.Ó., Dawar, A.: Game comonads & generalised quantifiers. In: Baier, C., Goubault-Larrecq, J. (eds.) 29th EACSL Annual Conference on Computer Science Logic (CSL 2021). Leibniz International Proceedings in Informatics (LIPIcs), vol. 183, pp. 16:1–16:17. Schloss Dagstuhl-Leibniz-Zentrum für Informatik, Dagstuhl, Germany (2021). https://doi.org/10.4230/LIPIcs.CSL.2021.16
13. Dawar, A., Jakl, T., Reggio, L.: Lovász-type theorems and game comonads. In: Proceedings of the 36th Annual ACM/IEEE Symposium on Logic in Computer Science (LICS), pp. 1–13. Association for Computing Machinery, New York, NY, USA (2021). https://doi.org/10.1109/LICS52264.2021.9470609
14. Diers, Yves: Categories of Boolean sheaves of simple algebras. In: Categories of Boolean Sheaves of Simple Algebras. LNM, vol. 1187, pp. 48–113. Springer, Heidelberg (1986). https://doi.org/10.1007/BFb0075890
15. Kan extensions in Enriched Category Theory. LNM, vol. 145. Springer, Heidelberg (1970). https://doi.org/10.1007/BFb0060485
16. Dvořák, Z.: On recognizing graphs by numbers of homomorphisms. J. Graph Theory **64**(4), 330–342 (2010)
17. Grohe, M.: Counting bounded tree depth homomorphisms. In: Proceedings of the 35th Annual ACM/IEEE Symposium on Logic in Computer Science (LICS), pp. 507–520 (2020)
18. Grohe, M., Rattan, G., Seppelt, T.: Homomorphism tensors and linear equations (2022). https://arxiv.org/abs/2111.11313, Sccepted at the 49th EATCS International Colloquium on Automata, Languages and Programming (ICALP)
19. Hammack, R.H., Imrich, W., Klavžar, S.: Handbook of Product Graphs, vol. 2. CRC Press, Taylor & Francis Group, New York (2011)
20. Jakl, T., Marsden, D., Shah, N.: A game comonadic account of Courcelle and Feferman-Vaught-Mostowski theorems (2022). in preparation, arXiv preprint. https://arxiv.org/abs/2205.05387
21. Kock, A.: Continuous Yoneda representation of a small category, University of Aarhus, Denmark (1966)
22. Leinster, T.: Codensity and the ultrafilter monad. Theory Appl. Categ. **28**(13), 332–370 (2013)

23. Lovász, L.: Operations with structures. Acta Math. Acad. Sci. Hungar. **18**, 321–328 (1967)
24. Lovász, L.: Large Networks and Graph Limits, vol. 60. American Mathematical Society (2012)
25. Mančinska, L., Roberson, D.E.: Quantum isomorphism is equivalent to equality of homomorphism counts from planar graphs. In: 2020 IEEE 61st Annual Symposium on Foundations of Computer Science (FOCS), pp. 661–672. IEEE (2020)
26. Montacute, Y., Shah, N.: The Pebble-Relation Comonad in Finite Model Theory (2021). https://arxiv.org/abs/2110.08196, Accepted at the 37th Annual ACM/IEEE Symposium on Logic in Computer Science (LICS)
27. Nešetřil, J., De Mendez, P.O.: On nowhere dense graphs. Eur. J. Comb. **32**(4), 600–617 (2011)
28. Ramana, M.V., Scheinerman, E.R., Ullman, D.: Fractional isomorphism of graphs. Discret. Math. **132**, 247–265 (1994)
29. Reggio, L.: Polyadic sets and homomorphism counting (2021). Submitted, arXiv preprint. https://arxiv.org/abs/2110.11061
30. Wisbauer, R.: Algebras versus coalgebras. Appl. Categ. Struct. **16**(1), 255–295 (2008)

Coalgebraic Semantics for Nominal Automata

Florian Frank, Stefan Milius(✉), and Henning Urbat

Friedrich-Alexander-Universität Erlangen-Nürnberg, Erlangen, Germany
stefan.milius@fau.de

Abstract. This paper provides a coalgebraic approach to the language semantics of two types of non-deterministic automata over nominal sets: non-deterministic orbit-finite automata (NOFAs) and regular nominal non-deterministic automata (RNNAs), which were introduced in previous work. While NOFAs are a straightforward nominal version of non-deterministic automata, RNNAs feature ordinary as well as name binding transitions. Correspondingly, words accepted by RNNAs are strings formed by ordinary letters and name binding letters. Bar languages are sets of such words modulo α-equivalence, and to every state of an RNNA one associates its accepted bar language. We show that the semantics of NOFAs and RNNAs, respectively, arise both as an instance of the Kleisli-style coalgebraic trace semantics as well as an instance of the coalgebraic language semantics obtained via generalized determinization. On the way we revisit coalgebraic trace semantics in general and give a new compact proof for the main result in that theory stating that an initial algebra for a functor yields the terminal coalgebra for the Kleisli extension of the functor. Our proof requires fewer assumptions on the functor than all previous ones.

1 Introduction

Classical automata and their language semantics have long been understood in the theory of coalgebras. For example, it is a well-known exercise [37] that standard deterministic automata over a fixed alphabet can be modelled as coalgebras, that the terminal coalgebra is formed by all formal languages over that alphabet, and the unique homomorphism into the terminal coalgebra assigns to each state of an automaton the language it accepts. Non-deterministic automata are also coalgebras for a functor extending the one for deterministic automata in order to accomodate non-deterministic branching. Their language semantics can be obtained coalgebraically in two different ways. First, in the *coalgebraic trace semantics* by Hasuo et al. [18] one considers coalgebras for composed functors TF where F is a set functor modelling the type of transitions and T is a set monad modelling the type of branching; for example, for non-deterministic branching one takes the

Funded by the Deutsche Forschungsgemeinschaft (DFG, German Research Foundation) – project number 419850228.

Published by Springer Nature Switzerland AG 2022
H. H. Hansen and F. Zanasi (Eds.): CMCS 2022, LNCS 13225, pp. 45–66, 2022.
https://doi.org/10.1007/978-3-031-10736-8_3

power-set monad. Under certain conditions on F and T, including that F has an extension $\bar{F}$ to the Kleisli category of T, an initial F-algebra is seen to lift to the terminal coalgebra for $\bar{F}$. Its universal property then yields the coalgebraic trace semantics. Among the instances of this is the standard language semantics of non-deterministic automata.

Second the *coalgebraic language semantics* [7] is based on generalized determinization by Silva et al. [40]. Here one considers coalgebras for composed functors GT where G models transition types and T again models the branching type. Assuming that G has a lifting to the Eilenberg-Moore category for T, generalized determinization turns such a coalgebra into a G-coalgebra by taking the unique extension of the coalgebra structure to the free Eilenberg-Moore algebra on the set of states. Moreover, taking the unique homomorphism from that coalgebra into the terminal G-coalgebra yields the coalgebraic language semantics. In the leading instance of non-deterministic automata, generalized determinization is the well-known power-set construction and coalgebraic language semantics the standard automata-theoretic language semantics once again.

These two approaches were brought together by Jacobs et al. [21] who study those species of systems which can be modelled as coalgebras in both of the above ways. They show that whenever there exists an *extension* natural transformation $TF \to GT$ satisfying two natural equational laws, then the two above semantics are canonically related, and they agree in the instances studied in op. cit.

It is our aim in this paper to draw a similar picture for non-deterministic automata for languages over infinite alphabets. Such alphabets allow to model *data*, such as nonces [28], object identities [16], or abstract resources [8], and the ensuing languages are therefore called *data languages*. There are several species of automata for data languages in the literature. We focus on two types which are known to have a presentation as coalgebras over the category of nominal sets: non-deterministic orbit-finite automata (NOFA) [4] and regular non-deterministic nominal automata (RNNA) [39]. For both of these types of automata one works with the category of nominal sets and takes the set of *names* as the alphabet. While NOFAs are a straightforward nominal version of standard non-deterministic automata, RNNAs feature *binding transitions*, which can be thought as storing an input name in a 'register' for comparison with future input names. Correspondingly, they accept words including name binding letters and which are taken modulo α-equivalence; such words form *bar languages* (the name stems from the bar in front of name binding letters $|a$). However, while these automata are understood as coalgebras, their semantics has not been studied from a coalgebraic perspective so far.

We fill this gap here and prove that the data language accepted by a NOFA and the bar language accepted by an RNNA arise as instances of both coalgebraic trace semantics (Theorems 3.18 and 3.21) and coalgebraic language semantics (Corollaries 4.16 and 4.22). The latter result is obtained by using canonical extension natural transformations obtained from the result by Jacobs et al. [21].

While these results will perhaps hardly surprise the cognoscenti, and the treatment of NOFAs indeed appears as an(other) exercise in coalgebra, we should like to point out that there are a number of technical subtleties arising in the treatment of RNNAs. Essentially, what causes some trouble is the presence of the abstraction functor in their type. We solve all these difficulties by working with the uniformly finitely supported power-set monad $\mathcal{P}_{\mathsf{ufs}}$ on nominal sets in lieu of the more common finitely supported power-set monad $\mathcal{P}_{\mathsf{fs}}$ (which provides the power objects of the topos of nominal sets). Note also that for a nominal set X, neither $\mathcal{P}_{\mathsf{fs}}X$ nor $\mathcal{P}_{\mathsf{ufs}}X$ form cpos (so, in particular, they do not form complete lattices). Hence, it may come as a bit of a surprise that the Kleisli categories of both monads are nevertheless enriched over complete lattices (Proposition 3.7), one of the key requirements for coalgebraic trace semantics.

We present our results in a modular way so that they may be reusable for the study of coalgebraic semantics for other types of nominal systems, such as nominal tree automata. For example, we show that all *binding polynominal functors*, e.g. those functors arising from a binding signature in the sense of Fiore et al. [10] have a canonical extension to the Kleisli category of $\mathcal{P}_{\mathsf{ufs}}$ (Corollary 3.13). Analogously, we show a lifting result for terminal coalgebras to the Eilenberg-Moore category for a subclass of these functors (Corollary 4.11).

Last but not least, on the way to the coalgebraic semantics of NOFAs and RNNAs we take a fresh look at coalgebraic trace semantics in general. We provide a new compact proof for the main theorem of that theory. It states that for a functor F and a monad T satisfying certain conditions, including that F has an extension $\bar{F}$ to the Kleisli category of T, the initial F-algebra extends to a terminal coalgebra for $\bar{F}$ (Theorem 3.5). We obtain this essentially as a combination of Hermida and Jacobs' adjoint lifting theorem [19, Theorem 2.14] and an argument originally given by Freyd [11] that for locally continuous endofunctors on categories enriched in cpos an initial algebra yields a terminal coalgebra. Here we adjust this argument to work for locally monotone endofunctors on categories enriched in directed-complete partial orders. As a consequence, our proof does not require the existence of a zero object in the Kleisli category of T and, notably, we only need the mere existence of the initial algebra for F and not that it is obtained after ω steps of the initial-algebra chain given by $F^n 0$ $(n < \omega)$.

2 Preliminaries

2.1 Nominal Sets

Nominal sets form a convenient formalism for dealing with names and freshness; for our present purposes, names play the role of data. We briefly recall basic notions and facts and refer to Pitts' book [34] for a comprehensive introduction. Fix a countably infinite set $\mathbb{A}$ of *names*, and let $\mathrm{Perm}(\mathbb{A})$ denote the group of finite permutations on $\mathbb{A}$, which is generated by the *transpositions* $(a\,b)$ for $a \neq b \in \mathbb{A}$ (recall that $(a\,b)$ just swaps a and b). A *nominal set* is a set X equipped with a (left) group action $\mathrm{Perm}(\mathbb{A}) \times X \to X$, denoted $(\pi, x) \mapsto \pi \cdot x$, such that every element $x \in X$ has a finite *support* $S \subseteq \mathbb{A}$, i.e. $\pi \cdot x = x$ for every $\pi \in \mathrm{Perm}(\mathbb{A})$ such that $\pi(a) = a$ for all $a \in S$. Every element x of a nominal set X has a least finite support, denoted $\mathsf{supp}(x)$. Intuitively, one should think of X as a set of syntactic objects (e.g. strings, λ-terms, programs), and of $\mathsf{supp}(x)$

as the set of names needed to describe an element $x \in X$. A name $a \in \mathbb{A}$ is *fresh* for x, denoted $a \# x$, if $a \notin \mathsf{supp}(x)$. The *orbit* of an element $x \in X$ is given by $\{\pi \cdot x : \pi \in \mathrm{Perm}(\mathbb{A})\}$. The orbits form a partition of X. The nominal set X is *orbit-finite* if it has only finitely many orbits.

A map $f\colon X \to Y$ between nominal sets is *equivariant* if $f(\pi\cdot x) = \pi\cdot f(x)$ for all $x \in X$ and $\pi \in \mathrm{Perm}(\mathbb{A})$. Equivariance implies $\mathsf{supp}(f(x)) \subseteq \mathsf{supp}(x)$ for all $x \in X$. We denote by Nom the category of nominal sets and equivariant maps.

Putting $\pi \cdot a = \pi(a)$ makes $\mathbb{A}$ into a nominal set. Moreover, $\mathrm{Perm}(\mathbb{A})$ acts on subsets $A \subseteq X$ of a nominal set X by $\pi \cdot A = \{\pi \cdot x : x \in A\}$. A subset $A \subseteq X$ is *equivariant* if $\pi \cdot A = A$ for all $\pi \in \mathrm{Perm}(\mathbb{A})$. More generally, it is *finitely supported* if it has finite support w.r.t. this action, i.e. there exists a finite set $S \subseteq \mathbb{A}$ such that $\pi \cdot A = A$ for all $\pi \in \mathrm{Perm}(\mathbb{A})$ such that $\pi(a) = a$ for all $a \in S$. The set A is *uniformly finitely supported* if $\bigcup_{x\in A} \mathsf{supp}(x)$ is a finite set. This implies that A is finitely supported, with least support $\mathsf{supp}(A) = \bigcup_{x\in A} \mathsf{supp}(x)$ [12, Theorem 2.29]. (The converse does not hold, e.g. the set $\mathbb{A}$ is finitely supported but not uniformly finitely supported.) Uniformly finitely supported orbit-finite sets are always finite (since an orbit-finite set contains only finitely many elements with a given finite support). We denote by $\mathcal{P}_{\mathsf{ufs}}\colon \mathsf{Nom} \to \mathsf{Nom}$ and $\mathcal{P}_{\mathsf{fs}}\colon \mathsf{Nom} \to \mathsf{Nom}$ the endofunctors sending a nominal set X to its set of (uniformly) finitely supported subsets and an equivariant map $f\colon X \to Y$ to the map $A \mapsto f[A]$.

The coproduct $X + Y$ of nominal sets X and Y is given by their disjoint union with the group action inherited from the two summands. Similarly, the product $X \times Y$ is given by the cartesian product with the componentwise group action; we have $\mathsf{supp}(x, y) = \mathsf{supp}(x) \cup \mathsf{supp}(y)$. Given a nominal set X equipped with an equivariant equivalence relation, i.e. an equivalence relation $\sim$ that is equivariant as a subset $\sim \subseteq X \times X$, the quotient $X/{\sim}$ is a nominal set under the expected group action defined by $\pi \cdot [x]_\sim = [\pi \cdot x]_\sim$.

A key role in the theory of nominal sets is played by *abstraction sets*, which provide a semantics for binding mechanisms [13]. Given a nominal set X, an equivariant equivalence relation $\sim$ on $\mathbb{A} \times X$ is defined by $(a, x) \sim (b, y)$ iff $(a\,c)\cdot x = (b\,c)\cdot y$ for some (equivalently, all) fresh c. The *abstraction set* $[\mathbb{A}]X$ is the quotient set $(\mathbb{A}\times X)/{\sim}$. The $\sim$-equivalence class of $(a, x) \in \mathbb{A}\times X$ is denoted by $\langle a\rangle x \in [\mathbb{A}]X$. We may think of $\sim$ as an abstract notion of α-equivalence, and of $\langle a\rangle$ as binding the name a. Indeed we have $\mathsf{supp}(\langle a\rangle x) = \mathsf{supp}(x) \setminus \{a\}$ (while $\mathsf{supp}(a, x) = \{a\} \cup \mathsf{supp}(x)$), as expected in binding constructs.

The object map $X \mapsto [\mathbb{A}]X$ extends to an endofunctor $[\mathbb{A}]\colon \mathsf{Nom} \to \mathsf{Nom}$ sending an equivariant map $f\colon X \to Y$ to the equivariant map $[\mathbb{A}]f\colon [\mathbb{A}]X \to [\mathbb{A}]Y$ given by $\langle a\rangle x \mapsto \langle a\rangle f(x)$ for $a \in \mathbb{A}$ and $x \in X$.

2.2 Nominal Automata

In this section, we recall two notions of nominal automata earlier introduced in the literature: non-deterministic orbit-finite automata (NOFAs) [4] and regular non-deterministic nominal automata (RNNAs) [39]. The former accept *data languages* (consisting of finite words over an infinite alphabet) while the latter accept *bar languages* (consisting of finite words formed by ordinary letters and name binding ones, taken modulo α-equivalence).

Definition 2.1 [4]. (1) A *NOFA* $A = (Q, R, F)$ is given by an orbit-finite nominal set Q of *states*, an equivariant relation $R \subseteq Q \times \mathbb{A} \times Q$ specifying *transitions*, and an equivariant set $F \subseteq Q$ of *final states*. We write $q \xrightarrow{a} q'$ in lieu of $(q, a, q') \in R$.
(2) Given a string $w = a_1 a_2 \cdots a_n \in \mathbb{A}^*$ and a state $q \in Q$, a *run* for w from q is a sequence of transitions $q \xrightarrow{a_1} q_1 \xrightarrow{a_2} \cdots \xrightarrow{a_n} q_n$. The run is *accepting* if q_n is final. The state q *accepts* w if there exists an accepting run for w from q. The data language *accepted* by q is given by $\{w \in \mathbb{A}^* : q \text{ accepts } w\}$.

NOFAs are known to be expressively equivalent to *finite memory automata* [25]. We note that in contrast to [4] we do not require NOFAs to have an initial state $q_0 \in Q$; this is more natural from a coalgebraic point of view. Moreover, the orbit-finiteness of the states is not relevant for our results and could be dropped.

Remark 2.2. (1) Given an endofunctor F on a category $\mathscr{C}$, an *F-coalgebra* is a pair (C, c) of an object C and a morphism $c\colon C \to FC$ on $\mathscr{C}$. A *homomorphism* of F-coalgebras from (C, c) to (D, d) is a morphism $h\colon C \to D$ with $d \cdot h = Fh \cdot c$.
(2) A NOFA corresponds precisely to an orbit-finite coalgebra $\langle f, \delta\rangle\colon Q \longrightarrow 2 \times \mathcal{P}_{\mathsf{fs}}(\mathbb{A} \times Q)$ for the functor on Nom given by

$$Q \mapsto \mathcal{P}_{\mathsf{fs}}(1 + \mathbb{A} \times Q) \cong 2 \times \mathcal{P}_{\mathsf{fs}}(\mathbb{A} \times Q).$$

In fact, $f\colon Q \to 2$ defines the equivariant set $F \subseteq Q$ of final states and $\delta\colon Q \to \mathcal{P}_{\mathsf{fs}}(\mathbb{A} \times Q)$ defines the transitions via $q \xrightarrow{a} q'$ iff $(a, q') \in \delta(q)$.

In order to incorporate explicit name binding into the automata-theoretic setting, we work with *bar strings*, i.e. finite words over the infinite alphabet

$$\bar{\mathbb{A}} := \mathbb{A} \cup \{|a : a \in \mathbb{A}\}.$$

We denote the nominal set of all bar strings by $\bar{\mathbb{A}}^*$, and we equip it with the group action defined pointwise. The letter $|a$ is interpreted as binding the name a to the right. Accordingly, a name $a \in \mathbb{A}$ is said to be *free* in a bar string $w \in \bar{\mathbb{A}}^*$ if (1) the letter a occurs in w, and (2) the first occurrence of a is not preceded by any occurrence of $|a$. For instance, the name a is free in $a|aba$ but not free in $|aaba$, while the name b is free in both bar strings. This yields a natural notion of α-equivalence:

Definition 2.3 (α-equivalence). Let $=_\alpha$ be the least equivalence relation on $\bar{\mathbb{A}}^*$ such that $x|av =_\alpha x|bw$ for all $a, b \in \mathbb{A}$ and $x, v, w \in \bar{\mathbb{A}}^*$ such that $\langle a\rangle v = \langle b\rangle w$. We denote by $\bar{\mathbb{A}}^*/{=_\alpha}$ the sets of α-equivalence classes of bar strings, and we write $[w]_\alpha$ for the α-equivalence class of $w \in \bar{\mathbb{A}}^*$.

Remark 2.4. (1) Pitts [34, Lemma 4.3], for every pair $v, w \in \bar{\mathbb{A}}^*$ the condition $\langle a\rangle v = \langle b\rangle w$ holds if and only if

$$a = b \text{ and } v = w, \qquad \text{or} \qquad b \,\#\, v \text{ and } (a\,b) \cdot v = w.$$

(2) The equivalence relation $=_\alpha$ is equivariant. Therefore, $\bar{\mathbb{A}}^*/{=_\alpha}$ forms a nominal set with the group action $\pi \cdot [w]_\alpha = [\pi \cdot w]_\alpha$ for $\pi \in \mathrm{Perm}(\mathbb{A})$ and $w \in \bar{\mathbb{A}}^*$. The least support of $[w]_\alpha$ is the set of free names of w.

Definition 2.5 [39]. (1) An *RNNA* $A = (Q, R, F)$ is given by an orbit-finite nominal set Q of *states*, an equivariant relation $R \subseteq Q \times \bar{\mathbb{A}} \times Q$ specifying *transitions*, and an equivariant set $F \subseteq Q$ of *final states*. We write $q \xrightarrow{\sigma} q'$ if $(q, \sigma, q') \in R$. The transitions are subject to two conditions:

(a) *α-invariance*: if $q \xrightarrow{\mathsf{l}a} q'$ and $\langle a \rangle q' = \langle b \rangle q''$, then $q \xrightarrow{\mathsf{l}b} q''$.
(b) *Finite branching up to α-invariance:* For every $q \in Q$ the sets

$$\{(a, q') : q \xrightarrow{a} q'\} \qquad \text{and} \qquad \{\langle a \rangle q' : q \xrightarrow{\mathsf{l}a} q'\}$$

are finite (equivalently, uniformly finitely supported).

(2) Given a bar string $w = \sigma_1\sigma_2 \cdots \sigma_n \in \bar{\mathbb{A}}^*$ and a state $q \in Q$, a *run* for w from q is a sequence of transitions $q \xrightarrow{\sigma_1} q_1 \xrightarrow{\sigma_2} \cdots \xrightarrow{\sigma_n} q_n$. The run is *accepting* if q_n is final. The state q *accepts* w if there exists an accepting run for w from q. The bar language *accepted* by q is given by $\{[w]_\alpha : w \in \bar{\mathbb{A}}^*,\ A \text{ accepts } w\}$.

Remark 2.6. As for NOFAs, we do not equip RNNAs with explicit initial states. Similar to Remark 2.2, RNNAs are seen to correspond to coalgebras $\langle f, \delta, \tau \rangle \colon Q \longrightarrow 2 \times \mathcal{P}_{\mathsf{ufs}}(\mathbb{A} \times Q) \times \mathcal{P}_{\mathsf{ufs}}([\mathbb{A}]Q)$ for the functor on Nom given by

$$Q \mapsto \mathcal{P}_{\mathsf{ufs}}(1 + \mathbb{A} \times Q + [\mathbb{A}]Q) \cong 2 \times \mathcal{P}_{\mathsf{ufs}}(\mathbb{A} \times Q) \times \mathcal{P}_{\mathsf{ufs}}([\mathbb{A}]Q).$$

Here f and δ correspond to final states and free transitions, and the equivariant map $\tau \colon Q \to \mathcal{P}_{\mathsf{ufs}}([\mathbb{A}]Q)$ defines the α-invariant bound transitions via $q \xrightarrow{\mathsf{l}a} q'$ iff $\langle a \rangle q' \in \tau(q)$. The use of $\mathcal{P}_{\mathsf{ufs}}$ (in lieu of $\mathcal{P}_{\mathsf{fs}}$) ensures that if Q is orbit-finite, then the finiteness conditions in the definition of an RNNA are met.

However, we note that while our results on coalgebraic semantics are stated for RNNAs they actually hold without orbit-finiteness assumptions.

Our goal is to interpret the above ad-hoc definition of the data languages of a NOFA and the bar languages of an RNNA within the coalgebraic framework.

2.3 Initial Algebras in $\mathsf{DCPO}_\perp$-enriched Categories

For the Kleisli-style coalgebraic trace semantics we shall make use of a result which shows that in categories where the hom-sets are enriched over directed-complete partial orders, the initial algebra and terminal coalgebra coincide.

Recall that a subset $D \subseteq P$ of a poset P is *directed* if every finite subset of D has an upper bound in D; equivalently, D is nonempty and for every $x, y \in D$, there exists a $z \in D$ with $x, y \leq z$. The poset P is a *dcpo with bottom* if it has a least element and *directed joins*, that is, every directed subset has a join in P. We write $\mathsf{DCPO}_\perp$ for the category of dcpos with bottom and continuous maps between them; a map is *continuous* if it is monotone and preserves directed joins.

Definition 2.7. (1) A category $\mathscr{C}$ is *left strictly* $\mathsf{DCPO}_\perp$*-enriched* provided that each hom-set is equipped with the structure of a dcpo with bottom, and composition preserves bottom on the left and is *continuous*: for every morphism f and appropriate directed sets of morphisms g_i ($i \in D$) we have

$$\perp \cdot f = \perp, \qquad f \cdot \bigvee_{i \in D} g_i = \bigvee_{i \in D} f \cdot g_i, \qquad \left(\bigvee_{i \in D} g_i\right) \cdot f = \bigvee_{i \in D} g_i \cdot f.$$

(2) A functor on $\mathscr{C}$ is *locally monotone* if its restrictions $\mathscr{C}(A, B) \to \mathscr{C}(FA, FB)$ to the hom-sets are monotone.

Theorem 2.8 [1, **Prop. 5.6**]. *Let F be a locally monotone functor on a left strictly* $\mathsf{DCPO}_\perp$*-enriched category. If an initial algebra $(\mu F, \iota)$ exists, then $(\mu F, \iota^{-1})$ is a terminal coalgebra.*

(This uses that the structure $\iota\colon F(\mu F) \to \mu F$ of the initial algebra is an isomorphism by Lambek's Lemma [29].) This result is an adaptation of an earlier related result proved by Freyd [11] for locally continuous functors on ω-cpo-enriched categories. Note that preservation of bottom on the right ($f \cdot \perp = \perp$) is not needed for this result.

3 Coalgebraic Trace Semantics

In this section we shall see that the (bar) language semantics of NOFAs and RNNAs is an instance of coalgebraic trace semantics. To this end we first adapt and generalize the coalgebraic trace semantics for set functors by Hasuo et al. [18] to arbitrary categories. Here one considers coalgebras for composed functors TF, where T is a monad modelling a branching type like non-determinism or probabilistic branching, and F models the type of transitions of systems. We then instantiate this to coalgebras in Nom for functors TF, where T is $\mathcal{P}_{\mathsf{fs}}$ and F a polynominal functor or $T = \mathcal{P}_{\mathsf{ufs}}$ and F a binding polynomial functor. Specifically, we obtain the two desired types of nominal automata as instances.

3.1 General Coalgebraic Trace Semantics Revisited

We begin by recalling a few facts about extensions of functors to Kleisli categories.

Remark 3.1. Let F be a functor and (T, η, μ) a monad, both on the category $\mathscr{C}$.
(1) The *Kleisli category* $\mathsf{Kl}(T)$ has the same objects as $\mathscr{C}$ and a morphisms f from X to Y is a morphism $f\colon X \to TY$ of $\mathscr{C}$. The composition of f with $g\colon Y \to TZ$ is defined by $\mu_Z \cdot Tg \cdot f$ and the identity on X is $\eta_X\colon X \to TX$. We have the identity-on-objects functor $J\colon \mathscr{C} \to \mathsf{Kl}(T)$ defined by $J(f\colon X \to Y) = \eta_Y \cdot f$.
(2) An endofunctor $\bar{F}\colon \mathsf{Kl}(T) \to \mathsf{Kl}(T)$ *extends* the functor F if $\bar{F}J = JF$. It is well known and easy to prove (see Mulry [33]) that extensions of F to $\mathsf{Kl}(T)$ are

in bijective correspondence with *distributive laws* of F over T; these are natural transformations $\lambda\colon FT \to TF$ compatible with the monad structure of T:

$$\begin{array}{ccc} F & \xrightarrow{F\eta} & FT \\ & \searrow_{\eta F} & \downarrow \lambda \\ & & TF \end{array} \qquad \begin{array}{ccccc} FTT & \xrightarrow{\lambda T} & TFT & \xrightarrow{T\lambda} & TTF \\ F\mu \downarrow & & & & \downarrow \mu F \\ FT & & \xrightarrow{\lambda} & & TF \end{array}$$

(3) Let G be a quotient functor of F, which means that we have a natural transformation with epimorphic components $q\colon F \twoheadrightarrow G$. Suppose that F extends to $\mathsf{Kl}(T)$ via a distributive law $\lambda\colon FT \to TF$. Then an object-indexed family of morphisms $\varrho_X\colon GTX \to TGX$ is a distributive law of G over T provided that the following squares commute

$$\begin{array}{ccc} FTX & \xrightarrow{\lambda_X} & TFX \\ q_{TX} \downarrow & & \downarrow Tq_X \\ GTX & \xrightarrow{\varrho_X} & TGX \end{array} \qquad \text{for every object } X \text{ of } \mathscr{C}.$$

Example 3.2. (1) Constant functors and the identity functor on $\mathscr{C}$ obviously extend to $\mathsf{Kl}(T)$.
(2) For a pair F, G of endofunctors which extend to $\mathsf{Kl}(T)$, their composition extends, too, and we have $\overline{GF} = \bar{G}\bar{F}$.
(3) Suppose that $\mathscr{C}$ has coproducts. Then $\overline{F+G} = \bar{F} + \bar{G}$, for a pair F, G of endofunctors which extend to $\mathsf{Kl}(\mathscr{C})$. Indeed, for a coproduct $F+G$ one uses that $J\colon \mathscr{C} \to \mathsf{Kl}(T)$, being a left adjoint, preserves coproducts. Given extensions $\bar{F}$ and $\bar{G}$, it is then clear that $\bar{F}+\bar{G}$ extends $F+G$: for every morphism $f\colon X \to TY$ in $\mathsf{Kl}(T)$ one has

$$\overline{F+G}(f) = \big(FX + FY \xrightarrow{\bar{F}f + \bar{G}f} TFY + TGY \xrightarrow{[T\mathsf{inl}, T\mathsf{inr}]} T(FY + GY)\big),$$

where $FY \xrightarrow{\mathsf{inl}} FY + GY \xleftarrow{\mathsf{inr}} GY$ are the coproduct injections. This works similarly for arbitrary coproducts.
(4) Suppose that $\mathscr{C}$ has finite products. Then finite products of functors with an extension can be extended when the monad T is *commutative*; this notion was introduced by Kock [26, Definition 3.1]. It is based on the notion of a *strong* monad, that is a monad T equipped with a natural transformation $s_{X,Y}\colon X \times TY \to T(X \times Y)$ (called *strength*) satisfying four natural equational laws (two w.r.t. 1 and $\times$ on $\mathscr{C}$ and two w.r.t. the monad structure).
We do not recall these laws explicitly since they are not needed for our exposition. A strength gives rise to a *costrength* $t_{X,Y}\colon TX \times Y \to T(X \times Y)$ defined by

$$t_{X,Y} = \big(TX \times Y \cong Y \times TX \xrightarrow{s_{Y,X}} T(Y \times X) \xrightarrow{T(\cong)} T(X \times Y)\big).$$

The monad T is *commutative* if the following diagram commutes:

$$\begin{array}{ccccc} & \overset{s_{TX,Y}}{\nearrow} & T(TX \times Y) \xrightarrow{Tt_{X,Y}} TT(X \times Y) & \overset{\mu_{X\times Y}}{\searrow} & \\ TX \times TY & & \xrightarrow{\quad\quad\quad d_{X,Y} \quad\quad\quad} & & T(X \times Y) \\ & \underset{t_{X,TY}}{\searrow} & T(X \times TY) \xrightarrow{Ts_{X,Y}} TT(X \times Y) & \underset{\mu_{X\times Y}}{\nearrow} & \end{array}$$

The ensuing natural transformation d in the middle is used to extend the product $F \times G$ of endofunctors on $\mathscr{C}$ having extensions $\bar{F}$ and $\bar{G}$ on $\mathsf{Kl}(T)$: for every morphism $f\colon X \to TY$ in $\mathsf{Kl}(T)$ one puts

$$\overline{F \times G}(f) = \big(FX \times GX \xrightarrow{\bar{F}f \times \bar{G}f} TFY \times TGY \xrightarrow{d_{FY,GY}} T(FY \times GY)\big).$$

Remark 3.3. (1) Every set monad is strong via a canonical strength; this follows, for example, from Moggi's result [32, Theorem 3.4]. For example, the power-set functor $\mathcal{P}\colon \mathbf{Set} \to \mathbf{Set}$ is commutative via its canonical strength

$$s_{X,Y}\colon X \times \mathcal{P}Y \to \mathcal{P}(X \times Y) \quad \text{defined by} \quad (x, S) \mapsto \{(x, s) : s \in S\}. \tag{3.1}$$

(2) As a consequence of what we saw in Example 3.2 every polynomial set functor has a canonical extension to the Kleisli category of any commutative set monad (cf. [18, Lemma 2.4]).
(3) More generally, this results holds for analytic set functors [30, Theorem 2.9]. That notion was introduced by Joyal [23,24], and he proved that analytic set functors are precisely those set functors which weakly preserve wide pullbacks.

With the help of Hermida and Jacobs' result [19, Theorem 2.14] on extending adjunctions to categories of algebras one easily obtains the following extension result for initial algebras:

Proposition 3.4. *Let T be a monad on the category $\mathscr{C}$ and let $F\colon \mathscr{C} \to \mathscr{C}$ have an extension $\bar{F}$ on $\mathsf{Kl}(T)$. If $(\mu F, \iota)$ is an initial F-algebra, then μF is an initial $\bar{F}$-algebra with the structure $J\iota = \eta_{\mu F} \cdot \iota\colon F(\mu F) \to T(\mu F)$.*

Coalgebraic trace semantics can be defined when the extended initial algebra above is also a terminal coalgebra for $\bar{F}$.

Theorem 3.5. *Let F be a functor and T a monad on the category $\mathscr{C}$. Assume that $\mathsf{Kl}(T)$ is left strictly $\mathsf{DCPO}_\perp$-enriched and that F has a locally monotone extension $\bar{F}$ on $\mathsf{Kl}(T)$ and an initial algebra $(\mu F, \iota)$. Then $(\mu F, J\iota^{-1})$ is a terminal coalgebra for $\bar{F}$.*

Proof. Immediate from Proposition 3.4 and Theorem 2.8. □

Compared to the previous result for **Set** [18, Theorem 3.3] our assumption on the enrichment of the Kleisli category is slightly stronger; in op. cit. only enrichment in ω-cpos is required. A related result [20, Theorem 5.3.4] for general base categories uses enrichment in directed-complete partial orders. However, in contrast to both of these results, we do not require that $\mathsf{Kl}(T)$ has a zero object and, most notably, we only need the mere existence of μF and not that the initial algebra for F is obtained by the first ω steps of the initial-algebra chain, that is, as the colimit of the ω-chain given by $F^n 0$ $(n < \omega)$. The technical reason for this is that the proof of Theorem 2.8 does not make use of the classical limit-colimit coincidence technique used e.g. by Smyth and Plotkin in their seminal work [42]. Consequently, our proof is easier and shorter than the previous ones.

Definition 3.6 (Coalgebraic Trace Semantics). Given F and T on $\mathscr{C}$ satisfying the assumptions in Theorem 3.5 and a coalgebra $c\colon X \to TFX$. The *coalgebraic trace map* is the unique coalgebra homomorphism tr_c from (X, c) to $(\mu F, J\iota^{-1})$; that is, the following diagram commutes in $\mathsf{Kl}(T)$:

$$\begin{array}{ccc} X & \xrightarrow{\mathsf{tr}_c} & \mu F \\ {\scriptstyle c}\downarrow & & \downarrow{\scriptstyle J\iota^{-1}} \\ \bar{F}X & \xrightarrow{\bar{F}\mathsf{tr}_c} & \bar{F}(\mu F) \end{array} \tag{3.2}$$

Among the instances of coalgebraic trace semantics are the trace semantics of labelled transition systems with explicit termination [18], which are the coalgebras for the set functor $\mathcal{P}(1 + \Sigma \times X)$, and that of probabilistic labelled transitions systems [17, Chapter 4], which are the coalgebras for the set functor $\mathcal{D}_{\leq}(1 + \Sigma \times X)$, where $\mathcal{D}_{\leq}$ denotes the subdistribution monad.

3.2 Coalgebraic Trace Semantics of Non-deterministic Nominal Systems

We will now work towards showing that the semantics of nominal automata is an instance of the coalgebraic trace semantics. To this end we will instantiate Theorem 3.5 to $\mathscr{C} = \mathsf{Nom}$, $FX = 1 + \mathbb{A} \times X$ and $T = \mathcal{P}_{\mathsf{fs}}$ (for NOFAs), or to $FX = 1 + \mathbb{A} \times X + [\mathbb{A}]X$ and $T = \mathcal{P}_{\mathsf{ufs}}$ (for RNNAs), cf. Remarks 2.2 and 2.6. More generally, we show that every endofunctor arising from a nominal algebraic signature in the sense of Pitts [34, Definition 8.2] has a locally monotone extension to $\mathsf{Kl}(\mathcal{P}_{\mathsf{ufs}})$. For $T = \mathcal{P}_{\mathsf{fs}}$ most of the development also works out, as we shall see. However, the distributive law for the abstraction functor in the proof of Proposition 3.11 is not well-defined for $\mathcal{P}_{\mathsf{fs}}$.

But the first obstacle is that the nominal sets $\mathcal{P}_{\mathsf{fs}}X$ and $\mathcal{P}_{\mathsf{ufs}}X$ are in general no complete lattices (and not even ω-cpos) since the union of a chain of (uniformly) finitely supported sets may fail to be (uniformly) finitely supported.In this light, the following result is slightly surprising.

Proposition 3.7. *For every pair X, Y of nominal sets, the sets $\mathsf{Kl}(\mathcal{P}_{\mathsf{fs}})(X, Y)$ and $\mathsf{Kl}(\mathcal{P}_{\mathsf{ufs}})(X, Y)$ form complete lattices (whence dcpos with bottom).*

Corollary 3.8. *If a locally monotone endofunctor H on $\mathsf{Kl}(\mathcal{P}_{\mathsf{fs}})$ or $\mathsf{Kl}(\mathcal{P}_{\mathsf{ufs}})$ has an initial algebra $(\mu H, \iota)$, then $(\mu H, \iota^{-1})$ is its terminal coalgebra.*

This is a consequence of Theorem 2.8 since the composition in $\mathsf{Kl}(\mathcal{P}_{\mathsf{fs}})$ and $\mathsf{Kl}(\mathcal{P}_{\mathsf{ufs}})$ is easily seen to preserve the bottom (empty set) on the left and all joins (unions).

Extending functors to $\mathsf{Kl}(\mathcal{P}_{\mathsf{fs}})$ and $\mathsf{Kl}(\mathcal{P}_{\mathsf{ufs}})$. We now show that endofunctors arising from a nominal algebraic signature (with one name and one data sort) [34, Definition 8.2] have a canonical locally monotone extension to $\mathsf{Kl}(\mathcal{P}_{\mathsf{ufs}})$. For instance, the functor F used for RNNAs has a locally monotone extension $\bar{F}$ on $\mathsf{Kl}(\mathcal{P}_{\mathsf{ufs}})$.

Definition 3.9. The class of *binding polynomial functors* on Nom is the smallest class of functors containing the constant and identity and abstraction functors and being closed under coproducts, finite products and composition.

In other words, binding polynomial functors are formed according to the grammar:

$$F ::= C \mid \mathsf{Id} \mid [\mathbb{A}](-) \mid F \times F \mid \textstyle\coprod_{i\in I} F_i \mid FF, \tag{3.3}$$

where C ranges over all constant functors on Nom and I is an arbitrary index set. Functors arising from a binding signature in the sense of Fiore et al. [10] and those associated to a nominal algebraic signature with one name sort and one data sort (see Pitts [34, Definition 8.12]) are instances of binding polynomial functors.

Proposition 3.10. *The monads $\mathcal{P}_{\mathsf{fs}}$ and $\mathcal{P}_{\mathsf{ufs}}$ are commutative w.r.t. to the strengths obtained by restricting the one in* (3.1).

Proposition 3.11. *The abstraction functor $[\mathbb{A}](-)$ has a locally monotone extension on $\mathsf{Kl}(\mathcal{P}_{\mathsf{ufs}})$.*

Proof (Sketch). One uses Remark 3.1(3): the abstraction functor is a quotient of the functor $FX = \mathbb{A} \times X$ which is equipped with the canonical distributive law $\lambda_X\colon \mathbb{A} \times \mathcal{P}_{\mathsf{ufs}}X \to \mathcal{P}_{\mathsf{ufs}}(\mathbb{A} \times X)$ obtained using the strength of $\mathcal{P}_{\mathsf{ufs}}$ (Example 3.2(4) and cf. (3.1)). The maps $\varrho_X\colon [\mathbb{A}](\mathcal{P}_{\mathsf{ufs}}X) \to \mathcal{P}_{\mathsf{ufs}}([\mathbb{A}]X)$ are defined by $\varrho_X(\langle a\rangle S) = \{\langle a\rangle s : s \in S\}$. □

Remark 3.12. For the monad $\mathcal{P}_{\mathsf{fs}}$ our proof does not work. The problem is that ϱ_X above is not well-defined in general if S is not uniformly finitely supported. For example, for $\mathbb{A} \in \mathcal{P}_{\mathsf{fs}}\mathbb{A}$ we have $\langle a\rangle\mathbb{A} = \langle b\rangle\mathbb{A}$ for every pair a, b of names. However, if $a \neq b$, then the sets $\{\langle a\rangle c : c \in \mathbb{A}\}$ and $\{\langle b\rangle c : c \in \mathbb{A}\}$ differ: $\langle a\rangle b$ is contained in the former but not in the latter set. In fact, since $a \neq b$, $\langle a\rangle b = \langle b\rangle c$ can hold only if $a \# \{b, c\}$ and $b = (a\, b)\cdot c$ (see Pitts [34, Lemma 4.3]). The latter means that $c = a$ contradicting freshness of a.

Corollary 3.13. *Every binding polynomial functor has a canonical locally monotone extension to $\mathsf{Kl}(\mathcal{P}_{\mathsf{ufs}})$.*

Unsurprisingly, an analogous result holds for polynomial functors and $\mathcal{P}_{\mathsf{fs}}$ by the same reasoning applied to a grammar as in (3.3) that does not include the abstraction functor:

Corollary 3.14. *Every polynomial functor has a canonical locally monotone extension to* $\mathsf{Kl}(\mathcal{P}_{\mathsf{fs}})$.

Nominal Coalgebraic Trace Semantics. Every binding polynomial functor F is finitary and therefore has an initial algebra. In particular, if F arises from a nominal algebraic signature, we know from Pitts [34, Theorem 8.15] its initial algebra μF is carried by the nominal set of terms modulo α-equivalence (defined in Definition 8.6 of op. cit.) of the nominal algebraic signature. If F is polynomial, then α-equivalence is trivial and μF the usual set of terms. By Corollary 3.8 we have

Corollary 3.15. (1) *For every polynomial functor F the terminal coalgebra of its canonical extension $\bar{F}$ on $\mathsf{Kl}(\mathcal{P}_{\mathsf{fs}})$ is carried by the nominal set μF.* (2) *For every binding polynomial functor F the terminal coalgebra of its canonical extension $\bar{F}$ on $\mathsf{Kl}(\mathcal{P}_{\mathsf{ufs}})$ is carried by the nominal set μF.*

According to Definition 3.6 we can thus define a coalgebraic trace semantics for every coalgebra $X \to \mathcal{P}_{\mathsf{fs}} F X$ with F a polynomial functor, as well as for every coalgebra $X \to \mathcal{P}_{\mathsf{ufs}} F X$ with F a binding polynomial functor. We now instantiate this to the two types of nominal automata introduced in Sect. 2.2.

Coalgebraic Trace Semantics of NOFAs. Recall from Remark 2.2 that NOFAs are coalgebras $X \to \mathcal{P}_{\mathsf{fs}} F X$ where $FX = 1 + \mathbb{A} \times X$ on Nom.

Proposition 3.16. *The initial algebra for F is the nominal set $\mathbb{A}^*$ with structure $\iota\colon 1 + \mathbb{A} \times \mathbb{A}^* \to \mathbb{A}^*$ defined by $\iota(*) = \varepsilon$ and $\iota(a, w) = aw$.*

Indeed, the functor F arises from from the algebraic signature with a constant ε and unary operations $a(-)$ for every $a \in \mathbb{A}$, and clearly the corresponding term algebra is isomorphic to the algebra $\mathbb{A}^*$.

Corollary 3.17. *The terminal coalgebra for the extension $\bar{F}\colon \mathsf{Kl}(\mathcal{P}_{\mathsf{fs}}) \to \mathsf{Kl}(\mathcal{P}_{\mathsf{fs}})$ is $(\mathbb{A}^*, J\iota^{-1})$ for ι from Proposition 3.16.*

Theorem 3.18. *For every NOFA $c\colon X \to \mathcal{P}_{\mathsf{fs}} F X$ its coalgebraic trace map $\mathsf{tr}_c\colon X \to \mathcal{P}_{\mathsf{fs}}(\mathbb{A}^*)$ assigns to every state of X its accepted data language.*

Indeed, one readily works out that assigning to every state of X its data language is a coalgebra homomorphism from (X, c) to $(\mu F, J\iota^{-1})$ in $\mathsf{Kl}(\mathcal{P}_{\mathsf{fs}})$.

Coalgebraic Trace Semantics of RNNAs. Recall from Remark 2.6 that RNNAs are coalgebras $X \to \mathcal{P}_{\mathsf{ufs}} F X$ where $FX = 1 + \mathbb{A} \times X + [\mathbb{A}]X$ on Nom.

Proposition 3.19. *The initial algebra for $F\colon \mathsf{Nom} \to \mathsf{Nom}$ is the nominal set $\bar{\mathbb{A}}^*/{=_\alpha}$ of all bar strings modulo α-equivalence with the algebra structure $\iota\colon 1 + \mathbb{A} \times (\bar{\mathbb{A}}^*/{=_\alpha}) + [\mathbb{A}](\bar{\mathbb{A}}^*/{=_\alpha}) \to \bar{\mathbb{A}}^*/{=_\alpha}$ defined by*

$$\iota(*) = [\varepsilon]_\alpha, \qquad \iota(a, [w]_\alpha) = [aw]_\alpha, \qquad \iota(\langle a \rangle [w]_\alpha) = [|aw]_\alpha. \tag{3.4}$$

Indeed, the functor F arises from a nominal algebraic signature with a constant ε, unary operations $a(-)$ for every $a \in \mathbb{A}$ and one unary name binding operation I. Terms over this signature are obviously the same as bar strings. Moreover, it is not difficult to show that Pitts' notion of α-equivalence for terms [34, Definition 8.6] is equivalent to α-equivalence for bar strings in Definition 2.3. Finally, the algebra structure in (3.4) above corresponds to the one given by term formation by Pitts [34, Theorem 8.15]. Using Theorem 3.5 we thus obtain the following result.

Corollary 3.20. *The terminal coalgebra for the extension* $\bar{F}\colon \mathsf{Kl}(\mathcal{P}_{\mathsf{ufs}}) \to \mathsf{Kl}(\mathcal{P}_{\mathsf{ufs}})$ *is* $(\bar{\mathbb{A}}^*/{=_\alpha}, J\iota^{-1})$ *for* ι *from* (3.4).

Theorem 3.21. *For every RNNA* $c\colon X \to \mathcal{P}_{\mathsf{ufs}}FX$ *its coalgebraic trace map* $\mathsf{tr}_c\colon X \to \mathcal{P}_{\mathsf{ufs}}(\bar{\mathbb{A}}^*/{=_\alpha})$ *assigns to every state of* X *its accepted bar language.*

Indeed, one readily works out that assigning to every state of X its bar language is a coalgebra homomorphism from (X, c) to $(\mu F, J\iota^{-1})$ in $\mathsf{Kl}(\mathcal{P}_{\mathsf{ufs}})$.

4 Coalgebraic Language Semantics

In this section we shall see that the language semantics of NOFAs and RNNAs is an instance of coalgebraic language semantics [7]. The latter is based on the generalized determinization construction by Silva et al. [40]. Here one considers coalgebras for a functor GT, where T models a branching type and G models the type of transition of a system (similarly as before in the coalgebraic trace semantics, but this time the order of composition is reversed). Again, we will apply this to coalgebras in Nom for functors GT, where $T = \mathcal{P}_{\mathsf{fs}}$ and G is functor composed of products and exponentials, or to $T = \mathcal{P}_{\mathsf{ufs}}$ and G composed of products, exponentials and binding functors. Specifically, we obtain the two desired types of nominal automata as instances.

4.1 A Recap of General Coalgebraic Language Semantics

We begin by recalling a few facts about liftings of functors to Eilenberg-Moore categories.

Remark 4.1. Let G be a functor and (T, η, μ) be a monad on the category $\mathscr{C}$.
(1) The *Eilenberg-Moore* category $\mathsf{EM}(T)$ consists of algebras (A, a) for T, that is, pairs formed by an object A and a morphism $a\colon TA \to A$ such that $a \cdot \eta_A = id_A$ and $a \cdot \mu_A = a \cdot Ta$. A morphism in $\mathsf{EM}(T)$ from (A, a) to (B, b) is a morphism $h\colon A \to B$ of $\mathscr{C}$ such that $h \cdot a = b \cdot Th$. We write $U\colon \mathsf{EM}(T) \to \mathscr{C}$ for the forgetful functor mapping an algebra (A, a) to its underlying object A.
(2) A *lifting* of G is an endofunctor $\widehat{G}\colon \mathsf{EM}(T) \to \mathsf{EM}(T)$ such that $GU = U\widehat{G}$. As shown by Applegate [2] (see also Johnstone [22]), liftings of G to $\mathsf{EM}(T)$ are in bijective correspondence with distributive laws of T over G. The latter are natural transformations $\lambda\colon TG \to GT$ compatible with the monad structure:

$$G\eta = \lambda \cdot \eta G, \qquad \lambda \cdot \mu G = G\mu \cdot \lambda T \cdot T\lambda.$$

(3) Suppose that G has a terminal coalgebra $(\nu G, \tau)$ and the lifting $\widehat{G}$ on $\mathsf{EM}(T)$ via the distributive law λ. It follows from the work of Turi and Plotkin [35] (see also Bartels [3, Theorem 3.2.3]) that the terminal coalgebra for G lifts to a terminal coalgebra for $\widehat{G}$. In fact, one obtains a canonical structure of a T-algebra on νG by taking the unique coalgebra homomorphism α in the diagram below:

$$\begin{array}{ccccc} T(\nu G) & \xrightarrow{T\tau} & TG(\nu G) & \xrightarrow{\lambda_{\nu G}} & GT(\nu G) \\ {\scriptstyle\alpha}\downarrow & & & & \downarrow{\scriptstyle G\alpha} \\ \nu G & & \xrightarrow{\quad\tau\quad} & & G(\nu G) \end{array}$$

It is then easy to prove that α is indeed the structure of an algebra for T and that $\tau\colon \nu G \to G(\nu G)$ is a homomorphism of Eilenberg-Moore algebras (in fact, this is expressed by the commutativity of the above diagram). Moreover, $(\nu G, \tau)$ is the terminal $\widehat{G}$-coalgebra.

Proposition 4.2. *Let $T\colon \mathscr{C} \to \mathscr{C}$ be a monad and $L \dashv R\colon \mathscr{C} \to \mathscr{C}$ an adjunction with the counit $\varepsilon\colon LR \to \mathsf{Id}$. Given a distributive law $\lambda\colon LT \to TL$, we obtain a distributive law $\varrho\colon TR \to RT$ as the adjoint transpose of $LTR \xrightarrow{\lambda R} TLR \xrightarrow{T\varepsilon} T$.*

Recall that the *adjoint transpose* of a morphism $LX \to Y$ is the corresponding morphism $X \to RY$ under the natural isomorphism $\mathscr{C}(LX, Y) \cong \mathscr{C}(X, RY)$.

Example 4.3. (1) The identity functor on $\mathscr{C}$ obviously lifts to $\mathsf{EM}(T)$, and so does a constant functor on the carrier object of an Eilenberg-Moore algebra for T.
(2) Suppose that $\mathscr{C}$ has products. Then for a product $F \times G$ of functors one uses that U preserves products. Given liftings $\widehat{F}$ and $\widehat{G}$, it is clear that $\widehat{F} \times \widehat{G}$ is a lifting of $F \times G$. This works similarly for arbitrary products.
(3) Suppose that $\mathscr{C}$ is cartesian closed and that the monad T is strong (cf. Example 3.2(4)). Then the exponentiation functor $(-)^A$ lifts to $\mathsf{EM}(T)$ for every object A of $\mathscr{C}$. In fact, we apply Proposition 4.2 to the adjunction $A \times (-) \dashv (-)^A$ and use that two of the axioms of the strength $A \times TX \to T(A \times X)$ state that it is a distributive law of $A \times (-)$ over T.

Remark 4.4. Recall that, for every monad (T, η, μ) on $\mathscr{C}$, the pair (TX, μ_X) is the free algebra for T on X with the universal morphism $\eta_X\colon X \to TX$. Given an Eilenberg-Moore algebra (A, a) for T and a morphism $f\colon X \to A$ in $\mathscr{C}$, we have a unique morphism $f^\sharp\colon (TX, \mu_X) \to (A, a)$ in $\mathsf{EM}(T)$ such that $f^\sharp \cdot \eta_X = f$. We call $f^\sharp$ the *homomorphic extension* of f.

Construction 4.5 (Generalized Determinization [40]**).** Let T be a monad on the category $\mathscr{C}$ and G an endofunctor on $\mathscr{C}$ having a lifting $\widehat{G}$ on $\mathsf{EM}(T)$. Given a coalgebra $c\colon X \to GTX$ its *(generalized) determinization* is the G-coalgebra obtained by taking the homomorphic extension $c^\sharp\colon TX \to GTX$ using that $\widehat{G}(TX, \mu_X)$ is an algebra for T carried by GTX.

Among the instances of this construction are the well-known power-set construction of deterministic automata [40] as well as the non-determinization of alternating automata and that of Simple Segala systems [21].

Definition 4.6 (Coalgebraic Language Semantics [7]**).** Given T, G and a coalgebra $c\colon X \to GTX$ as in Construction 4.5, the *coalgebraic language morphism* $\ddagger c\colon X \to \nu G$ is the composite of the unique coalgebra homomorphism h from the determinization of (X, c) to νG with the unit η_X of the monad T, which is summarized in the diagram on the left below:

$$
\begin{array}{ccccc}
 & & \overset{\ddagger c}{\frown} & & \\
X & \xrightarrow{\eta_X} & TX & \xrightarrow{h} & \nu G \\
{\scriptstyle c}\downarrow & \swarrow{\scriptstyle c^\sharp} & & & \downarrow{\scriptstyle \tau} \\
GTX & & \xrightarrow{\quad Gh \quad} & & G(\nu G)
\end{array}
$$

Among the instances of coalgebraic language semantics are the language semantics of non-deterministic [21,40], weighted and probabilistic automata, but also the languages generated by context-free grammars [31,43], constructively $-algebraic formal power series for a semiring $ (the 'context-free' weighted languages) [31,44]. Less direct instances are the languages accepted by machines with extra memory such as (deterministic) push-down automata and Turing machines [15].

Relation of Coalgebraic Trace and Language Semantics. Jacobs et al. [21] show how the coalgebraic trace semantics and coalgebraic language semantics are connected in cases where both are applicable. We now give a terse review of this.

Assumption 4.7. We assume that T is a monad and F, G are endofunctors, all on the category $\mathscr{C}$, such that F has the extension $\bar{F}$ on $\mathsf{Kl}(T)$ via the distributive law $\lambda\colon FT \to TF$ and G has the lifting $\widehat{G}$ on $\mathsf{EM}(T)$ via the distributive law $\varrho\colon TG \to GT$. Moreover, we assume that we have an *extension* natural transformation $\varepsilon\colon TF \to GT$ compatible with the two distributive laws:

$$
\begin{array}{ccccc}
TFT & \xrightarrow{T\lambda} & TTF & \xrightarrow{\mu F} & TF \\
{\scriptstyle \varepsilon T}\downarrow & & & & \downarrow{\scriptstyle \varepsilon} \\
GTT & & \xrightarrow{\quad G\mu \quad} & & GT
\end{array}
\qquad
\begin{array}{ccccc}
TTF & & \xrightarrow{\quad \mu F \quad} & & TF \\
{\scriptstyle T\varepsilon}\downarrow & & & & \downarrow{\scriptstyle \varepsilon} \\
TGT & \xrightarrow{\varrho T} & GTT & \xrightarrow{G\mu} & GT
\end{array}
\tag{4.1}
$$

Remark 4.8. (1) For every object X of $\mathscr{C}$ the morphism ε_X is a homomorphism of Eilenberg-Moore algebras for T from (TFX, μ_{FX}) to $\widehat{G}(TX, \mu_X)$. Indeed, this is precisely what the commutativity of the diagram on the right in (4.1) expresses. (2) For every coalgebra $c\colon X \to TFX$ the extension natural transformation yields a coalgebra $\varepsilon_X \cdot c\colon X \to GTX$, and we take its determinization $(TX, (\varepsilon_X \cdot c)^\sharp)$. This is the object assignment of the functor $E\colon \mathsf{Coalg}(\bar{F}) \to \mathsf{Coalg}(\widehat{G})$ which maps an $\bar{F}$-coalgebra homomorphism $h\colon (X, c) \to (Y, d)$ to $Eh = h^\sharp\colon TX \to TY$, the homomorphic extension of $h\colon X \to TY$ (in $\mathscr{C}$). One readily proves that $h^\sharp$ is a $\widehat{G}$-coalgebra homomorphism using the naturality of ε as well as the

laws in (4.1). Functoriality follows since E is clearly a lifting of the canonical comparison functor $\mathsf{Kl}(T) \to \mathsf{EM}(T)$; see Jacobs et al. [21, Theorem 2] for the proof.
(3) We obtain a canonical morphism $e\colon T(\mu F) \to \nu G$ by applying the functor E to the coalgebra $J\iota^{-1}\colon \mu F \to TF(\mu F)$ (cf. Proposition 3.4) and taking the unique coalgebra homomorphism from it to the terminal $\widehat{G}$-coalgebra (Remark 4.1(3)).

Now recall the coalgebraic trace semantics from Definition 3.6. The following result follows from Jacobs et al.'s result [21, Prop. 5].

Proposition 4.9. *For every coalgebra $c\colon X \to TFX$ we have*

$$‡(\varepsilon_X \cdot c) = \big(X \xrightarrow{\mathsf{tr}_c} T(\mu F) \xrightarrow{e} \nu G\big).$$

4.2 Coalgebraic Language Semantics of Nominal Systems

We will now work towards that the language semantics of nominal automata is an instance of coalgebraic language semantics. To this end we will instantiate the results of Sect. 4.1 to $\mathscr{C} = \mathsf{Nom}$, $GX = 2 \times X^{\mathbb{A}}$ and $T = \mathcal{P}_{\mathsf{fs}}$ (for NOFAs), or to $GX = 2 \times X^{\mathbb{A}} \times [\mathbb{A}]X$ and $T = \mathcal{P}_{\mathsf{ufs}}$ (for RNNAs). More generally, in the former case we show that certain polynomial functors G with exponentiation lift to $\mathsf{EM}(\mathcal{P}_{\mathsf{fs}})$, and in the latter case, certain binding polynomial functors with exponentation lift to $\mathsf{EM}(\mathcal{P}_{\mathsf{ufs}})$. For our specific instances of interest we show that the terminal coalgebra νG is given by (data or bar) languages. The desired end result then follows by an application of Proposition 4.9.

The class of functors G we consider are formed according to the grammar

$$G ::= A \mid \mathsf{Id} \mid [\mathbb{A}](-) \mid \textstyle\prod_{i\in I} G_i \mid G^N, \tag{4.2}$$

where A ranges over all nominal sets equipped with the structure $a\colon \mathcal{P}_{\mathsf{ufs}}A \to A$ of an algebra for the monad $\mathcal{P}_{\mathsf{ufs}}$, I is an arbitrary index set, and N ranges over all nominal sets. Every such functor G has a canonical lifting to $\mathsf{EM}(\mathcal{P}_{\mathsf{ufs}})$. This can be proved by induction over the grammar using Example 4.3 and the following result.

Proposition 4.10. *The abstraction functor has a canonical lifting to* $\mathsf{EM}(\mathcal{P}_{\mathsf{ufs}})$.

Proof. The abstraction functor $[\mathbb{A}](-)$ has a left-adjoint $\mathbb{A}*(-)$, where $*$ denotes the *fresh product* defined for two nominal sets X and Y by

$$X * Y = \{(x, y) : x \in X,\ y \in Y,\ \mathsf{supp}(x) \cap \mathsf{supp}(y) = \emptyset\},$$

see [34, Theorem 4.12]. The strength of $\mathcal{P}_{\mathsf{ufs}}$ restricts to the fresh product; we have

$$s_{X,Y}\colon X * \mathcal{P}_{\mathsf{ufs}}Y \to \mathcal{P}_{\mathsf{ufs}}(X * Y) \qquad (x, S) \mapsto \{(x, s) : s \in S\}.$$

Indeed, if $\mathsf{supp}(x) \cap \mathsf{supp}(S) = \emptyset$, then $\mathsf{supp}(x) \cap \mathsf{supp}(s) = \emptyset$ for every $s \in S$ because S is uniformly finitely supported and thus $\mathsf{supp}(s) \subseteq \mathsf{supp}(S)$. It follows that $s_{\mathbb{A},X}\colon \mathbb{A} * \mathcal{P}_{\mathsf{ufs}}X \to \mathcal{P}_{\mathsf{ufs}}(\mathbb{A} * X)$ yields a distributive law of $\mathbb{A} * (-)$ over $\mathcal{P}_{\mathsf{ufs}}$. By Proposition 4.2 we thus obtain a distributive law of $\mathcal{P}_{\mathsf{ufs}}$ over $[\mathbb{A}](-)$. □

Corollary 4.11. *For every functor G according to the grammar in* (4.2) *the terminal coalgebra νG lifts to a terminal coalgebra of $\widehat{G}$ on* $\mathsf{EM}(\mathcal{P}_{\mathsf{ufs}})$.

The terminal coalgebra νG exists since every such G is an accessible functor on Nom. This can be shown by induction on the structure of G; for exponentiation in the induction step one argues similarly as Wißmann [45, Cor. 3.7.4] has done for orbit-finite sets: an exponentiation functor $(-)^N$ is λ-accessible iff the set of orbits of N has cardinality less than λ. Now use Remark 4.1(3).

Consequently, one can define a coalgebraic language semantics for every functor G according to the grammar (4.2).

Remark 4.12. (1) For $T = \mathcal{P}_{\mathsf{fs}}$ one has the same results for functors G on Nom according to the grammar obtained from the one in (4.2) by dropping the abstraction functor $[\mathbb{A}](-)$ and letting A range over all nominal sets equipped with the structure $a\colon \mathcal{P}_{\mathsf{fs}}A \to A$ of an algebra for the monad $\mathcal{P}_{\mathsf{fs}}$. Every functor according to the changed grammar has a canonical lifting to $\mathsf{EM}(\mathcal{P}_{\mathsf{fs}})$. More generally, this works whenever T is a strong monad on a cartesian closed category (by Example 4.3).
(2) We have dropped the abstraction functor in the previous item because our proof of Proposition 4.10 does not work for $\mathcal{P}_{\mathsf{fs}}$. The problem is that the strength in (3.1) does not restrict to the fresh product for all finitely supported subsets. Indeed, even if $\mathsf{supp}(x)$ and $\mathsf{supp}(S)$ are disjoint, the support of x may not be disjoint from that of every element $s \in S$, whence (x, s) does not lie in $X * Y$. For example, take $X = Y = \mathbb{A}$ and $S = \mathbb{A} \setminus \{a\}$ for some $a \in \mathbb{A}$. Clearly, $\mathsf{supp}(S) = \{a\}$. Thus, for every $b \neq a$, we see that (b, S) lies in $\mathbb{A} * \mathcal{P}_{\mathsf{fs}}\mathbb{A}$. However, while $b \in S$ we do not have that $(b, b) \in \mathbb{A} * \mathbb{A} = \{(a, a') : a, a' \in \mathbb{A}, a \neq a'\}$, which means that $s_{\mathbb{A},\mathbb{A}}(b, S)$ does not lie in $\mathcal{P}_{\mathsf{fs}}(\mathbb{A} * \mathbb{A})$.

Coalgebraic Language Semantics of NOFAs. We now apply the previous results to $T = \mathcal{P}_{\mathsf{fs}}$ and $GX = 2 \times X^{\mathbb{A}}$.

Remark 4.13. We have a canonical isomorphism $\mathcal{P}_{\mathsf{fs}}(\mathbb{A} \times X) \cong (\mathcal{P}_{\mathsf{fs}}X)^{\mathbb{A}}$ given by $S \mapsto (a \mapsto \{x : (a, x) \in S\})$. This follows from the fact that $\mathcal{P}_{\mathsf{fs}}$ is the power object functor on the topos Nom and so we have $\mathcal{P}_{\mathsf{fs}}X \cong 2^X$.

Consequently, a NOFA may be regarded as a coalgebra for $G\mathcal{P}_{\mathsf{fs}}$:

$$X \to \mathcal{P}_{\mathsf{fs}}(1 + \mathbb{A} \times X) \cong 2 \times (\mathcal{P}_{\mathsf{fs}}X)^{\mathbb{A}} = G\mathcal{P}_{\mathsf{fs}}X.$$

Proposition 4.14. *The terminal coalgebra for G is the nominal set $\mathcal{P}_{\mathsf{fs}}(\mathbb{A}^*)$ of all data languages with the structure*

$$\mathcal{P}_{\mathsf{fs}}(\mathbb{A}^*) \xrightarrow{\tau} 2 \times \mathcal{P}_{\mathsf{fs}}(\mathbb{A}^*)^{\mathbb{A}}, \quad L \mapsto (b, a \mapsto a^{-1}L),$$

where $b = 1$ if $\varepsilon \in L$ and 0 else, and $a^{-1}L = \{w \in \mathbb{A}^* : aw \in L\}$.

The proof is analogous to the one that for every alphabet A the set functor $X \to 2 \times X^A$ has the terminal coalgebra $\mathcal{P}(A^*)$, see e.g. Rutten [38].

We may thus define the coalgebraic language semantics for NOFAs as in Definition 4.6.

Remark 4.15. We take $FX = 1 + \mathbb{A} \times X$ as in Theorem 3.18 and obtain $\mu F = \mathbb{A}^*$ (Proposition 3.16) and $\nu G = \mathcal{P}_{\mathsf{fs}}(\mathbb{A}^*)$ (Proposition 4.14). Moreover, analogous to ordinary non-deterministic automata [21, Sec. 7.1], we have an extension natural transformation $\varepsilon_X : \mathcal{P}_{\mathsf{fs}}(1 + \mathbb{A} \times X) \to 2 \times (\mathcal{P}_{\mathsf{fs}} X)^{\mathbb{A}}$ given by

$$\varepsilon_X(S) = (b, a \mapsto S_a),$$

where $b = 1$ iff the element $*$ of 1 lies S and $S_a = \{x : (a, x) \in S\}$. The ensuing canonical morphism $e\colon \mathcal{P}_{\mathsf{fs}}(\mu F) \to \nu G$ from Remark 4.8 is then easily seen to be just the identity map on $\mathcal{P}_{\mathsf{fs}}(\mathbb{A}^*)$.

Corollary 4.16. *The coalgebraic language semantics assigns to each state of a NOFA the data language it accepts.*

Indeed, this follows from Theorem 3.18 and Proposition 4.9 using that in the latter result e is the identity map on $\mathcal{P}_{\mathsf{fs}}(\mathbb{A}^*)$.

Coalgebraic Language Semantics of RNNAs. We now apply the previous results to $T = \mathcal{P}_{\mathsf{ufs}}$ and $GX = 2 \times X^{\mathbb{A}} \times [\mathbb{A}]X$.

Remark 4.17. (1) The canonical isomorphism from Remark 4.13 restricts to an injection $i\colon \mathcal{P}_{\mathsf{ufs}}(\mathbb{A} \times X) \rightarrowtail (\mathcal{P}_{\mathsf{ufs}} X)^{\mathbb{A}}$. Indeed, take a uniformly finitely supported subset $S \subseteq \mathbb{A} \times X$. Then for every $a \in \mathbb{A}$, every element x of the set $i(S)(a) = \{x : (a, x) \in S\}$ satisfies $\mathsf{supp}(x) \subseteq \{a\} \cup \mathsf{supp}(x) = \mathsf{supp}(a, x) \subseteq \mathsf{supp}(S)$ and therefore that set lies in $\mathcal{P}_{\mathsf{ufs}} X$. However, note that the inverse of the isomorphism from Remark 4.13 does not restrict to uniformly finitely supported subsets.
(2) The components $\varrho_X : [\mathbb{A}]\mathcal{P}_{\mathsf{ufs}} X \to \mathcal{P}_{\mathsf{ufs}}([\mathbb{A}]X)$ of the distributive law from the proof of Proposition 3.11 are in fact isomorphisms with inverses $\psi_X : \mathcal{P}_{\mathsf{ufs}}([\mathbb{A}]X) \to [\mathbb{A}]\mathcal{P}_{\mathsf{ufs}} X$ defined by $\psi_X(S) = \langle a \rangle \{x : \langle a \rangle x \in S\}$, where a is fresh for S. These inverses can also be gleaned from Pitts' result [34, Prop. 4.14] which shows that the abstraction functor preserves exponentials specializing to $\mathcal{P}_{\mathsf{fs}}([\mathbb{A}]X) \cong [\mathbb{A}]\mathcal{P}_{\mathsf{fs}} X$. However, note that ϱ_X has a more involved description in the case of $\mathcal{P}_{\mathsf{fs}}$.

It follows that for every nominal set X we have an injection

$$m_X : 2 \times \mathcal{P}_{\mathsf{ufs}}(\mathbb{A} \times X) \times \mathcal{P}_{\mathsf{ufs}}([\mathbb{A}]X) \rightarrowtail 2 \times (\mathcal{P}_{\mathsf{ufs}} X)^{\mathbb{A}} \times [\mathbb{A}](\mathcal{P}_{\mathsf{ufs}} X). \tag{4.3}$$

Thus every RNNA (Remark 2.6) may be regarded as a coalgebra for $G\mathcal{P}_{\mathsf{ufs}}$.

A description of the terminal coalgebra for G has previously been given by Kozen et al. [27, Theorem 4.10]. We provide a different (of course, isomorphic) description as a final ingredient for our desired result.

Proposition 4.18. *The terminal coalgebra for G is the nominal set $\mathcal{P}_{\mathsf{fs}}(\bar{\mathbb{A}}^*/{=_\alpha})$ of all bar languages with the structure*

$$\mathcal{P}_{\mathsf{fs}}(\bar{\mathbb{A}}^*/{=_\alpha}) \xrightarrow{\tau} 2 \times (\mathcal{P}_{\mathsf{fs}}(\bar{\mathbb{A}}^*/{=_\alpha}))^{\mathbb{A}} \times [\mathbb{A}]\mathcal{P}_{\mathsf{fs}}(\bar{\mathbb{A}}^*/{=_\alpha}), \quad S \mapsto (b, a \mapsto S_a, S_{\mathsf{l}a}),$$

where $b = 1$ if $[\varepsilon]_\alpha \in S$ and 0 else, $S_a = \{[w]_\alpha : [aw]_\alpha \in S\}$ and $S_{\mathsf{l}a} = \langle a \rangle \{[w]_\alpha : [\mathsf{l}aw]_\alpha \in S\}$ for any a which is fresh for S.

We may thus define the coalgebraic language semantics for RNNAs as in Definition 4.6.

Remark 4.19. We take $FX = 1 + \mathbb{A} \times X + [\mathbb{A}]X$ as in Theorem 3.21 and obtain $\mu F = \bar{\mathbb{A}}^*/{=_\alpha}$ (Proposition 3.19) and $\nu G = \mathcal{P}_{\mathsf{fs}}(\bar{\mathbb{A}}^*/{=_\alpha})$ (Proposition 4.18). We also define a natural transformation $\varepsilon\colon \mathcal{P}_{\mathsf{ufs}}F \to G\mathcal{P}_{\mathsf{ufs}}$ by composing the canonical isomorphism $\mathcal{P}_{\mathsf{ufs}}(1 + \mathbb{A} \times X + [\mathbb{A}]X) \cong 2 \times \mathcal{P}_{\mathsf{ufs}}(\mathbb{A} \times X) \times \mathcal{P}_{\mathsf{ufs}}([\mathbb{A}]X)$ with the injection m_X from (4.3). For every uniformly finitely supported subset $S \subseteq 1 + \mathbb{A} \times X + [\mathbb{A}]X$ we have $\varepsilon_X(S) = (b, a \mapsto S_a, S_{\mathsf{I}a})$, where $b = 1$ iff the element $*$ of 1 lies in S, $S_a = \{s : (a, s) \in S\}$ and $S_{\mathsf{I}a} = \langle a \rangle \{s : \langle a \rangle s \in S\}$, where a is fresh for (all elements $\langle b \rangle s$ in) S.

Lemma 4.20. *The natural transformation* $\varepsilon\colon \mathcal{P}_{\mathsf{ufs}}F \to G\mathcal{P}_{\mathsf{ufs}}$ *is an extension.*

Lemma 4.21. *The canonical morphism* $e\colon \mathcal{P}_{\mathsf{ufs}}(\mu F) \to \nu G$ *from Remark 4.8 is the inclusion map* $\mathcal{P}_{\mathsf{ufs}}(\bar{\mathbb{A}}^*/{=_\alpha}) \hookrightarrow \mathcal{P}_{\mathsf{fs}}(\bar{\mathbb{A}}^*/{=_\alpha})$.

Corollary 4.22. *The coalgebraic language semantics assigns to each state of an RNNA the bar language it accepts.*

Indeed, this follows from Theorem 3.21 and Proposition 4.9 using that in the latter result $e\colon \mathcal{P}_{\mathsf{ufs}}(\bar{\mathbb{A}}^*/{=_\alpha}) \hookrightarrow \mathcal{P}_{\mathsf{fs}}(\bar{\mathbb{A}}^*/{=_\alpha})$ is the inclusion map by Lemma 4.21.

5 Conclusions and Future Work

We have worked out coalgebraic semantics for two species of non-deterministic automata for data languages: NOFAs [4] and RNNAs [39]. We have seen that their semantics arises both as an instance of the Kleisli style coalgebraic trace semantics and from the Eilenberg-Moore style coalgebraic language semantics, which is based on generalized determinization. To see that both semantics coincide we have employed the results by Jacobs et al. [21].

We have also revisited coalgebraic trace semantics in general and given a new compact proof of the main extension result for initial algebras in that theory. Our proof avoids assumptions on the convergence of the initial algebra chain; mere existence of an initial algebra suffices.

Having provided coalgebraic semantics for non-deterministic nominal systems makes the powerful toolbox of coalgebraic methods fully available to those systems. For example, generic constructions like coalgebraic ε-elimination [5,41] can be instantiated to them. Or coalgebraic up-to techniques starting with the work by Rot et al. [36] might lead to new proof principles and algorithms, cf. [6].

Our general extension and lifting results for nominal systems may be applied to related kinds of systems, e.g. nominal transition systems and the coalgebraic study of equivalences for them. Going a step beyond the standard coalgebraic trace and language semantics, graded semantics [9] should lead to a nominal spectrum of equivalences generalizing van Glabbeek's famous linear time – branching time spectrum [14].

References

1. Adámek, J., Milius, S., Moss, L.S.: Initial algebras without iteration. In: Gaducci, F., Silva, A. (eds.) 9th Conference on Algebra and Coalgebra in Computer Science (CALCO). LIPIcs, vol. 211, pp. 5:1–5:20. Schloss Dagstuhl (2021)
2. Applegate, H.: Acyclic models and resolvent functors. Ph.D. thesis, Columbia University (1965)
3. Bartels, F.: On generalized coinduction and probabilistic specification formats. Ph.D. thesis, Vrije Universiteit Amsterdam (2004)
4. Bojańczyk, M., Klin, B., Lasota, S.: Automata theory in nominal sets. Log. Methods Comput. Sci. **10**(3), 1–8 (2014)
5. Bonchi, F., Milius, S., Silva, A., Zanasi, F.: Killing epsilons with a dagger: a coalgebraic study of systems with algebraic label structure. Theoret. Comput. Sci. **604**, 102–126 (2015)
6. Bonchi, F., Pous, D.: Checking NFA equivalence with bisimulations up to congruence. In: Giacobazzi, R., Cousot, R. (eds.) Proceedings of 40th ACM SIGPLAN-SIGACT Symposium on Principles of Programming Languages (POPL 2013), pp. 457–468. ACM (2013)
7. Bonsangue, M.M., Milius, S., Silva, A.: Sound and complete axiomatizations of coalgebraic language equivalence. ACM Trans. Comput. Logic **14**(1), 7:1–7:52 (2013)
8. Ciancia, V., Sammartino, M.: A class of automata for the verification of infinite, resource-allocating behaviours. In: Maffei, M., Tuosto, E. (eds.) TGC 2014. LNCS, vol. 8902, pp. 97–111. Springer, Heidelberg (2014). https://doi.org/10.1007/978-3-662-45917-1_7
9. Dorsch, U., Milius, S., Schröder, L.: Graded monads and graded logics for the linear time - branching time spectrum. In: Fokkink, W.J., van Glabbeek, R. (eds.) Proceedings of 30th International Conference on Concurrency Theory (CONCUR). LIPIcs, vol. 140, pp. 36:1–36:16. Schloss Dagstuhl (2019)
10. Fiore, M., Plotkin, G.D., Turi, D.: Abstract syntax and variable binding. In: Proceedings of Logic in Computer Science (LICS), pp. 193–202. IEEE Computer Society (1999)
11. Freyd, P.: Remarks on algebraically compact categories. In: Fourman, M.P., Johnstone, P.T., Pitts, A.M. (eds.) Applications of category theory in computer science: Proceedings of the London Mathematical Society Symposium, Durham 1991. London Mathematical Society Lecture Note Series, vol. 177, pp. 95–106. Cambridge University Press (1992)
12. Gabbay, M.J.: Foundations of nominal techniques: logic and semantics of variables in abstract syntax. Bull. Symb. Log. **17**(2), 161–229 (2011)
13. Gabbay, M.J., Pitts, A.M.: A new approach to abstract syntax involving binders. In: Logic in Computer Science, LICS 1999, pp. 214–224. IEEE Computer Society (1999)
14. van Glabbeek, R.: The linear time - branching time spectrum i; the semantics of concrete, sequential processes. In: Bergstra, J., Ponse, A., Smolka, S. (eds.) Handbook of Process Algebra, pp. 3–99. Elsevier (2001)
15. Goncharov, S., Milius, S., Silva, A.: Towards a uniform theory of effectful state machines. ACM Trans. Comput. Log.**21**(3), 63 (2020). (article 23)
16. Grigore, R., Distefano, D., Petersen, R.L., Tzevelekos, N.: Runtime verification based on register automata. In: Piterman, N., Smolka, S.A. (eds.) TACAS 2013. LNCS, vol. 7795, pp. 260–276. Springer, Heidelberg (2013). https://doi.org/10.1007/978-3-642-36742-7_19

17. Hasuo, I.: Tracing Anonymity with Coalgebras. Ph.D. thesis, Radboud University Nijmegen (2008)
18. Hasuo, I., Jacobs, B., Sokolova, A.: Generic trace semantics via coinduction. Log. Methods Comput. Sci. **3**(4:11), 1–36 (2007)
19. Hermida, C., Jacobs, B.: Structural induction and conduction in a fibrational setting. Inform. Comput. **145**, 107–152 (1998)
20. Jacobs, B.: Introduction to Coalgebra. Towards Mathematics of States and Observation, Cambridge University Press, Cambridge (2016)
21. Jacobs, B., Silva, A., Sokolova, A.: Trace semantics via determinization. J. Comput. System Sci. **81**, 859–879 (2015)
22. Johnstone, P.T.: Adjoint lifting theorems for categories of algebras. Bull. London Math. Soc. **7**, 294–297 (1975)
23. Joyal, A.: Une théorie combinatoire des séries formelles. Adv. Math. **42**, 1–82 (1981)
24. Joyal, A.: Foncteurs analytiques et espèces de structures. Lect. Notes Math. **1234**, 126–159 (1986)
25. Kaminski, M., Francez, N.: Finite-memory automata. Theor. Comput. Sci. **134**(2), 329–363 (1994)
26. Kock, A.: Monads on symmetric monoidal closed categories. Arch. Math. (Basel) **21**, 1–10 (1970)
27. Kozen, D., Mamouras, K., Petrişan, D., Silva, A.: Nominal Kleene coalgebra. In: Halldórsson, M.M., Iwama, K., Kobayashi, N., Speckmann, B. (eds.) ICALP 2015. LNCS, vol. 9135, pp. 286–298. Springer, Heidelberg (2015). https://doi.org/10.1007/978-3-662-47666-6_23
28. Kürtz, K., Küsters, R., Wilke, T.: Selecting theories and nonce generation for recursive protocols. In: Formal methods in security engineering, FMSE 2007, pp. 61–70. ACM (2007)
29. Lambek, J.: A fixpoint theorem for complete categories. Math. Z. **103**, 151–161 (1968)
30. Milius, S., Palm, T., Schwencke, D.: Complete iterativity for algebras with effects. In: Kurz, A., Lenisa, M., Tarlecki, A. (eds.) CALCO 2009. LNCS, vol. 5728, pp. 34–48. Springer, Heidelberg (2009). https://doi.org/10.1007/978-3-642-03741-2_4
31. Milius, S., Pattinson, D., Wißmann, T.: A new foundation for finitary corecursion and iterative algebras. Inform. and Comput. **271**, 104456 (2020)
32. Moggi, E.: Notions of computations and monads. Inform. Comput. **93**(1), 55–92 (1991)
33. Mulry, P.S.: Lifting theorems for Kleisli categories. In: Brookes, S., Main, M., Melton, A., Mislove, M., Schmidt, D. (eds.) MFPS 1993. LNCS, vol. 802, pp. 304–319. Springer, Heidelberg (1994). https://doi.org/10.1007/3-540-58027-1_15
34. Pitts, A.M.: Nominal Sets: Names and Symmetry in Computer Science. Cambridge University Press (2013)
35. Plotkin, G.D., Turi, D.: Towards a mathematical operational semantics. In: Proceedings of Logic in Computer Science (LICS) (1997)
36. Rot, J., Bonsangue, M., Rutten, J.: Coalgebraic bisimulation-up-to. In: van Emde Boas, P., Groen, F.C.A., Italiano, G.F., Nawrocki, J., Sack, H. (eds.) SOFSEM 2013. LNCS, vol. 7741, pp. 369–381. Springer, Heidelberg (2013). https://doi.org/10.1007/978-3-642-35843-2_32
37. Rutten, J.J.M.M.: Automata and coinduction (an exercise in coalgebra). In: Sangiorgi, D., de Simone, R. (eds.) CONCUR 1998. LNCS, vol. 1466, pp. 194–218. Springer, Heidelberg (1998). https://doi.org/10.1007/BFb0055624

38. Rutten, J.: Universal coalgebra: a theory of systems. Theoret. Comput. Sci. **249**(1), 3–80 (2000)
39. Schröder, L., Kozen, D., Milius, S., Wißmann, T.: Nominal automata with name binding. In: Esparza, J., Murawski, A.S. (eds.) FoSSaCS 2017. LNCS, vol. 10203, pp. 124–142. Springer, Heidelberg (2017). https://doi.org/10.1007/978-3-662-54458-7_8
40. Silva, A., Bonchi, F., Bonsangue, M.M., Rutten, J.J.M.M.: Generalizing determinization from automata to coalgebras. Log. Methods Comput. Sci. **9**(1:9), 1–22 (2013)
41. Silva, A., Westerbaan, B.: A coalgebraic view of ε-transitions. In: Heckel, R., Milius, S. (eds.) CALCO 2013. LNCS, vol. 8089, pp. 267–281. Springer, Heidelberg (2013). https://doi.org/10.1007/978-3-642-40206-7_20
42. Smyth, M.B., Plotkin, G.D.: The category-theoretic solution of recursive domain equations. SIAM J. Comput. **11**(4), 761–783 (1982)
43. Winter, J., Bonsangue, M., Rutten, J.: Coalgebraic characterizations of context-free languages. Log. Methods Comput. Sci. **9**(3:14), 39 (2013)
44. Winter, J., Bonsangue, M., Rutten, J.: Context-free coalgebras. J. Comput. System Sci. **81**, 911–939 (2015)
45. Wißmann, T.: Coalgebraic Semantics and Minimization in Sets and Beyond. Ph.D. thesis, Friedrich-Alexander-Universität Erlangen-Nürnberg (FAU) (2020). https://opus4.kobv.de/opus4-fau/frontdoor/index/index/docId/14222

A Categorical Framework for Learning Generalised Tree Automata

Gerco van Heerdt[1(✉)], Tobias Kappé[2], Jurriaan Rot[3], Matteo Sammartino[1,4], and Alexandra Silva[1,5]

[1] University College London, London, UK
gerco.heerdt@ucl.ac.uk
[2] ILLC, University of Amsterdam, Amsterdam, The Netherlands
t.kappe@uva.nl
[3] Radboud University, Nijmegen, The Netherlands
jrot@cs.ru.nl
[4] Royal Holloway, University of London, Egham, UK
matteo.sammartino@rhul.ac.uk
[5] Cornell University, Ithaca, USA
alexandra.silva@cornell.edu

Abstract. Automata learning is a popular technique used to automatically construct an automaton model from queries. Much research went into devising ad hoc adaptations of algorithms for different types of automata. The CALF project seeks to unify these using category theory in order to ease correctness proofs and guide the design of new algorithms. In this paper, we extend CALF to cover learning of algebraic structures that may not have a coalgebraic presentation. Furthermore, we provide a detailed algorithmic account of an abstract version of the popular $L^\star$ algorithm, which was missing from CALF. We instantiate the abstract theory to a large class of **Set** functors, by which we recover for the first time practical tree automata learning algorithms from an abstract framework and at the same time obtain new algorithms to learn algebras of quotiented polynomial functors.

1 Introduction

Automata learning—automated discovery of automata models from system observations—is emerging as a highly effective bug-finding technique with applications in verification of passports [3], bank cards [2], and network protocols [19]. The design of algorithms for automata learning of different models is a fundamental research problem, and in the last years much progress has been made in developing and understanding new algorithms. The roots of the area go back to the 50s, when Moore studied the problem of inferring deterministic finite automata. Later, the same problem, albeit under different names, was studied by control theorists [21] and computational linguists [17]. The algorithm that caught the attention of the verification community is the one presented in Dana

Published by Springer Nature Switzerland AG 2022
H. H. Hansen and F. Zanasi (Eds.): CMCS 2022, LNCS 13225, pp. 67–87, 2022.
https://doi.org/10.1007/978-3-031-10736-8_4

Angluin's seminal paper in 1987 [8]. She proves that it is possible to infer minimal deterministic automata in polynomial time using only so-called membership and equivalence queries. Vaandrager's CACM article [42] provides an extensive review of the literature in automata learning and its applications to verification.

Angluin's algorithm, called $L^\star$, has served as a basis for many extensions that work for more expressive models than plain deterministic automata: I/O automata [4], weighted automata [13,28], register automata [1,31,37], nominal automata [36], and Büchi automata [9]. Many of these extensions were developed independently and, though they bear close resemblance to the original algorithm, arguments of correctness and termination had to be repeated every time. This motivated Silva and Jacobs to provide a categorical understanding of $L^\star$ [32] and capture essential data structures abstractly, in the hope of developing a generic, modular, and parametric framework for automata learning based on (co)algebra. Their early work was taken much further in Van Heerdt's master thesis [23], which then formed the basis of a wider project on developing a *Categorical Automata Learning Framework*—CALF.[1] CALF was described in the 2017 paper [29], but several problems were left open:

1. An abstract treatment of counterexamples: in the original $L^\star$ algorithm, counterexamples are a core component, as they enable refinement of the state space of the learned automaton to ensure progress towards termination.
2. The development of a full abstract learning algorithm that could readily be instantiated for a given model: in essence, CALF provided only the abstract data structures needed in the learning process, but no direct algorithm.
3. Finding suitable constraints on the abstract framework to cover interesting examples, such as tree automata [16], that did not fit the constraints in [29].

In this paper, we resolve the open problems above, and develop CALF further to provide concrete learning algorithms for models that are algebras for a given functor, which notably include tree automata. In a nutshell, the contributions and technical roadmap of the paper are as follows. After recalling some categorical notions, the basics of $L^\star$ (Sect. 2), and CALF (Sect. 3), we provide:

1. A general treatment of counterexamples (Sect. 4), together with an abstract analysis of progress, that enables termination analysis of a generic algorithm.
2. A step-by-step generalisation of all components of $L^\star$ for models that are algebras of a given functor (Sect. 5).
3. An instantiation of the abstract algorithm to concrete categories (Sect. 6), providing the first abstractly derived learning algorithm for tree automata.

The present paper complements other recent work on abstract automata learning algorithms: Barlocco, Kupke, and Rot [12] gave an algorithm for coalgebras of a functor, whereas Urbat and Schröder [41] provided an algorithm for structures that can be represented as both algebras and coalgebras. More recently, Colcombet, Petrisan, and Stabile [15] gave an abstract learning algorithm based on modelling automata as functors. Our focus is on algebras, such as tree automata,

[1] http://www.calf-project.org.

that cannot be covered by the aforementioned frameworks. A detailed comparison is given in Sect. 7. We conclude with directions for future work in Sect. 8. Proofs can be found in the extended version [26].

2 Preliminaries

We now introduce some categorical notions that we will need later in our technical development, and describe Angluin's original $\mathtt{L}^\star$ algorithm. We assume some prior knowledge of category theory (categories, functors); see e.g., [11,33].

An $(\mathcal{E},\mathcal{M})$-*factorisation system* on a category $\mathbf{C}$ consists of classes of morphisms $\mathcal{E}$ and $\mathcal{M}$, closed under composition with isos, such that for every morphism f in $\mathbf{C}$ there exist $e \in \mathcal{E}$ and $m \in \mathcal{M}$ with $f = m \circ e$, and we have a unique diagonal fill-in property. Given a morphism f, we write $f^{\triangleright}$ and $f^{\triangleleft}$ for the $\mathcal{E}$-part and $\mathcal{M}$-part of its factorisation, respectively.

We work in a category $\mathbf{C}$ with finite products and coproducts. When $f: X \to Z$ and $g: Y \to Z$, we write $[f,g]$ for the unique arrow from $X+Y$ to Z induced by the coproduct. We assume that $\mathbf{C}$ admits a fixed factorisation system $(\mathcal{E},\mathcal{M})$, where $\mathcal{E}$ consists of epis and $\mathcal{M}$ consists of monos. We fix a *varietor* F in $\mathbf{C}$, that is, an endofunctor such that there is a free F-algebra monad (T,η,μ). We write γ_X for the F-algebra structure $FTX \to TX$, which is natural in X. Given an F-algebra (Y,y), we write $f^\sharp\colon (TX,\mu_X) \to (Y,y)$ for the extension of $f\colon X \to Y$ and denote $y^* = \mathrm{id}_Y^\sharp\colon (TY,\mu_Y) \to (Y,y)$. We often implicitly apply forgetful functors. We fix an input object I and an output object O and write F_I for the functor $I + F(-)$. Lastly, we assume F preserves $\mathcal{E}$.

2.1 Abstract Automata

We recall the automaton definition from Arbib and Manes [10], which we will use in this paper, and its basic properties of accepted language and minimality.

Definition 1 (Automaton). *An* automaton *is a tuple $\mathcal{A} = (Q,\delta,i,o)$ consisting of a state space object Q, dynamics $\delta\colon FQ \to Q$, initial states $i\colon I \to Q$, and an output $o\colon Q \to O$. A* homomorphism *from $\mathcal{A}$ to $\mathcal{A}' = (Q',\delta',i',o')$ is an F-algebra morphism h from (Q,δ) to (Q',δ')—that is to say, a function $h\colon Q \to Q'$ with $\delta' \circ Fh = h \circ \delta$—such that $h \circ i = i'$ and $o' \circ h = o$.*

We will use the case of deterministic automata as a running example.

Example 2. If $\mathbf{C} = \mathbf{Set}$ with the (surjective, injective) factorisation system, $F = (-) \times A$ for a finite set A, $I = 1 = \{*\}$, and $O = 2 = \{0,1\}$, we recover deterministic automata (DAs) as automata: the state space is a set Q, the transition function is the dynamics, the initial state is represented as a function $1 \to Q$, and the classification of states into accepting and rejecting ones is represented by a function $Q \to 2$. In this case we obtain the monad $T = (-) \times A^*$, with its unit pairing an element with the empty word ε and the multiplication concatenating words. The extension of $\delta\colon Q \times A \to Q$ to $\delta^*\colon Q \times A^* \to Q$ is the usual one that lets the automaton read a word starting from a given state.

Algorithm 1. Make table closed and consistent

function $\textsc{Fix}(S, E)$
  while T is not closed or not consistent **do**
    if T is not closed **then**
      find $t \in S, a \in A$ such that $\forall s \in S.\ \mathsf{T}(ta) \neq \mathsf{T}(s)$
      $S \leftarrow S \cup \{sa\}$
    else if T is not consistent **then**
      find $s_1, s_2 \in S$, $a \in A$ and $e \in E$ such that
        $\mathsf{T}(s_1) = \mathsf{T}(s_2)$ and $\mathsf{T}(s_1 a)(e) \neq \mathsf{T}(s_2 a)(e)$
      $E \leftarrow E \cup \{ae\}$
  return S, E

Algorithm 2. $\mathtt{L}^\star$ algorithm

$S \leftarrow \{\varepsilon\}$
$E \leftarrow \{\varepsilon\}$
$S, E \leftarrow \textsc{Fix}(S, E)$
while $\mathsf{EQ}(\mathcal{H}_\mathsf{T}) = c$ **do**
  $S \leftarrow S \cup \mathsf{prefixes}(c)$
  $S, E \leftarrow \textsc{Fix}(S, E)$
return $\mathcal{H}_\mathsf{T}$

Definition 3 (Language). *A* language *is a morphism $TI \to O$. The language accepted by an automaton $\mathcal{A} = (Q, \delta, i, o)$ is given by $\mathcal{L}_\mathcal{A} = TI \xrightarrow{\mathsf{reach}_\mathcal{A}} Q \xrightarrow{o} O$, where $\mathsf{reach}_\mathcal{A} \colon TI \to Q$ is the* reachability map *of $\mathcal{A}$ given by $i^\sharp$.*

Definition 4 (Minimality [10]). *An automaton $\mathcal{A}$ is said to be* reachable *if $\mathsf{reach}_\mathcal{A} \in \mathcal{E}$. $\mathcal{A}$ is* minimal *if it is reachable and every reachable automaton $\mathcal{A}'$ s.t. $\mathcal{L}_\mathcal{A} = \mathcal{L}_{\mathcal{A}'}$ admits a (necessarily unique) homomorphism to $\mathcal{A}$.*

Example 5. Recall the setting from Example 2. The reachability map $\mathsf{reach}_\mathcal{A} \colon 1 \times A^* \to Q$ for a DA $\mathcal{A} = (Q, \delta, i, o)$ assigns to each word the state reached after reading that word from the initial state. The language $\mathcal{L}_\mathcal{A} \colon 1 \times A^* \to 2$ accepted by $\mathcal{A}$ is precisely the language accepted by $\mathcal{A}$ in the traditional sense. Reachability of $\mathcal{A}$ means that for every state $q \in Q$ there exists a word that leads to q from the initial state. If this is the case, the unique homomorphism into a language-equivalent minimal automaton identifies states that accept the same language. Here, minimality is equivalent to having a minimal number of states.

A general study of existence of minimal automata in this setting is given in [7]; see also [25].

2.2 The $\mathtt{L}^\star$ Algorithm

In this section, we recall Angluin's algorithm $\mathtt{L}^\star$, which learns the *minimal* DFA accepting a given unknown regular language $\mathcal{L}$. The algorithm can be seen as a game between two players: a *learner* and a *teacher*. The learner can ask two types of *queries* to the teacher:

1. **Membership queries**: is a word $w \in A^*$ in $\mathcal{L}$?
2. **Equivalence queries**: is a *hypothesis* DFA $\mathcal{H}$ correct? That is, is $\mathcal{L}_\mathcal{H} = \mathcal{L}$?

The teacher answers *yes* or *no* to these queries. Moreover, negative answers to equivalence queries are witnessed by a *counterexample*—a word classified incorrectly by $\mathcal{H}$. The learner gathers the results of queries into an *observation table*: a function $\mathsf{T}\colon S \cup S \cdot A \to 2^E$, where $S, E \subseteq A^*$ are finite and $\mathsf{T}(s)(e) = \mathcal{L}(se)$. This function can be depicted as a table where elements of $S \cup S \cdot A$ label rows ($\cdot$ is pointwise concatenation) and elements of E label columns.

As an example, consider the table on the right, over the alphabet $A = \{a, b\}$, where $S = \{\varepsilon\}$ and $E = \{\varepsilon, b, ab\}$. This table approximates a language that contains ε, ab, but not a, b, bb, aab, bab. Following the visual intuition, we will refer to the part of the table indexed by S as the *top* part of the table, and the one indexed by $S \cdot A$ as the *bottom* part.

		E: ε	b	ab
S	ε	1	0	1
$S \cdot A$	a	0	1	0
	b	0	0	0

Intuitively, the content of each row labelled by a word s approximates the Myhill–Nerode equivalence class of s. This is in fact the main idea behind the construction of a hypothesis DFA $\mathcal{H}_{\mathsf{T}}$ from T: states of $\mathcal{H}_{\mathsf{T}}$ are distinct rows of T, corresponding to distinct Myhill–Nerode equivalence classes. Formally, $\mathcal{H}_{\mathsf{T}} = (Q, q_0, \delta, F)$ is defined as follows:

- $Q = \{\mathsf{T}(s) \mid s \in S\}$ is the set of states;
- $F = \{\mathsf{T}(s) \mid s \in S, \mathsf{T}(s)(\varepsilon) = 1\}$ is the set of final states;
- $q_0 = \mathsf{T}(\varepsilon)$ is the initial state;
- $\delta\colon Q \times A \to Q, (\mathsf{T}(s), a) \mapsto \mathsf{T}(sa)$ is the transition function.

For F and q_0 to be well-defined we need ε in E and S respectively. Moreover, for δ to be well-defined we need $\mathsf{T}(sa) \in Q$ for all $sa \in S \cdot A$, and we must ensure that the choice of s to represent a row does not affect the transition. These constraints are captured in the following two properties.

Definition 6 (Closedness and consistency). *A table* T *is* closed *if for all* $t \in S$ *and* $a \in A$ *there exists* $s \in S$ *such that* $\mathsf{T}(s) = \mathsf{T}(ta)$. *A table is* consistent *if for all* $s_1, s_2 \in S$ *with* $\mathsf{T}(s_1) = \mathsf{T}(s_2)$ *we have* $\mathsf{T}(s_1 a) = \mathsf{T}(s_2 a)$ *for any* $a \in A$.

Closedness and consistency form the core of L*, described in Algorithm 2. The sets S and E are initialised with the empty word ε (lines 1 and 2), and extended as a closed and consistent table is built using the subroutine Fix, given in Algorithm 1. The main loop uses an equivalence query, denoted EQ, to ask the teacher whether the hypothesis induced by the table is correct. If the result is a counterexample c, the table is updated by adding all prefixes of c to S (line 5) and made closed and consistent again (line 6). Otherwise, the algorithm returns with the correct hypothesis (line 7). See Appendix A of the extended version [26] for an example.

3 The Abstract Data Structures in CALF

We recall the basic notions underpinning CALF [29]: generalisations of the observation table, closedness, consistency and hypothesis. The generalised table is called a *wrapper*:

Definition 7 (Wrapper). *A* wrapper *for an object Q is a pair of morphisms*

$$\mathcal{W} = \left(S \xrightarrow{\alpha} Q \;,\; Q \xrightarrow{\beta} P\right)$$

We denote the factorisation of $\beta \circ \alpha$ by $S \xrightarrow{e_{\mathcal{W}}} H_{\mathcal{W}} \xrightarrow{m_{\mathcal{W}}} P$.

This will be instantiated with Q the state space of the target automaton, S a collection of row labels of an observation table, and P a collection of possible values of the rows. Then α *selects* states in Q, and β *classifies* them into P. We note that although such α and β underly the learning algorithm, they are not actually known to the learner, as they explicitly involve the unknown target automaton. However, we will see that we only need to represent certain compositions involving these morphisms, and that when α and β are chosen appropriately it will be possible to compute these compositions.

Example 8 (Observation table wrapper). Recall the DA setting from Example 2 and consider a DA $\mathcal{A} = (Q, \delta, i, o)$. For $S \subseteq A^*$ and $E \subseteq A^*$, we can define a wrapper $\mathcal{W} = \left(S \xrightarrow{\alpha_S} Q \;,\; Q \xrightarrow{\beta_E} 2^E\right)$ for Q as follows:

$$\alpha_S(w) = \mathsf{reach}_{\mathcal{A}}(*, w) \qquad\qquad \beta_E(q)(e) = (o \circ \delta^*)(q, e).$$

The composition $\beta_E \circ \alpha_S \colon S \to 2^E$ is precisely the top part of the observation table of L*, with rows S and columns E. In fact, we have $(\beta_E \circ \alpha_S)(s)(e) = \mathcal{L}_{\mathcal{A}}(*, se)$. The image of $\beta_E \circ \alpha_S$ is the set of rows that appear in the table. In L*, this set is used as states of the hypothesis, and in our setting can be obtained as $H_{\mathcal{W}}$, recalling that the (surjective, injective) factorisation system in **Set** gives factorisation through the image.

Before we define hypotheses in this abstract framework, we need generalised notions of closedness and consistency.

Definition 9 (Closedness and consistency). *Given a wrapper*

$$\mathcal{W} = \left(S \xrightarrow{\alpha} Q \;,\; Q \xrightarrow{\beta} P\right),$$

where Q is the state space of an automaton (Q, δ, i, o), we say that $\mathcal{W}$ is closed *if there exist morphisms $i_{\mathcal{W}} \colon I \to H_{\mathcal{W}}$ and $\mathsf{close}_{\mathcal{W}} \colon FS \to H_{\mathcal{W}}$ making the diagrams below commute.*

$$\begin{array}{ccc} I & \xrightarrow{i} & Q \\ {\scriptstyle i_{\mathcal{W}}}\downarrow & & \downarrow{\scriptstyle \beta} \\ H_{\mathcal{W}} & \xrightarrow{m_{\mathcal{W}}} & P \end{array} \qquad\qquad \begin{array}{ccccc} FS & \xrightarrow{F\alpha} & FQ & \xrightarrow{\delta} & Q \\ {\scriptstyle \mathsf{close}_{\mathcal{W}}}\downarrow & & & & \downarrow{\scriptstyle \beta} \\ H_{\mathcal{W}} & & \xrightarrow{m_{\mathcal{W}}} & & P \end{array}$$

Furthermore, we say that $\mathcal{W}$ *is* consistent *if there exist morphisms* $o_{\mathcal{W}} \colon H_{\mathcal{W}} \to O$ *and* $\mathsf{cons}_{\mathcal{W}} \colon FH_{\mathcal{W}} \to P$ *making the diagrams below commute.*

$$\begin{array}{ccc} S & \xrightarrow{e_{\mathcal{W}}} & H_{\mathcal{W}} \\ {\scriptstyle\alpha}\downarrow & & \downarrow{\scriptstyle o_{\mathcal{W}}} \\ Q & \xrightarrow{o} & O \end{array} \qquad \begin{array}{ccccc} FS & & \xrightarrow{Fe_{\mathcal{W}}} & & FH_{\mathcal{W}} \\ {\scriptstyle F\alpha}\downarrow & & & & \downarrow{\scriptstyle \mathsf{cons}_{\mathcal{W}}} \\ FQ & \xrightarrow{\delta} & Q & \xrightarrow{\beta} & P \end{array}$$

Example 10. In the DA case, generalised closedness and consistency instantiate to the conditions allowing the hypothesis to be well-defined in L* (see Sect. 2.2):

Closedness: The wrapper (α_S, β_E) is closed if: (i) there exists $s \in S$ such that $(\beta_E \circ \alpha_S)(s) = (\beta_E \circ i)(*)$ and; (ii) for all $s \in S$ and $a \in A$ there exists $s_a \in S$ such that $(\beta_E \circ \alpha_S)(s_a) = (\beta_E \circ \delta)(\alpha_S(s), a)$. Condition (i) holds immediately if $\varepsilon \in S$—the function $(\beta_E \circ i)(*) \colon E \to 2$ maps $e \in E$ to $\mathcal{L}_{\mathcal{A}}(*, e)$. Condition (ii) corresponds to closedness in Definition 6. In fact, $\beta_E \circ \delta \circ (\alpha_S \times \mathrm{id}_A) \colon S \times A \to 2^E$ represents the lower part of the observation table associated with S and E.

Consistency: The wrapper (α_S, β_E) is consistent if: (iii) for all $s_1, s_2 \in S$ such that $(\beta_E \circ \alpha_S)(s_1) = (\beta_E \circ \alpha_S)(s_2)$ we have $(o \circ \alpha_S)(s_1) = (o \circ \alpha_S)(s_2)$ and; (iv) for all $a \in A$ we have $(\beta_E \circ \delta)(\alpha_S(s_1), a) = (\beta_E \circ \delta)(\alpha_S(s_2), a)$. Condition (iii) holds immediately if $\varepsilon \in E$—the function $o \circ \alpha_S \colon S \to 2$ maps $s \in S$ to $\mathcal{L}_{\mathcal{A}}(*, s)$. Condition (iv) corresponds to consistency in Definition 6.

To determine these properties, we do not need the individual descriptions of α_S and β_E, which refer to the target automaton and are thus not available to the learner; we just need the compositions $\beta_E \circ \alpha_S$, $\beta_E \circ i$, $\beta_E \circ \delta \circ (\alpha_S \times \mathrm{id}_A)$, and $o \circ \alpha_S$, which can be determined using membership queries in this case. In general, for any instantiation of our abstract algorithm it will be important to show that these compositions (adapted to the wrapper and functor involved) can be determined and used concretely by the instantiated algorithm.

So far, we have used the wrapper to obtain the state space $\mathcal{H}_{\mathcal{W}}$ of the hypothesis. When a wrapper is closed and consistent, we can equip $\mathcal{H}_{\mathcal{W}}$ with a full automaton structure, leveraging the unique diagonal fill-in property of the factorisation.

Definition 11 (Hypothesis). *A closed and consistent wrapper*

$$\mathcal{W} = \left(S \xrightarrow{\alpha} Q\,,\, Q \xrightarrow{\beta} P\right)$$

for (Q, δ, i, o) *induces a* hypothesis *automaton* $\mathcal{H}_{\mathcal{W}} = (H_{\mathcal{W}}, \delta_{\mathcal{W}}, i_{\mathcal{W}}, o_{\mathcal{W}})$, *where* $\delta_{\mathcal{W}}$ *is the unique diagonal in the commutative square below.*

$$\begin{array}{ccc} FS & \xrightarrow{Fe_{\mathcal{W}}} & FH_{\mathcal{W}} \\ {\scriptstyle \mathsf{close}_{\mathcal{W}}}\downarrow & \swarrow{\scriptstyle \delta_{\mathcal{W}}} & \downarrow{\scriptstyle \mathsf{cons}_{\mathcal{W}}} \\ H_{\mathcal{W}} & \xrightarrow{m_{\mathcal{W}}} & P \end{array}$$

4 Counterexamples, Generalised

We now provide a key missing element for the development and analysis of an abstract learning algorithm in CALF: counterexamples. In the original L* algorithm, counterexamples are used to refine the state space of the hypothesis—namely the representations of the Myhill–Nerode classes of the language being learned. A crucial property for termination, which we prove at a high level of generality in this section, is that adding counterexamples to a closed and consistent table results in a table which is either not closed or not consistent, and hence needs to be extended. Such an extension, in turn, results in progress being made in the algorithm. We show how we can use *recursive coalgebras* [38,40] as witnesses for discrepancies—i.e., as counterexamples—between a hypothesis and the target language in our abstract approach.[2] Here, and throughout the paper, we fix a *target* automaton $\mathcal{A}_t = (Q_t, \delta_t, i_t, o_t)$ whose language we want to learn.

Definition 12 (Recursive coalgebras). *An F-coalgebra $\rho\colon S \to FS$ is* recursive *if for every algebra $x\colon FX \to X$ there is a unique morphism $x^\rho\colon S \to X$ making the diagram below commute.*

$$\begin{array}{ccc} FS & \xrightarrow{Fx^\rho} & FX \\ {\scriptstyle\rho}\uparrow & & \downarrow{\scriptstyle x} \\ S & \xrightarrow{x^\rho} & X \end{array}$$

Example 13. A *prefix-closed* subset $S \subseteq A^*$ is easily equipped with a coalgebra structure $\rho\colon S \to 1+S\times A$ that detaches the last letter from each non-empty word and assigns $*$ to the empty one. Such a coalgebra is recursive, with the unique map into an algebra being defined as a restricted reachability map. In fact, under certain conditions that are satisfied in the DA setting, recursivity of a coalgebra is equivalent to having a coalgebra homomorphism into the initial algebra [5, Corollary 5.6]. This means that every recursive coalgebra is isomorphic to one given by a prefix-closed multiset of words. If the unique morphism into the initial algebra is injective, then the multiset becomes a set.

Given an automaton $\mathcal{A} = (Q, \delta, i, o)$ and a recursive coalgebra $\rho\colon S \to F_I S$, the map $S \xrightarrow{[i,\delta]^\rho} Q$ can be seen as a generalised reachability map, allowing states in Q to be reached from S. We use this map to derive a notion of generalised language induced by a recursive coalgebra. This will be used to compare languages of the hypothesis and of the target automaton with respect to a specific recursive coalgebra, i.e., a specific counterexample.

Definition 14 (ρ-languages). *Given a recursive coalgebra $\rho\colon S \to F_I S$ and an automaton $\mathcal{A} = (Q, \delta, i, o)$, the ρ-language of $\mathcal{A}$ is $\mathcal{L}^\rho_\mathcal{A} = S \xrightarrow{[i,\delta]^\rho} Q \xrightarrow{o} O$.*

[2] Recursive coalgebras have been used to generalise prefix-closedness in an automata learning context in earlier work [24], as well as to generalise counterexamples [12,41].

For instance, in the case of a DA $\mathcal{A}$ and a recursive coalgebra as in Example 13, $\mathcal{L}^{\rho}_{\mathcal{A}}$ is the restriction of the language of $\mathcal{A}$ to the prefix-closed set of words S.

In Algorithm 2, a counterexample is produced by the teacher (line 4) when the hypothesis does not agree with the target automaton. We now generalise counterexamples to wrappers: counterexamples are recursive coalgebras on which the languages of the hypothesis and of the target automaton disagree.

Definition 15 (Counterexample). *A closed and consistent wrapper $\mathcal{W}$ is said to be* correct up to *a recursive $\rho\colon S \to F_I S$ if $\mathcal{L}^{\rho}_{\mathcal{H}_\mathcal{W}} = \mathcal{L}^{\rho}_{\mathcal{A}_\mathrm{t}}$. A* counterexample *for $\mathcal{W}$ (or $\mathcal{H}_\mathcal{W}$) is a recursive $\rho\colon S \to F_I S$ such that $\mathcal{W}$ is not correct up to ρ.*

The following guarantees incorrect hypotheses yield counterexamples.

Proposition 16 (Language equivalence via recursion). *Given an automaton $\mathcal{A} = (Q, \delta, i, o)$, we have $\mathcal{L}_{\mathcal{A}_\mathrm{t}} = \mathcal{L}_\mathcal{A}$ if and only if $\mathcal{L}^{\rho}_{\mathcal{A}_\mathrm{t}} = \mathcal{L}^{\rho}_{\mathcal{A}}$ for every recursive coalgebra $\rho\colon S \to F_I S$.*

Corollary 17 (Counterexample existence). *Given a closed and consistent wrapper $\mathcal{W}$ for Q_t, we have $\mathcal{L}_{\mathcal{H}_\mathcal{W}} \neq \mathcal{L}_{\mathcal{A}_\mathrm{t}}$ iff there exists a counterexample for $\mathcal{W}$.*

The next step in Algorithm 2 is to fix the table by adding all prefixes of the counterexample to S (line 5). We generalise this step by incorporating the counterexample given by a recursive coalgebra into the wrapper. In the DA case, this precisely corresponds to adding a prefix-closed subset to S. The following results say that doing so will lead to either a closedness or a consistency defect. In other words, we give theoretical guarantees that resolving counterexamples results in progress being made towards convergence.

Theorem 18 (Resolving counterexamples). *Given a closed and consistent wrapper $\mathcal{W} = \left(S \xrightarrow{\alpha} Q_\mathrm{t}\,,\ Q_\mathrm{t} \xrightarrow{\beta} P \right)$ and a recursive coalgebra $\rho\colon S' \to F_I S'$, the following holds. If the wrapper $\mathcal{W}' = ([\alpha, [i_\mathrm{t}, \delta_\mathrm{t}]^{\rho}], \beta)$ is closed and consistent, then $\mathcal{W}$ is correct up to ρ.*

This theorem is used contrapositively: given a closed and consistent wrapper, adding a counterexample yields a wrapper that is either not closed or inconsistent.

5 Generalised Learning Algorithm

We are now in a position to describe our general algorithm. Similarly to L* (Sect. 2.2) it is organised into two procedures: Algorithm 3, which contains the abstract procedure for making a wrapper closed and consistent, and Algorithm 4, containing the learning iterations. These generalise the analogous procedures in L*, Algorithm 1 and Algorithm 2, respectively. We note again that although the algorithmic description operates on a wrapper (α, β), these individual morphisms will not be known to the learner. In fact, at this level of abstraction the descriptions should be seen as algorithmic templates rather than concrete algorithms.

Algorithm 3. Make wrapper closed and consistent

```
function Fix(α, β)
    while (α, β) not closed
            or not consistent do
        if (α, β) not closed then
            α ← α′ such that (α′, β) is
                locally closed w.r.t. α
        else if (α, β) not consistent then
            β ← β′ such that (α, β′) is
                locally consistent w.r.t. β
    return α, β
```

Algorithm 4. Abstract automata learning algorithm

1: $\alpha, \beta \leftarrow \text{Fix}(!\colon 0 \to Q_\mathrm{t}, !\colon Q_\mathrm{t} \to 1)$
2: **while** $\mathsf{EQ}(\mathcal{H}_{(\alpha,\beta)}) = \rho\colon S \to F_I S$ **do**
3: $\quad \alpha \leftarrow \alpha'$ s.t. $\alpha'^{\triangleleft} = [\alpha, [i_\mathrm{t}, \delta_\mathrm{t}]^{\rho}]^{\triangleleft}$
4: $\quad \alpha, \beta \leftarrow \text{Fix}(\alpha, \beta)$
5: **return** $\mathcal{H}_{(\alpha,\beta)}$

An instantiation must ensure that at least the compositions required to determine the closedness and consistency conditions and to construct the hypothesis can be maintained. These compositions are $\beta \circ \alpha$, $\beta \circ i_\mathrm{t}$, $\beta \circ \delta_\mathrm{t} \circ F\alpha$, and $o_\mathrm{t} \circ \alpha$. We have previously shown how these instantiate to recover $\mathtt{L}^\star$, and in Sect. 6 we will discuss the class of examples given by generalised tree automata.

In Algorithm 4, the wrapper is initialised with trivial maps and extended to be closed and consistent using the subroutine Fix (line 1). The equivalence query for the main loop (line 2) returns a counterexample in the form of a recursive coalgebra, which is used to update the wrapper (line 3, which will be explained in more detail later, when we define runs). The updated wrapper is passed on to the subroutine Fix (line 4) to be made closed and consistent.

A crucial point for Algorithm 3 is defining what it means to resolve the "current" closedness and consistency defects. We call these *local* defects, meaning the ones that can be directly detected in the current wrapper. For DAs, local closedness defects are rows from the bottom part missing in the top part, and the empty word row if it is missing. Local consistency defects are pairs of row labels which are distinguished by the target language, or with differing rows when the labels are extended with a single symbol.

We first introduce additional notions to formalise these ideas. We partially order the subobjects and quotients of the target automaton's state space Q_t in the usual way. Given two subobjects $j\colon J \hookrightarrow Q_\mathrm{t}$ and $k\colon K \hookrightarrow Q_\mathrm{t}$, we say $j \leq k$ if there is $f\colon J \to K$ such that $k \circ f = j$. Intuitively, j is "contained" in k. Given two quotients $x\colon Q_\mathrm{t} \twoheadrightarrow X$ and $y\colon Q_\mathrm{t} \twoheadrightarrow Y$, we say $x \leq y$ if there exists $g\colon X \to Y$ such that $y = g \circ x$. Intuitively, x is "finer" than y.

Now, consider a wrapper (α, β) for Q_t. We have that $\alpha^{\triangleleft}$ and $\beta^{\triangleright}$ are a subobject and a quotient of Q_t, respectively. For instance, in the DA case, $\alpha^{\triangleleft}$ is the set of states in Q_t currently represented by the table, and $\beta^{\triangleright}$ is the equivalence relation on states induced by the rows. We can now say another wrapper (α', β') is a locally closed extension of (α, β) if (a) it represents at least the same states of the target automaton as α, formalised $\alpha^{\triangleleft} \leq \alpha'^{\triangleleft}$, and; (b) it solves the closedness defects present in α. Local consistency is analogous: it requires the extended wrapper to distinguish at least the same states of Q_t as the original one.

Definition 19 (Local closedness and consistency). *Consider a wrapper*

$$\mathcal{W} = (\, S' \xrightarrow{\alpha'} Q_{\mathsf{t}} \,,\, Q_{\mathsf{t}} \xrightarrow{\beta'} P' \,)$$

We call $\mathcal{W}$ locally closed *w.r.t. a morphism* $\alpha\colon S \to Q_{\mathsf{t}}$ *if* $\alpha^{\triangleleft} \leq \alpha'^{\triangleleft}$ *and there are morphisms* $i_{\mathcal{W}}\colon I \to H_{\mathcal{W}}$ *and* $\mathsf{lclose}_{\mathcal{W},\alpha}\colon FS \to H_{\mathcal{W}}$ *s.t. these diagrams commute:*

$$\begin{array}{ccc} I & \xrightarrow{i_{\mathsf{t}}} & Q_{\mathsf{t}} \\ {\scriptstyle i_{\mathcal{W}}}\downarrow & & \downarrow{\scriptstyle \beta'} \\ H_{\mathcal{W}} & \xrightarrow{m_{\mathcal{W}}} & P' \end{array} \qquad \begin{array}{ccccc} FS & \xrightarrow{F\alpha} & FQ_{\mathsf{t}} & \xrightarrow{\delta_{\mathsf{t}}} & Q_{\mathsf{t}} \\ {\scriptstyle \mathsf{lclose}_{\mathcal{W},\alpha}}\downarrow & & & & \downarrow{\scriptstyle \beta'} \\ H_{\mathcal{W}} & & \xrightarrow{m_{\mathcal{W}}} & & P' \end{array}$$

Given $\beta\colon Q \to P$, *we say that* $\mathcal{W}$ *is* locally consistent *w.r.t.* β *if* $\beta'^{\triangleright} \leq \beta^{\triangleright}$ *and there exist morphisms* $o_{\mathcal{W}}\colon H_{\mathcal{W}} \to O$ *and* $\mathsf{lcons}_{\mathcal{W},\beta}\colon FH_{\mathcal{W}} \to P$ *making the diagrams below commute.*

$$\begin{array}{ccc} S' & \xrightarrow{e_{\mathcal{W}}} & H_{\mathcal{W}} \\ {\scriptstyle \alpha'}\downarrow & & \downarrow{\scriptstyle o_{\mathcal{W}}} \\ Q_{\mathsf{t}} & \xrightarrow{o_{\mathsf{t}}} & O \end{array} \qquad \begin{array}{ccccc} FS' & & \xrightarrow{Fe_{\mathcal{W}}} & & FH_{\mathcal{W}} \\ {\scriptstyle F\alpha'}\downarrow & & & & \downarrow{\scriptstyle \mathsf{lcons}_{\mathcal{W},\beta}} \\ FQ_{\mathsf{t}} & \xrightarrow{\delta_{\mathsf{t}}} & Q_{\mathsf{t}} & \xrightarrow{\beta} & P \end{array}$$

A wrapper (α, β) is closed if and only if it is locally closed w.r.t. α and consistent if and only if it is locally consistent w.r.t. β.

Example 20. For the case of DAs, consider a wrapper $(\alpha_{S'}, \beta_{E'})$ representing an observation table (S', E') for the target DA $\mathcal{A}_{\mathsf{t}}$ (see Example 8):

Local closedness: Given $\alpha_S\colon S \to Q_{\mathsf{t}}$, $(\alpha_{S'}, \beta_{E'})$ is locally closed w.r.t. α_S if (1) $S \subseteq S'$ (to ensure $\alpha_S^{\triangleleft} \leq \alpha_{S'}^{\triangleleft}$); (2) S' contains the empty word (left diagram); and (3) any row in the bottom part of the table (S, E') occurs in the top part of (S', E') (right diagram).

Local consistency: Similarly, given $\beta_E\colon Q_{\mathsf{t}} \to 2^E$, $(\alpha_{S'}, \beta_{E'})$ is locally consistent w.r.t. β_E if (1) $E \subseteq E'$ (to ensure $\beta_{E'}^{\triangleright} \leq \beta_E^{\triangleright}$); (2) E' contains the empty word (left diagram); and (3) for all $s, s' \in S'$ and $a \in A$, if s and s' map to the same row in the top part of (S', E'), then the rows for sa and $s'a$ are the same in the bottom part of (S', E) (right diagram).

In Algorithm 3 we assume that we can always find locally closed and consistent wrappers (lines 4 and 6 respectively). This assumption holds in general for local closedness: for each wrapper (α, β) for Q_{t} we can always find α' such that (α', β) is locally closed w.r.t. α.

Lemma 21. *Given a wrapper* (α, β) *for* Q_{t}, $([\alpha, [i_{\mathsf{t}}, \delta_{\mathsf{t}}] \circ F_I\alpha], \beta)$ *is locally closed w.r.t.* α.

This result is enabled by the algebraic nature of automata. Local consistency is not inherently algebraic, so ensuring it takes more effort. We shall see in Sect. 6 that existence of locally closed/consistent extensions can be proved constructively for a broad class of automata.

Termination. To analyse termination of Algorithm 4, we introduce its *runs.*

Definition 22 (Run of the algorithm). *A* run *of the algorithm is a stream of wrappers $\mathcal{W}_n = (\alpha_n, \beta_n)$ satisfying the following conditions:*

1. $\alpha_0 \colon 0 \to Q_t$ *and* $\beta_0 \colon Q_t \to 1$ *are the unique morphisms;*
2. *if* $\mathcal{W}_n$ *is not closed, then* $\beta_{n+1} = \beta_n$ *and* α_{n+1} *is s.t.* (α_{n+1}, β_n) *is locally closed w.r.t.* α_n*;*
3. *if* $\mathcal{W}_n$ *is closed but not consistent, then* $\alpha_{n+1} = \alpha_n$ *and* β_{n+1} *is s.t.* (α_n, β_{n+1}) *is locally consistent w.r.t.* β_n*;*
4. *if* $\mathcal{W}_n$ *is closed and consistent and we obtain a counterexample* $\rho \colon S \to F_I S$ *for* $\mathcal{W}_n$*, then* $\alpha_{n+1}^{\triangleleft} = [\alpha_n, [i_t, \delta_t]^{\rho}]^{\triangleleft}$ *and* $\beta_{n+1} = \beta_n$*; and*
5. *if* $\mathcal{W}_n$ *is closed and consistent and correct up to all recursive* F_I*-coalgebras, then* $\mathcal{W}_{n+1} = \mathcal{W}_n$*.*

Note that, in point 4 above and in line 3 of Algorithm 4, we admit a more general counterexample resolution than Theorem 18: we only require that α_{n+1} and $[\alpha_n, [i_t, \delta_t]^{\rho}]$ represent the same states of the target automaton. This captures how observation tables are updated in practice; for instance in $\mathtt{L}^\star$ a counterexample prefix already in the table is discarded.

Proposition 23. *Algorithm 4 halts if and only if for all runs* $\{\mathcal{W}_n\}_{n \in \mathbb{N}}$ *there is* n *with* $\mathcal{W}_{n+1} = \mathcal{W}_n$*.*

We can establish an invariant on the order of subsequent wrappers in runs.

Lemma 24. *Let* $\{\mathcal{W}_n = (\alpha_n, \beta_n)\}_{n \in \mathbb{N}}$ *be a run. For all* $n \in \mathbb{N}$*, we have* $\alpha_n^{\triangleleft} \leq \alpha_{n+1}^{\triangleleft}$ *and* $\beta_{n+1}^{\triangleright} \leq \beta_n^{\triangleright}$*. Moreover, if* $\alpha_{n+1}^{\triangleleft} \leq \alpha_n^{\triangleleft}$*, then* $\alpha_{n+1} = \alpha_n$*; if* $\beta_n^{\triangleright} \leq \beta_{n+1}^{\triangleright}$*, then* $\beta_{n+1} = \beta_n$*.*

Putting these results together, we conclude that the algorithm terminates with a correct automaton, which is minimal under certain conditions. Satisfaction of the requirement on recursive coalgebras ρ_k depends on the implementation of counterexamples and closing of wrappers; for DAs, it suffices to keep the set of row labels S prefix-closed.

Theorem 25 (Termination). *If* Q_t *has finitely many subobject and quotient isomorphism classes, then for all runs* $\{\mathcal{W}_n = (\alpha_n, \beta_n)\}_{n \in \mathbb{N}}$ *there exists* $n \in \mathbb{N}$ *such that* $\mathcal{W}_n$ *is closed and consistent and its hypothesis is correct. If* $\mathcal{A}_t$ *is minimal and for all* $k \in \mathbb{N}$ *there exists a recursive* $\rho_k \colon S_k \to F_I S_k$ *such that* $\alpha_k = [i_t, \delta_t]^{\rho_k}$*, then the final hypothesis is minimal.*

6 Generalised Tree Automata

We now instantiate the above development to a wide class of **Set** endofunctors. This yields an abstract algorithm for *generalised* tree automata—i.e., automata accepting sets of trees, possibly subject to equations—including bottom-up tree automata and unordered tree automata. We first introduce the running examples.

Example 26 (Tree automata). Let Γ be a ranked alphabet, i.e., a finite set where $\gamma \in \Gamma$ comes with $\mathsf{arity}(\gamma) \in \mathbb{N}$. The set of Γ-trees over a finite set of leaf symbols I is the smallest set $T_\Gamma(I)$ such that $I \subseteq T_\Gamma(I)$, and for all $\gamma \in \Gamma$ we have that $t_1, \ldots, t_{\mathsf{arity}(\gamma)} \in T_\Gamma(I)$ implies $(\gamma, t_1, \ldots, t_{\mathsf{arity}(\gamma)}) \in T_\Gamma(I)$. The alphabet Γ gives rise to the polynomial functor $FX = \coprod_{\gamma \in \Gamma} X^{\mathsf{arity}(\gamma)}$. The free F-algebra monad is precisely T_Γ, where the unit turns elements into leaves, and the multiplication flattens nested trees into a tree. A bottom-up deterministic tree automaton [16] is then an automaton over F, with finite I and $O = 2$.

Example 27 (Unordered tree automata). Consider the finite powerset functor $\mathcal{P}_\mathsf{f} \colon \mathbf{Set} \to \mathbf{Set}$, mapping a set to its finite subsets. The corresponding free $\mathcal{P}_\mathsf{f}$-monad maps a set X to the set of finitely-branching unordered trees with nodes in X. Automata over $\mathcal{P}_\mathsf{f}$, with output set $O = 2$ and finite I, accept sets of such trees. Note that unordered trees can be seen as trees over a ranked alphabet $\Gamma = \{\mathfrak{s}_i \mid i \in \mathbb{N}\}$, where $\mathsf{arity}(\mathfrak{s}_i) = i$, satisfying equations that collapse duplicate branches and identify lists of branches up to permutations.

Automata in these examples are algebras for endofunctors with the following properties: they are *strongly finitary* [6]—i.e., they are finitary and preserve finite sets—and they preserve weak pullbacks. We turn these into a global assumption, used in several places; in particular, that F is strongly finitary is used to guarantee the existence of finite counterexamples.

Assumption 28. *In the remainder we take* $\mathbf{C} = \mathbf{Set}$ *with the (surjective, injective) factorisation system, assume F is strongly finitary and preserves weak pullbacks, and I finite.*

If the target automaton $\mathcal{A}_\mathsf{t}$ is finite, the algorithm terminates by Theorem 25.

We start with the central notion of *contextual wrapper*, a specific form of wrapper using contexts to generalise string concatenation to trees. We then show that contextual wrappers enable effective procedures for local closedness and consistency, and for computing hypotheses. Moreover, they can always be updated via finite counterexamples. Altogether, this makes the ingredients of our abstract algorithm concrete for generalised tree automata.

6.1 Contextual Wrappers

Denote by 1 the set $\{\square\}$. Given $x \in X$ for any set X, we write $\mathfrak{e}_x$ for the function $1 \to X$ that assigns x to $\square$. We use the set 1 to define the set of *contexts* $T(I+1)$, where the *holes* $\square$ occurring in a context $c \in T(I+1)$ can be used to plug in further data such as another context or a tree, e.g., in the case of Examples 26 and 27. In fact, it is well known that $T(I + (-))$ forms a monad with unit $\hat{\eta}_X$ turning each hole from X into a context, and multiplication $\hat{\mu}_X$ plugging a context into another context [35].

Definition 29 (Contextual wrapper). *Let $S \subseteq TI$ and $E \subseteq T(I+1)$. Now:*

- *$\alpha_S \colon S \to Q_\mathsf{t}$ is defined as the restriction of the reachability map of $\mathcal{A}_\mathsf{t}$ to S;*

– $\beta_E : Q_t \to O^E$ *is defined as the function given by* $\beta_E(q)(e) = (o_t \circ [i_t, \mathfrak{e}_q]^\sharp)(e)$.

A wrapper is called contextual *if it is of the form* (α_S, β_E) *for some* S *and* E.

Intuitively, β_E classifies states by plugging them into every context in E and comparing the resulting outputs. In the DA case, contextual wrappers are equivalent to those of Example 8, where row labels are plugged into word contexts—i.e., words of the form $\square \cdot e$, with $e \in A^*$—to achieve string concatenation.

We now show how to compute several morphisms induced by a wrapper. These morphisms, intuitively, correspond to different parts of an observation table, and are used for (local) closedness and consistency, and to construct the hypothesis. In particular, we show that they can be computed concretely by querying the language $\mathcal{L}_{\mathcal{A}_t}$, i.e., via membership queries.

Proposition 30 (Computing wrapper morphisms). *Given* $S \subseteq TI$ *with inclusion* $j : S \to TI$ *and* $E \subseteq T(I+1)$ *with inclusion* $k : E \to T(I+1)$*, we have:*

– *The* top observation table $\beta_E \circ \alpha_S : S \to O^E$, $s \mapsto \mathcal{L}_{\mathcal{A}_t} \circ \mu_I \circ T[\eta_I, j \circ \mathfrak{e}_s] \circ k$*;*
– *The* bottom observation table $\beta_E \circ \delta_t \circ F\alpha_S : FS \to O^E$, $t \mapsto \mathcal{L}_{\mathcal{A}_t} \circ \mu_I \circ T[\eta_I, \gamma_I \circ Fj \circ \mathfrak{e}_t] \circ k$*;*
– *The* input rows $\beta_E \circ i_t : I \to O^E$ *given by* $(\beta_E \circ i_t)(x) = \mathcal{L}_{\mathcal{A}_t} \circ T[\mathrm{id}_I, \mathfrak{e}_x] \circ k$*;*
– *The* row output $o_t \circ \alpha_S : S \to O$ *given by* $(o_t \circ \alpha_S)(s) = \mathcal{L}_{\mathcal{A}_t}(s)$.

Example 31 For tree automata, a contextual wrapper is as follows: $S \subseteq T_\Gamma(I)$ is a set of Γ-trees over I, and $E \subseteq T_\Gamma(I+1)$ is formed by contexts, i.e., Γ-trees where a special leaf $\square$ may occur, or equivalently $\Gamma + \square$-trees, where $\Gamma + \square$ is the signature Γ extended with an additional constant $\square$. Plugging into a context intuitively amounts to replacing this leaf with a tree.

We now give the intuition behind the maps of Proposition 30:

– The top part of the observation table has rows labelled by trees in S, columns by contexts in E, and rows are computed by plugging their tree labels into each column context and querying the language. When E contains only contexts with exactly one instance of $\square$, this corresponds precisely to the observation tables of [14,18].
– The bottom part contains rows labelled over elements of $FS = \coprod_{\gamma \in \Gamma} S^{\mathsf{arity}(\gamma)}$, i.e., trees obtained by adding a new root symbol to those from S. This generalises adding an alphabet symbol to row labels, as done in the bottom observation table of L*. Rows are computed as in the top part, by plugging their tree labels into contexts E and querying.
– The input rows are those for the leaves I, and the row output function queries the language for each row label.

The case of unordered trees is analogous, with a key difference: wrapper maps are now up to equations, as both S and E are sets of unordered trees. The corresponding observation table can be understood as containing equivalence classes of rows and columns. For instance, the bottom part has only one successor row for each set of trees in S, whereas in the previous case we have one successor row for each symbol $\gamma \in \Gamma$ and $\mathsf{arity}(\gamma)$-list of trees from S.

Hypotheses. Recall that, given a closed and consistent wrapper (α_S, β_E), the state space of the associated hypothesis is given by the image of $\beta_E \circ \alpha_S \colon S \to O^E$. Since S and E are finite sets, we can compute the image of this function. For bottom-up and unordered tree automata, as in the DA case (see Example 8), this image consists of distinct rows. The initial states, outputs and dynamics of the hypothesis automaton are defined as follows:

$$i_{\mathcal{H}_\mathcal{W}}(x) = (\beta_E \circ i_{\mathsf{t}})(x) \qquad\qquad o_{\mathcal{H}_\mathcal{W}}(e_\mathcal{W}(s)) = (o_{\mathsf{t}} \circ \alpha_S)(s)$$

$$\delta_{\mathcal{H}_\mathcal{W}}(F(e_\mathcal{W})(x)) = (\beta_E \circ \delta_{\mathsf{t}} \circ F\alpha_S)(x).$$

Closedness and consistency ensure well-definedness. We know from Proposition 30 how to compute those functions via membership queries.

6.2 Witnessing Local Closedness and Consistency

We now consider local closedness and consistency. In the current setting, these amount to equality checks on finite structures, which can be performed effectively.

Lemma 32 (Local closedness for Set automata). *Given $S, S' \subseteq TI$ and $E \subseteq T(I+1)$ such that $S \subseteq S'$, $(\alpha_{S'}, \beta_E)$ is locally closed w.r.t. α_S if there exist $k \colon I \to S'$ and $\ell \colon FS \to S'$ such that (1) $\alpha_{S'} \circ k = i_{\mathsf{t}}$ and (2) $\alpha_{S'} \circ \ell = \delta_{\mathsf{t}} \circ F\alpha_S$.*

Example 33. For bottom-up tree automata, local closedness holds if the table (S', E) already contains each leaf row (Eq. 1), and it contains every successor row for S, namely $FS = \coprod_{\gamma \in \Gamma} S^{\mathsf{arity}(\gamma)}$ (Eq. 2). For unordered tree automata the condition is similar, and now involves successor trees in $\mathcal{P}_{\mathsf{f}}(S)$.

Lemma 34 (Local consistency for Set automata). *Let $S \subseteq TI$ and $E \subseteq E' \subseteq T(I+1)$, with S finite. Furthermore, suppose that for $s, s' \in S$ with $(\beta_{E'} \circ \alpha_S)(s) = (\beta_{E'} \circ \alpha_S)(s')$ we have: (1) $(o_{\mathsf{t}} \circ \alpha_S)(s) = (o_{\mathsf{t}} \circ \alpha_S)(s')$; and (2) $\beta_E \circ \delta_{\mathsf{t}} \circ F(\alpha_S \circ [\mathrm{id}_S, \mathfrak{e}_s]) = \beta_E \circ \delta_{\mathsf{t}} \circ F(\alpha_S \circ [\mathrm{id}_S, \mathfrak{e}_{s'}])$. Then $\mathcal{W} = (\alpha_S, \beta_{E'})$ is locally consistent w.r.t. β_E.*

Example 35. For bottom-up tree automata, local consistency amounts to require the following for the table for (S, E'). For all $s, s' \in S$ corresponding to the same row we must have: (1) s and s' are both accepted/rejected; (2) successor rows obtained by plugging s and s' into the same one-level context from $F(S+1) = \coprod_{\gamma \in \Gamma} (S + \{\Box\})^{\mathsf{arity}(\gamma)}$ are equal.

For unordered-tree automata, we need to compare s and s' only when they are equationally inequivalent. Note that one-level contexts are also up to equations, which means that the position of the hole in the context is irrelevant for computing extensions of s and s'.

We now develop procedures for fixing local closedness and consistency defects. First, we show that we can always extend S to make the wrapper locally closed.

Proposition 36. *Given finite $S \subseteq TI$ and $E \subseteq T(I+1)$, there exists a finite $S' \subseteq TI$ such that $(\alpha_{S'}, \beta_E)$ is locally closed w.r.t. α_S. If there exists a recursive $\rho\colon S \to F_I S$ such that $[\eta_I, \gamma_I]^{\rho}\colon S \to TI$ is the inclusion, then there exists a recursive $\rho'\colon S' \to F_I S'$ such that $[\eta_I, \gamma_I]^{\rho'}\colon S' \to TI$ is the inclusion.*

The condition of $[\eta_I, \gamma_I]^{\rho}\colon S \to TI$ being the inclusion map in the above result amounts to prefix closedness of S in tree automata, see Example 37 below. Further, under this condition we have that $\alpha_S = [i_{\mathrm{t}}, \delta_{\mathrm{t}}]^{\rho}$, since the reachability map is an algebra morphism, and similarly for $\alpha_{S'}$. This is crucial to satisfy the requirements for minimality of the termination theorem.

Example 37. To better understand the above proposition, it is worth describing what recursive coalgebras are for the automata of Examples 26 and 27. For bottom-up tree automata, they are coalgebras $\rho\colon S \to \coprod_{\gamma \in \Gamma} S^{\mathsf{arity}(\gamma)} + I$ satisfying suitable conditions. Subtree-closed subsets of $T_\Gamma(I)$ are sets of trees closed under taking subtrees. Every subtree-closed S can be made into a recursive coalgebra that returns the root symbol and its arguments, if applied to a tree of non-zero depth, and a leaf otherwise. For unordered tree automata, $\rho\colon S \to \mathcal{P}_{\mathsf{f}} S + I$ will just return the set of subtrees or a leaf.

The proof of Proposition 36, which can be found in ??, is constructive and describes a naive procedure to make a table locally closed: adding all (finitely-many) successor rows to the table. For instance, in the case of tree automata, one adds rows obtained by adding a new root symbol to trees labelling rows in all possible ways, for each symbol in the alphabet. One may optimise the algorithm by instead adding only missing rows.

We now show how to fix local consistency, by extending a finite set of column labels E to a finite set E' such that the resulting wrapper is locally consistent.

Proposition 38. *Given finite $S \subseteq TI$ and $E \subseteq T(I+1)$, define $E' \subseteq T(I+1)$ by $E' = E \cup \{(\eta_{I+1} \circ \kappa_2)(\Box)\} \cup \{(\hat{\mu}_1 \circ T(\mathrm{id}_I + c_x))(e) \mid e \in E, x \in F(S+1)\}$ where $c_x\colon 1 \to T(I+1)$, with $c_x = \gamma_{I+1} \circ F[T\kappa_1 \circ j, \hat{\eta}_1] \circ \mathfrak{e}_x$, and $j\colon S \to TI$ is set inclusion. It holds that E' is finite and $(\alpha_S, \beta_{E'})$ is locally consistent w.r.t. β_E.*

For tree automata, E' is E plus the empty context $\Box$ and the trees obtained by plugging one-level contexts formed from the current row labels (see Example 35) into all contexts in E. This amounts to extending columns so that *all* consistency defects are fixed. One can optimise the procedure above by incrementally adding to E only those elements of E' that result in new pairs of rows being distinguished.

6.3 Finite Counterexamples

Finally, we show that the teacher can always supply a finite counterexample.

Proposition 39 (Language equivalence via finite recursion). *Given an automaton $\mathcal{A} = (Q, \delta, i, o)$, we have $\mathcal{L}_{\mathcal{A}_{\mathrm{t}}} = \mathcal{L}_{\mathcal{A}}$ iff $\mathcal{L}^{\rho}_{\mathcal{A}_{\mathrm{t}}} = \mathcal{L}^{\rho}_{\mathcal{A}}$ for all recursive $\rho\colon S \to F_I S$ such that S is finite.*

Corollary 40 (Finite counterexamples). *Given a closed and consistent wrapper $\mathcal{W}$ for Q_t, we have $\mathcal{L}_{\mathcal{H}_\mathcal{W}} \neq \mathcal{L}_{\mathcal{A}_t}$ iff there exists a counterexample $\rho\colon S \to F_I S$ for $\mathcal{W}$ such that S is finite.*

Example 41. Recall from Example 37 that finite recursive coalgebras for bottom-up (resp. unordered) tree automata are coalgebras $\rho\colon S \to \coprod_{\gamma \in \Gamma} S^{\mathsf{arity}(\gamma)} + I$ (resp. $\rho\colon S \to \mathcal{P}_\mathsf{f} S + I$). Thus, finite counterexamples are recursive coalgebras of this form where S is finite or, more concretely, a finite subtree-closed set of trees.

Given a finite counterexample, if α_S arises from a recursive coalgebra ρ (e.g., when S is prefix-closed), updating the wrapper in line 3 of Algorithm 4 can be done as follows: (1) combine ρ with the recursive coalgebra in Corollary 40 via a coproduct (which preserves recursiveness); (2) take a suitable factorisation to make sure that there is an inclusion of S into TI, and thus that the updated $\alpha_{S'}$ forms a contextual wrapper (see ?? for a formal justification). Concretely, the latter step amounts to removing multiple copies of rows with the same label. Altogether, these steps take the union of the current rows with the (prefix-closed) counterexample, and guarantee that $\alpha_{S'}$ again arises from a recursive coalgebra.

6.4 Minimality

Theorem 25 gives sufficient conditions for minimality of the automaton obtained from the algorithm, namely: each α arises from a recursive coalgebra, and the target automaton should be minimal. For the first condition to hold, there are two parts of the algorithm that need to be implemented appropriately, as they change α: closing the table and adding counterexamples. This can always be done: for closing the table, this follows from Proposition 36; for counterexamples, the strategy outlined in the previous section yields a wrapper of the desired form. As for the second condition, a minimal automaton exists if the functor F preserves arbitrary cointersections, which is the case iff F is finitary [7].

7 Related Work

This paper proceeds in the line of work on categorical automata learning started in [32], and further developed in the CALF framework [29,30]. CALF provides abstract definitions of closedness, consistency, and hypothesis and several techniques to analyse and guide the development of concrete learning algorithms. CALF operates at a high level of abstraction and previously did not include an explicit learning algorithm. We discuss two further recent categorical approaches to learning, which make stronger assumptions than CALF in order to allow for the definition of concrete algorithms. The present paper is a third such approach.

Barlocco et al. [12] proposed an abstract algorithm for learning coalgebras, where tests are formed by an abstract version of coalgebraic modal logic. On the one hand, the notion of wrapper and closedness from CALF essentially instantiate to that setting; on the other hand, the combination of logic and coalgebra is what enables to define an actual algorithm in [12]. The current work focuses on algebras rather than coalgebras, and is orthogonal. In particular, it covers (bottom-up) tree automata, which are outside the scope of [12].

Urbat and Schröder proposed another categorical approach to automata learning [41], which—similarly to the work of Barlocco et al.—makes stronger assumptions than CALF in order to define a learning algorithm. Their work focuses primarily on automata, assuming that the systems of interest can be viewed both as algebras and coalgebras, and the generality comes from allowing to instantiate these in various categories. Moreover, it allows covering algebraic recognisers in certain cases, through a reduction to automata over a carefully constructed alphabet; this (orthogonal) extension allows covering, e.g., ω-languages as well as tree languages. However, the reduction to automata makes this process quite different than the approach to tree learning in the present paper: it makes use of an automaton over all (flat) contexts, yielding an infinite alphabet, and therefore the algorithmic aspect is not clear. The extension to an actual algorithm for learning tree automata is mentioned as future work in [41]. In the present paper, this is achieved by learning algebras directly.

Yet another categorical approach to learning was proposed recently by Colcombet, Petrisan, and Stabile [15]. Here, the way automata are modelled is rather different: not as algebras or coalgebras within a category, but as functors from a structure category to an output category. So far this has led the authors to develop an abstract automata learning algorithm that generalises algorithms for DFAs, weighted automata, and subsequential transducers. However, as their structure category is built by generating morphisms representing words by starting with a morphism for each alphabet symbol and closing under composition, it is unclear whether this approach could cover tree automata.

Concrete algorithms for learning tree automata and languages have appeared in the literature. The inference of regular tree languages using membership and equivalence queries appeared in [18], extending earlier work of Sakakibara [39]. Later, [14] provided a learning algorithm for regular tree automata using only membership queries. The instantiated algorithm in our paper has elements (such as the use of contexts) close to the concrete algorithms. The focus of the present paper is on presenting an algebraic framework that can effectively be instantiated to recover such concrete algorithms in a modular and canonical fashion, with proofs of correctness and termination stemming from the general framework.

8 Future Work

This paper makes use of the free monad of a functor F in the formulation of the generalised learning algorithm, and hence can only deal with quotienting in a restricted setting, namely by flat equations in the presentation of F. It remains an open challenge to extend the present algorithm to a setting with more general equations. For the concrete case of pomset languages [20, 22] represented by *bimonoids* [34], we note that we have successfully instantiated the abstract algorithm described in this paper, and augmented it to include optimisations specific to the equations that hold in that setting [27]. In future work, we aim to extend the ideas behind these optimisations to the abstract setting, as well.

Another direction is to extend the framework with side-effects, encoded by a monad, in the style of [30]. This would enable learning more compact automata—albeit with richer, monadic, transitions—representing languages and, as a concrete instance, provide an active learning algorithm for weighted tree automata.

Acknowledgements. T. Kappé was partially supported by the European Union's Horizon 2020 research and innovation programme under the Marie Skłodowska-Curie grant agreement No. 101027412 (VERLAN), as well as ERC Starting Grant 679127 (ProFoundNet). Gerco van Heerdt, Matteo Sammartino, and Alexandra Silva were partially supported by the EPSRC Standard Grant CLeVer (EP/S028641/1).

References

1. Aarts, F., Fiterau-Brostean, P., Kuppens, H., Vaandrager, F.: Learning register automata with fresh value generation. In: Leucker, M., Rueda, C., Valencia, F.D. (eds.) ICTAC 2015. LNCS, vol. 9399, pp. 165–183. Springer, Cham (2015). https://doi.org/10.1007/978-3-319-25150-9_11
2. Aarts, F., de Ruiter, J., Poll, E.: Formal models of bank cards for free. In: ICST, pp. 461–468 (2013). https://doi.org/10.1109/ICSTW.2013.60
3. Aarts, F., Schmaltz, J., Vaandrager, F.: Inference and abstraction of the biometric passport. In: Margaria, T., Steffen, B. (eds.) ISoLA 2010. LNCS, vol. 6415, pp. 673–686. Springer, Heidelberg (2010). https://doi.org/10.1007/978-3-642-16558-0_54
4. Aarts, F., Vaandrager, F.: Learning I/O automata. In: Gastin, P., Laroussinie, F. (eds.) CONCUR 2010. LNCS, vol. 6269, pp. 71–85. Springer, Heidelberg (2010). https://doi.org/10.1007/978-3-642-15375-4_6
5. Adámek, J., Milius, S., Moss, L.S.: On well-founded and recursive coalgebras. In: FoSSaCS 2020. LNCS, vol. 12077, pp. 17–36. Springer, Cham (2020). https://doi.org/10.1007/978-3-030-45231-5_2
6. Adámek, J., Milius, S., Velebil, J.: Free iterative theories: a coalgebraic view. MFCS **13**(2), 259–320 (2003). https://doi.org/10.1017/S0960129502003924
7. Adámek, J., Trnková, V.: Automata and Algebras in Categories. Kluwer, Dordrecht (1989)
8. Angluin, D.: Learning regular sets from queries and counterexamples. Inf. Comput. **75**(2), 87–106 (1987). https://doi.org/10.1016/0890-5401(87)90052-6
9. Angluin, D., Fisman, D.: Learning regular omega languages. Theor. Comput. Sci. **650**, 57–72 (2016). https://doi.org/10.1016/j.tcs.2016.07.031
10. Arbib, M.A., Manes, E.G.: A categorist's view of automata and systems. In: Manes, E.G. (ed.) Category Theory Applied to Computation and Control. LNCS, vol. 25, pp. 51–64. Springer, Heidelberg (1975). https://doi.org/10.1007/3-540-07142-3_61
11. Awodey, S.: Category Theory. Oxford University Press, Oxford (2010)
12. Barlocco, S., Kupke, C., Rot, J.: Coalgebra learning via duality. In: Bojańczyk, M., Simpson, A. (eds.) FoSSaCS 2019. LNCS, vol. 11425, pp. 62–79. Springer, Cham (2019). https://doi.org/10.1007/978-3-030-17127-8_4
13. Bergadano, F., Varricchio, S.: Learning behaviors of automata from multiplicity and equivalence queries. SIAM J. Comput. **25**(6), 1268–1280 (1996). https://doi.org/10.1137/S009753979326091X

14. Besombes, J., Marion, J.: Learning tree languages from positive examples and membership queries. Theor. Comput. Sci. **382**(3), 183–197 (2007). https://doi.org/10.1016/j.tcs.2007.03.038
15. Colcombet, T., Petrisan, D., Stabile, R.: Learning automata and transducers: a categorical approach. In: CSL, vol. 183, pp. 15:1–15:17 (2021). https://doi.org/10.4230/LIPIcs.CSL.2021.15
16. Comon, H., et al.: Tree Automata Techniques and Applications (2008). https://hal.inria.fr/hal-03367725
17. de la Higuera, C.: Grammatical Inference: Learning Automata and Grammars. Cambridge University Press, Cambridge (2010)
18. Drewes, F., Högberg, J.: Learning a regular tree language from a teacher. In: Ésik, Z., Fülöp, Z. (eds.) DLT 2003. LNCS, vol. 2710, pp. 279–291. Springer, Heidelberg (2003). https://doi.org/10.1007/3-540-45007-6_22
19. Fiterău-Broştean, P., Janssen, R., Vaandrager, F.: Combining model learning and model checking to analyze TCP implementations. In: Chaudhuri, S., Farzan, A. (eds.) CAV 2016. LNCS, vol. 9780, pp. 454–471. Springer, Cham (2016). https://doi.org/10.1007/978-3-319-41540-6_25
20. Gischer, J.L.: The equational theory of pomsets. Theor. Comput. Sci. **61**, 199–224 (1988). https://doi.org/10.1016/0304-3975(88)90124-7
21. Gold, E.M.: System identification via state characterization. Automatica **8**(5), 621–636 (1972). https://doi.org/10.1016/0005-1098(72)90033-7
22. Grabowski, J.: On partial languages. Fundam. Inform. **4**(2), 427 (1981)
23. van Heerdt, G.: An abstract automata learning framework. Master's thesis, Radboud Universiteit Nijmegen (2016)
24. van Heerdt, G., Jacobs, B., Kappé, T., Silva, A.: Learning to coordinate. In: de Boer, F., Bonsangue, M., Rutten, J. (eds.) It's All About Coordination. LNCS, vol. 10865, pp. 139–159. Springer, Cham (2018). https://doi.org/10.1007/978-3-319-90089-6_10
25. van Heerdt, G., Kappé, T., Rot, J., Sammartino, M., Silva, A.: Tree automata as algebras: minimisation and determinisation. In: CALCO, vol. 139, pp. 6:1–6:22 (2019). https://doi.org/10.4230/LIPIcs.CALCO.2019.6
26. van Heerdt, G., Kappé, T., Rot, J., Sammartino, M., Silva, A.: A categorical framework for learning generalised tree automata. arXiv e-prints (2020). https://arxiv.org/abs/2001.05786
27. van Heerdt, G., Kappé, T., Rot, J., Silva, A.: Learning pomset automata. In: FOSSACS 2021. LNCS, vol. 12650, pp. 510–530. Springer, Cham (2021). https://doi.org/10.1007/978-3-030-71995-1_26
28. van Heerdt, G., Kupke, C., Rot, J., Silva, A.: Learning weighted automata over principal ideal domains. In: FoSSaCS 2020. LNCS, vol. 12077, pp. 602–621. Springer, Cham (2020). https://doi.org/10.1007/978-3-030-45231-5_31
29. van Heerdt, G., Sammartino, M., Silva, A.: CALF: categorical automata learning framework. In: CSL, pp. 29:1–29:24 (2017). https://doi.org/10.4230/LIPIcs.CSL.2017.29
30. van Heerdt, G., Sammartino, M., Silva, A.: Learning automata with side-effects. In: Petrişan, D., Rot, J. (eds.) CMCS 2020. LNCS, vol. 12094, pp. 68–89. Springer, Cham (2020). https://doi.org/10.1007/978-3-030-57201-3_5
31. Isberner, M., Howar, F., Steffen, B.: The open-source LearnLib. In: Kroening, D., Păsăreanu, C.S. (eds.) CAV 2015. LNCS, vol. 9206, pp. 487–495. Springer, Cham (2015). https://doi.org/10.1007/978-3-319-21690-4_32

32. Jacobs, B., Silva, A.: Automata learning: a categorical perspective. In: van Breugel, F., Kashefi, E., Palamidessi, C., Rutten, J. (eds.) Horizons of the Mind. A Tribute to Prakash Panangaden. LNCS, vol. 8464, pp. 384–406. Springer, Cham (2014). https://doi.org/10.1007/978-3-319-06880-0_20
33. Lane, S.M.: Categories for the Working Mathematician. Graduate Texts in Mathematics, Springer, New York (1998). https://doi.org/10.1007/978-1-4612-9839-7
34. Lodaya, K., Weil, P.: Series-parallel languages and the bounded-width property. Theoret. Comput. Sci. **237**(1), 347–380 (2000). https://doi.org/10.1016/S0304-3975(00)00031-1
35. Lüth, C., Ghani, N.: Composing monads using coproducts. In: ICFP, pp. 133–144 (2002). https://doi.org/10.1145/581478.581492
36. Moerman, J., Sammartino, M., Silva, A., Klin, B., Szynwelski, M.: Learning nominal automata. In: POPL, pp. 613–625 (2017). https://doi.org/10.1145/3009837.3009879
37. Mues, M., Howar, F., Luckow, K.S., Kahsai, T., Rakamaric, Z.: Releasing the PSYCO: using symbolic search in interface generation for Java. ACM SIGSOFT Softw. Eng. Notes **41**(6), 1–5 (2016). https://doi.org/10.1145/3011286.3011298
38. Osius, G.: Categorical set theory: a characterization of the category of sets. J. Pure Appl. Algebra **4**(1), 79–119 (1974)
39. Sakakibara, Y.: Learning context-free grammars from structural data in polynomial time. Theor. Comput. Sci. **76**(2–3), 223–242 (1990). https://doi.org/10.1016/0304-3975(90)90017-C
40. Taylor, P.: Practical Foundations of Mathematics. Cambridge University Press, Cambridge (1999)
41. Urbat, H., Schröder, L.: Automata learning: an algebraic approach. In: LICS, pp. 900–914 (2020). https://doi.org/10.1145/3373718.3394775
42. Vaandrager, F.W.: Model learning. Commun. ACM **60**(2), 86–95 (2017). https://doi.org/10.1145/2967606

Saturated Kripke Structures as Vietoris Coalgebras

H. Peter Gumm(✉) and Mona Taheri

FB Mathematik und Informatik, Philipps-Universität, Marburg, Germany
gumm@mathematik.uni-marburg.de

Abstract. We show that the category of coalgebras for the compact Vietoris endofunctor $\mathbb{V}$ on the category *Top* of topological spaces and continuous mappings is isomorphic to the category of all modally saturated Kripke structures. Extending a result of Bezhanishvili, Fontaine and Venema [4], we also show that Vietoris subcoalgebras as well as bisimulations admit topological closure and that the category of Vietoris coalgebras has a terminal object.

Keywords: Saturated Kripke structures · Vietoris functor · Vietoris coalgebras · Vietoris convergence · Bisimulations

1 Introduction

The theory of coalgebras has provided Computer Science with a much needed general framework for dealing with all sorts of state based systems, with their structure theories and their logics. The varied types of systems, be they deterministic or nondeterministic automata, transition systems, probabilistic or weighted systems, neighborhood systems or the like, are fixed by the choice of an appropriate endofunctor F on the category of sets. From there on, with hardly any further assumptions, a mathematically pleasing structure theory and corresponding modal logics can be developed, see e.g. [10,20].

A particularly well behaved situation arises when choosing for F the finite-powerset functor $\mathbb{P}_\omega(-)$, perhaps augmented with a constant component $\mathbb{P}(\Phi)$ representing sets of atomic formulas. Coalgebras for the functor $\mathbb{P}_\omega(-) \times \mathbb{P}(\Phi)$ are precisely all image finite Kripke structures. Their logic is the standard modal logic based on the atomic formulae in Φ, and they possess a terminal coalgebra T, even though its description is usually of an "indirect" nature (see [2,3,11]).

The well known Hennessy-Milner theorem [12], relating bisimulations and logical equivalence, is a consequence of image finiteness and will not continue to hold for arbitrary Kripke structures, i.e. for coalgebras of type $\mathbb{P}(-) \times \mathbb{P}(\Phi)$, see [14].

The theory of modal logic knows of a class of Kripke structures, which lies between image finite structures and arbitrary Kripke structures and which continues to enjoy the Hennessy-Milner theorem. These structures are called *modally*

Published by Springer Nature Switzerland AG 2022
H. H. Hansen and F. Zanasi (Eds.): CMCS 2022, LNCS 13225, pp. 88–109, 2022.
https://doi.org/10.1007/978-3-031-10736-8_5

saturated, m-saturated in [9], or simply *saturated*. Unfortunately, though, there seems to be no *Set*-functor F, somehow located in between $\mathbb{P}_\omega(-) \times \mathbb{P}(\Phi)$ and $\mathbb{P}(-) \times \mathbb{P}(\Phi)$, whose coalgebras would be just the saturated Kripke structures.

It is well known, that much of the theory of coalgebras can be generalized by turning to other categories than *Set*, provided they are co-complete and come with a reasonable factorization structure. Some of the examples studied in the literature replace the base category *Set* with the category *Rel* of sets and relations [15], with the category *Pos* of posets [1] or *Cpo* of complete partial orders, with the category *Meas* of measurable spaces [6,17], or the category *Stone* of Stone spaces. Relevant to this present work will be the works of Kupke, Kurz and Venema [16] as well as Bezhanishvili, Fontaine and Venema [4] regarding coalgebras for the Vietoris functor on the category of Stone spaces, i.e. compact zero-dimensional Hausdorff spaces with continuous mappings.

When extending the Vietoris functor from Stone spaces to arbitrary topological spaces $\mathcal{X}$, two natural choices offer themselves for the object map: either the collection of all closed subsets of $\mathcal{X}$ or the collection of all compact subsets of $\mathcal{X}$, both equipped with appropriate topologies. Each of these choices yields a functor, generalizing the mentioned Vietoris functor on Stone spaces. Named the *lower Vietoris functor*, resp. the *compact Vietoris functor*, these endofunctors on the category *Top* of topological spaces and continuous functions were explored in recent work by Hofmann, Neves and Nora [13].

For our investigation of saturated Kripke structures, the *compact Vietoris functor*, which we denote by $\mathbb{V}(-)$, turns out to be appropriate. To model saturated Kripke-Structures, we choose the endofunctor $\mathbb{V}(-) \times \mathbb{P}(\Phi)$ on the category *Top* of topological spaces and continuous mappings, where the $\mathbb{P}(\Phi)$-part is a constant component equipped with an appropriate topology, intuitively representing a set of atomic propositions, as above. We show that $\mathbb{V}(-) \times \mathbb{P}(\Phi)$ coalgebras precisely correspond to saturated Kripke structures, in fact there is an isomorphism of categories between the category of saturated Kripke structures and the category of all topological coalgebras for the compact Vietoris functor $\mathbb{V}(-) \times \mathbb{P}(\Phi)$.

This correspondence also yields a direct description of the terminal $\mathbb{V}(-) \times \mathbb{P}(\Phi)$ coalgebra, which seems to be simpler and more natural than the terminal $\mathbb{P}_\omega(-) \times \mathbb{P}(\Phi)$ coalgebra mentioned above: it is simply the Vietoris coalgebra corresponding to the canonical model of normal modal logic over Φ.

For Stone coalgebras we know from [4], that the topological closure $\bar{R}$ of a bisimulation R is itself a bisimulation, again. We extend this result to the more general case of arbitrary Vietoris coalgebras, and we show also that a corresponding result holds true for subcoalgebras in place of bisimulations. For this we need to prepare some topological tools which may be interesting in their own right, relating convergence in the Vietoris space $\mathbb{V}(\mathcal{X})$ to convergence in the base space $\mathcal{X}$. In particular, nets $(\kappa_i)_{i\in I}$ converging to κ in the Vietoris space $\mathbb{V}(\mathcal{X})$ are shown to correspond, up to subnet formation, to nets $(a_i)_{i\in I}$ with $a_i \in \kappa_i$, converging in the base space $\mathcal{X}$ to some $a \in \kappa$, and conversely.

2 Preliminaries

For the remainder of this article we shall fix a set Φ, the elements of which shall be called *propositional variables* or *atomic propositions.*

2.1 Kripke Structures

Definition 1. *A* Kripke structure *(also called* Kripke model*)* $\mathcal{K} = (X, R, v)$ *consists of a set* X *of states together with a relation* $R \subseteq X \times X$*, and a map* $v : X \to \mathbb{P}(\Phi)$*, where* $\mathbb{P}$ *denotes the powerset functor.*

In applications, X will typically be a set of possible states of a system, R is called the *transition relation*, describing the allowed transitions between states from X, and v is called the *valuation*, since $v(x)$ consists of all atomic propositions true in state x. Instead of $(x, y) \in R$ we write $x \twoheadrightarrow y$ (or $x \twoheadrightarrow_R y$, if necessary). The idea is that $x \twoheadrightarrow y$ expresses that it is possible for the system to move from state x to state y. Instead of a relation, we can alternatively consider R as a map $R : X \to \mathbb{P}(X)$. This justifies the notation

$$R(x) := \{y \in X \mid (x, y) \in R\},$$

so $R(x)$ denotes the *successors* of x, i.e. all states reachable from x in one step.

Definition 2. *For subsets* $V \subseteq X$ *define* $\langle R \rangle V := \{x \in X \mid \exists y \in V.(x, y) \in R\}$ *and* $[R]V := \{x \in X \mid \forall y \in X.(x, y) \in R \implies y \in V\}$*.*

Thus $x \in \langle R \rangle V$ if from x it is *possible* to reach an element of V in one step, and $x \in [R]V$ says that starting from x, each transition will *necessarily* take us to V. Obviously, $\langle R \rangle (X - V) = X - [R]V$, and $[R](X - V) = X - \langle R \rangle V$, so $\langle R \rangle$ and $[R]$ are mutually expressible if complements are available.

2.2 Modal Logic

Starting with the elements of Φ as *atomic formulae*, we obtain *modal formulae* by combining them with the standard boolean connectors $\wedge$, $\vee$, $\neg$ or prefixing with the unary modal operator $\Box$. We also allow the usual shorthands $\bigvee_{i \in I_0} \phi_i$ and $\bigwedge_{i \in I_0} \phi_i$, whenever I_0 is a finite indexing set and each ϕ_i is a formula. Let $\mathcal{L}_\Phi$ be the set of all modal formulae so definable.

Validity $x \Vdash \phi$, is defined for $x \in X$ and $\phi \in \mathcal{L}_\Phi$ in the usual way (see [5]):

$$x \Vdash p :\iff p \in v(x), \quad \text{whenever } p \in \Phi$$
$$x \Vdash \Box\phi :\iff \forall y \in X. (x \twoheadrightarrow y \implies y \Vdash \phi).$$

For the boolean connectives $\wedge, \vee, \neg$, validity is defined as expected. We extend it to sets of formulas $\Sigma \subseteq \mathcal{L}_\Phi$, by

$$x \Vdash \Sigma \;:\iff\; \forall \phi \in \Sigma. x \Vdash \phi.$$

For any $x \in X$ we put $[\![x]\!] := \{\phi \in \mathcal{L}_\Phi \mid x \Vdash \phi\}$ and, similarly, for any $\phi \in \mathcal{L}_\Phi$ we set $[\![\phi]\!] := \{x \in X \mid x \Vdash \phi\}$. Two elements x, y from (possibly different) Kripke structures are called *logically equivalent* (in symbols $x \approx y$), if for each formula $\phi \in \mathcal{L}_\Phi$ we have $x \models \phi \iff y \models \phi$. Restricted to a single Kripke structure, $\approx$ is the kernel of the *semantic map* $x \mapsto [\![x]\!]$, and hence an equivalence relation. Similarly, two modal formulae ϕ, ψ are *equivalent*, and we write $\phi \equiv \psi$, if for each element x in any Kripke structure we have $x \Vdash \phi \iff x \Vdash \psi$.

Adding a further modality $\Diamond$ to our logical language by defining $\Diamond\phi := \neg\Box\neg\phi$ provides more than only a convenient abbreviation. The resulting equivalences $\neg\Box\phi \equiv \Diamond\neg\phi$ and $\neg\Diamond\phi \equiv \Box\neg\phi$ allow one to push negations inside, just as De Morgan's laws permit to do so for $\vee$ and $\wedge$, so that each modal formula becomes equivalent to a modal formula in *negation normal form* (*nnf*), where negations may only occur only in front of an atomic formula. We state this here for later reference:

Lemma 1. *Every modal formula is equivalent to a modal formula in negation normal form (nnf).*

2.3 Bisimulations

Definition 3. *A* bisimulation *between two Kripke structures* $\mathcal{K}_1 = (X_1, R_1, v_1)$ *and* $\mathcal{K}_2 = (X_2, R_2, v_2)$ *is a relation* $B \subseteq X_1 \times X_2$ *such that for each* $(x, y) \in B$*:*

1. $v_1(x) = v_2(y)$,
2. $\forall x' \in X_1 . x \twoheadrightarrow_{R_1} x' \implies \exists y' \in X_2 . y \twoheadrightarrow_{R_2} y' \wedge x' B\, y'$,
3. $\forall y' \in X_2 . y \twoheadrightarrow_{R_2} y' \implies \exists x' \in X_1 . x \twoheadrightarrow_{R_1} x' \wedge x' B\, y'$.

The empty relation $\emptyset \subseteq X_1 \times X_2$ is clearly a bisimulation, and the union of a family of bisimulations between $\mathcal{K}_1$ and $\mathcal{K}_2$ is again a bisimulation, hence there is a largest bisimulation between $\mathcal{K}_1$ and $\mathcal{K}_2$, which we call $\sim_{\mathcal{K}_1,\mathcal{K}_2}$ or simply $\sim$, when $\mathcal{K}_1$ and $\mathcal{K}_2$ are clear from the context.

If $B_1 \subseteq X_1 \times X_2$ is a bisimulation between $\mathcal{K}_1$ and $\mathcal{K}_2$, then the converse relation $B_1^{-1} \subseteq X_2 \times X_1$ is a bisimulation between $\mathcal{K}_2$ and $\mathcal{K}_1$. Given another bisimulation B_2 between Kripke structures $\mathcal{K}_2$ and $\mathcal{K}_3$ then the relational composition $B_1 \circ B_2$ is a bisimulation between $\mathcal{K}_1$ and $\mathcal{K}_3$.

A bisimulation *on* a Kripke structure $\mathcal{K} = (X, R, v)$ is a bisimulation between $\mathcal{K}$ and itself. The identity $\Delta_X = \{(x, x) \mid x \in X\}$ is always a bisimulation on $\mathcal{K}$. Consequently, the largest bisimulation *on* $\mathcal{K}$ is an equivalence relation, denoted by $\sim_\mathcal{K}$ or simply $\sim$. We say that two points $x \in X_1$ and $y \in X_2$ are *bisimilar*, if there exists a bisimulation B with $x\, B\, y$, which is the same as saying $x \sim y$. It is well known and easy to check by induction:

Lemma 2. *Bisimilar points satisfy the same formulae* $\phi \in \mathcal{L}_\Phi$.

A converse to this lemma was shown by Hennessy and Milner for the case of *image finite* Kripke structures. Here, an element x in a Kripke structure $\mathcal{K}$ is called *image finite* if it has only finitely many successors, i.e. $\{x' \mid x \twoheadrightarrow x'\}$ is finite. $\mathcal{K}$ is called image finite if each x from $\mathcal{K}$ is image finite. Thus Hennessy and Milner proved in [12]:

Proposition 1. *If x and y are image finite elements of Kripke structures $\mathcal{K}_1$ and $\mathcal{K}_2$, then $x \sim y$ iff $x \approx y$.*

2.4 Homomorphisms and Congruences

Definition 4. *A homomorphism $\varphi : \mathcal{K}_1 \to \mathcal{K}_2$ between Kripke structures $\mathcal{K}_1 = (X_1, R_1, v_1)$ and $\mathcal{K}_2 = (X_2, R_2, v_2)$ is a map whose graph*

$$G(\varphi) := \{(x, \varphi(x)) \mid x \in X_1\}$$

is a bisimulation.[1]

We call $\mathcal{K}_1$ a homomorphic preimage *of $\mathcal{K}_2$, and if φ is surjective (which we indicate by writing $\varphi : \mathcal{K}_1 \twoheadrightarrow \mathcal{K}_2$) then we call $\mathcal{K}_2$ a* homomorphic image *of $\mathcal{K}_1$. If $X_1 \subseteq X_2$ and the inclusion map $\iota : \mathcal{K}_1 \to \mathcal{K}_2$ is a homomorphism, then $\mathcal{K}_1$ is called a* Kripke substructure *of $\mathcal{K}_2$.*

It is easy to check that a subset $X_1 \subseteq X_2$ with the restrictions of R_2 and v_2 to X_1 is a substructure of $\mathcal{K}_2$ if and only if $R_2(x) \subseteq X_1$ for each $x \in X_1$.

If $\varphi : \mathcal{K}_1 \to \mathcal{K}_2$ is a homomorphism, then its kernel

$$\ker \varphi := \{(x, x') \in X_1 \times X_1 \mid \varphi(x) = \varphi(x')\}$$

is called a *congruence relation*. This is clearly an equivalence relation and a bisimulation as well, since we can express it as a relation product of $G(\varphi)$, the graph of φ, with its converse $G(\varphi)^{-1}$ as

$$\ker \varphi = G(\varphi) \circ G(\varphi)^{-1}.$$

3 Saturated Structures

The notion of *saturation* goes back to a similar concept of Fine in [7]. The terminology *m-saturation* (or *modal saturation*) is used in [5] and [9].

Definition 5. *An element x is called* saturated, *if for each set Σ of formulas, such that each finite subset $\Sigma_0 \subseteq \Sigma$ is satisfied at some successor y_0 of x, there is a successor y of x satisfying all formulas in Σ. A Kripke structure is called* saturated, *if each of its elements is saturated.*

In the following we shall find it convenient to informally use infinitary disjunctions $\bigvee_{i \in I} \phi_i$ – not as as logical expressions but as shorthands. In particular we write

$$x \Vdash \Box \bigvee_{i \in I} \phi_i$$

as an abbreviation for

$$\forall y.(x \to y \implies \exists i \in I. y \models \phi_i).$$

With this shorthand, the above definition can be reformulated:

[1] In the literature on Modal Logic (see e.g. [5,9]), homomorphisms are usually called "bounded morphisms".

Lemma 3. *An element x in a Kripke model $\mathcal{K} = (X, R, v)$ is saturated, if for each family $(\phi_i)_{i \in I}$ such that $x \Vdash \Box \bigvee_{i \in I} \phi_i$ there exists a finite subset $I_0 \subseteq I$ with $x \Vdash \Box \bigvee_{i \in I_0} \phi_i$.*

Image finite elements are clearly saturated, but they are not the only ones. Below, we consider two examples of Kripke structures. In both cases, we assume $v(x) := \emptyset$ for each x:

Example 1. On the set $S := \{s\} \cup \{s_i \mid i \in \mathbb{N}\}$ consider the relation $R = \{(s, s_i) \mid i \in \mathbb{N}\} \cup \{(s_{i+1}, s_i) \mid i \in \mathbb{N}\}$. Then for each s_i we have $s_i \Vdash \Box^{i+1} \bot$, but $s_i \not\Vdash \Box^j \bot$ for $j \leq i$. Therefore (S, R, v) is not saturated, since $s \Vdash \Box \bigvee_{i \in \mathbb{N}} (\Box^{i+1} \bot)$, but for no finite $I_0 \subseteq \mathbb{N}$ do we have $s \Vdash \Box \bigvee_{i \in I_0} (\Box^{i+1} \bot)$.

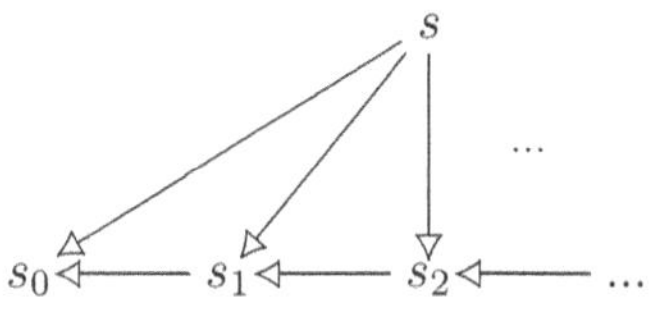

Next, we modify the above structure by adding a "point at infinity" s_∞ together with a self-loop $s_\infty \to s_\infty$ to obtain the following structure:

Example 2.

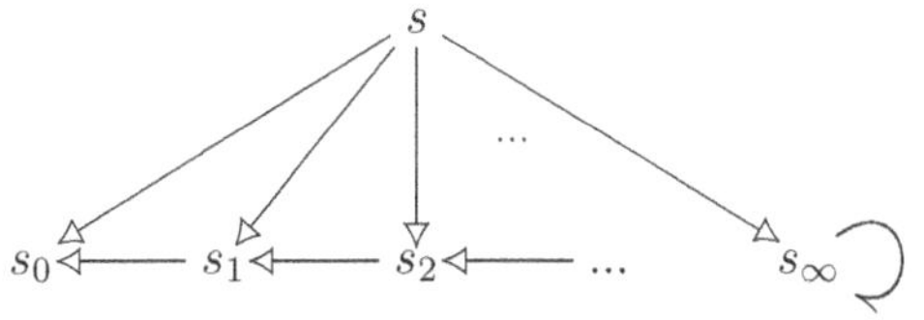

The point at infinity changes the situation. We claim:

Lemma 4. *The Kripke structure in Example 2 is saturated.*

Proof. We first observe that for s_∞ and any formula ϕ we have:

$$s_\infty \Vdash \Diamond \phi \iff s_\infty \Vdash \phi \iff s_\infty \Vdash \Box \phi.$$

Next we prove for each formula ϕ:

Claim. If $s_\infty \Vdash \phi$, then there is some $k \in \mathbb{N}$ such that $s_i \Vdash \phi$ for each $i \geq k$.

We prove this claim by induction over the construction of nnf-formulae:

- For $\phi = \bot$ and $\phi = \top$, the claim is vacuously true. For $\phi = \phi_1 \wedge \phi_2$, from $s_\infty \Vdash \phi_1 \wedge \phi_2$, the induction hypothesis yields k_1 and k_2 such that $s_i \Vdash \phi_1$ for each $i \geq k_1$ and $s_i \Vdash \phi_2$ for each $i \geq k_2$. With $k = max(k_1, k_2)$ we obtain $s_i \Vdash \phi_1 \wedge \phi_2$ for $i \geq k$. For $\phi = \phi_1 \vee \phi_2$ we could similarly choose $k = min(k_1, k_2)$.
- For $\phi = \Box \phi_1$ we have $s_\infty \Vdash \phi \iff s_\infty \Vdash \phi_1$. By assumption, there is some k such that $s_i \Vdash \phi_1$ for each $i \geq k$. It follows that $s_i \Vdash \Box \phi_1$ for $i \geq k + 1$. Similarly we argue for $\phi = \Diamond \phi_1$.

Now, to show that s in the structure of Example 2 is saturated, assume that $s \Vdash \Box \bigvee_{i\in I} \phi_i$, then there is some $i_\infty \in I$ such that $s_\infty \Vdash \phi_{i_\infty}$. The claim above provides a k such that for each $j \geq k$ we have $s_j \Vdash \phi_{i_\infty}$. Also, for each $j < k$ there is some $i_j \in I$ with $s_j \Vdash \phi_{i_j}$. Altogether then with $I_0 := \{i_0, i_1, ..., i_{k-1}\} \cup \{i_\infty\}$ we have $s \Vdash \Box \bigvee_{i\in I_0} \phi_i$.

Thus s is saturated, and all other points in the structure are image finite, hence they are saturated, too.

We can extend Lemma 2 to "infinitary formulas" in the following sense:

Lemma 5. *[Bisimulations preserve saturation] If $B \subseteq X_1 \times X_2$ is a bisimulation and $(x, y) \in B$, then x is saturated iff y is saturated.*

Proof. Assume that x is saturated and $(x, y) \in B$. Suppose $y \Vdash \Box \bigvee_{i\in I} \phi_i$, then each y' with $y \twoheadrightarrow y'$ satisfies one of the formulas ϕ_i. Each x' with $x \twoheadrightarrow x'$ is bisimilar to some y' with $y \twoheadrightarrow y'$, so by Lemma 2 each x' satisfies one of the ϕ_i. This means that $x \Vdash \Box \bigvee_{i\in I} \phi_i$. By saturation of x there is a finite subset $I_0 \subseteq I$ with $x \Vdash \Box \bigvee_{i\in I_0} \phi_i$. The latter, being an honest modal formula, is preserved by bisimulation, so $y \Vdash \Box \bigvee_{i\in I_0} \phi_i$.

If φ is a homomorphism, then Lemma 2 implies that for each element x and each formula ϕ we have

$$x \Vdash \phi \iff \varphi(x) \Vdash \phi \tag{3.1}$$

and Lemma 5 tells us that x is saturated iff $\varphi(x)$ is saturated, which we might combine to:

Corollary 1. *Homomorphisms preserve and reflect saturation.*

On the level of Kripke structures, rather than elements, this translates to:

Corollary 2. *Homomorphic images and homomorphic preimages of saturated Kripke structures are saturated.*

Let $\mathcal{K}_1 = (X_1, R_1, v_1)$ and $\mathcal{K}_2 = (X_2, R_2, v_2)$ be Kripke structures. Recall that for elements $x \in X_1$ and $y \in X_2$ we write $x \approx y$, if they are logically equivalent, i.e. they satisfy the same modal formulae. The following generalization of the Hennessy-Milner theorem [12] is credited in [5] to unpublished notes of Alfred Visser:

Proposition 2. *Let $\mathcal{K}_1 = (X_1, R_1, v_1)$ and $\mathcal{K}_2 = (X_2, R_2, v_2)$ be saturated Kripke structures. Then elements $x \in X_1$ and $y \in X_2$ are bisimilar if and only if they are logically equivalent. In short: $\sim_{\mathcal{K}_1,\mathcal{K}_2} = \approx_{\mathcal{K}_1,\mathcal{K}_2}$.*

We shall next show that saturation allows us to describe the minimal homomorphic image of a Kripke structure:

Lemma 6. *If a Kripke structure $\mathcal{K} = (X, R, v)$ is saturated, then $\approx$ is a congruence relation on $\mathcal{K}$.*

Proof. Clearly, $\approx$ is an equivalence relation and therefore it is the kernel of the map $\pi_\approx$ sending arbitrary elements x to $x/_\approx$, which denotes the equivalence class of $\approx$ containing x. To show that $\pi_\approx$ is a homomorphism, we need to exhibit a coalgebra structure on $X/_\approx$, the factor set of X. Put

$$x/_\approx \Vdash p :\iff \exists\, x' \approx x \,.\, x' \Vdash p.]$$
$$x/_\approx \to y/_\approx] :\iff \text{there exist } x' \approx x \text{ and } y' \approx y \text{ such that } x' \to y'.$$

We check that $\pi_\approx : \mathcal{K} \to \mathcal{K}/_\approx$ is indeed a Kripke homomorphism:

- Clearly, $x \Vdash p$ iff $x/_\approx \Vdash p$ by definition of $\Vdash$ on $X/_\approx$, and
- if $x \to y$, then $x/_\approx \to y/_\approx$ is also immediate by definition. Conversely, given $\pi_\approx(x) = x/_\approx \to y/_\approx$ for some y, we must find a y'' with $x \to y''$ and $\pi_\approx(y'') = y/_\approx$. Since $x/_\approx \to y/_\approx$, we know that there are $x' \approx x$ and $y' \approx y$ with $x' \to y'$. By Proposition 2, $\approx$ is a bisimulation, so it follows that there is some y'' with $x \to y''$ and $y'' \approx y'$. Consequently, $x \to y''$ and $\pi_\approx(y'') = \pi_\approx(y') = y/_\approx$, as required.

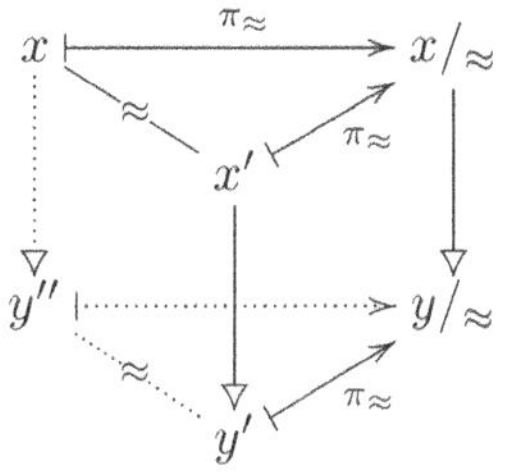

Thus, $\pi_\approx$ is a homomorphism with kernel $\approx$, which makes the latter a congruence relation.

Definition 6. *A Kripke structure is called* simple, *if it does not have a proper homomorphic image.*

Clearly, if $x \not\approx y$ then there cannot be a homomorphism φ with $\varphi(x) = \varphi(y)$, since $x \approx \varphi(x)$ and $y \approx \varphi(y)$. Thus, if $\approx$ is a congruence, $\mathcal{K}/_\approx$ must be simple. It follows:

Theorem 1. *A Kripke structure is saturated iff it has a simple and saturated homomorphic image.*

Observe that Example 2 is a Kripke structure, which is saturated and simple, but not image finite. In particular it does not have a homomorphism to an image finite Kripke structure.

4 *F*-coalgebras

Given a category $\mathscr{C}$ and an endofunctor $F : \mathscr{C} \to \mathscr{C}$, an *F-coalgebra* $\mathcal{A} = (A, \alpha)$ is an object A from $\mathscr{C}$ together with a morphism $\alpha : A \to F(A)$. The object A is called the *base object* and α is called the *structure morphism* of the F-coalgebra $\mathcal{A} = (A, \alpha)$.

Given a second coalgebra $\mathcal{B} = (B, \beta)$, a *homomorphism* $\varphi : \mathcal{A} \to \mathcal{B}$ is a $\mathscr{C}$-morphism $\varphi : A \to B$ which renders the following diagram commutative:

$$\begin{array}{ccc} A & \xrightarrow{\varphi} & B \\ {\scriptstyle\alpha}\downarrow & & \downarrow{\scriptstyle\beta} \\ F(A) & \xrightarrow[F(\varphi)]{} & F(B) \end{array}$$

F-coalgebras with homomorphisms, as defined above, form a category, which we shall call $\mathscr{C}_F$, or simply $Coalg_F$ when the base category is understood. When φ in the above figure is a monomorphism in the base category, then we call $\mathcal{A}$ a *subcoalgebra* of $\mathcal{B}$.

Kripke structures are prime examples of coalgebras. Indeed, the successor relation $R \subseteq X \times X$ can be understood as a map $R : X \to \mathbb{P}(X)$ and the valuation v as a map $v : X \to \mathbb{P}(\Phi)$, where $\mathbb{P}$ is the powerset functor and Φ is the fixed set of propositional atoms. Thus a Kripke structure is simply an F-coalgebra for the combined functor $\mathbb{P}(-) \times \mathbb{P}(\Phi)$, that is a map

$$\alpha : X \to \mathbb{P}(X) \times \mathbb{P}(\Phi),$$

whose first component models the successor relation R and whose second component is the valuation v.

It is easy to check (see [19]), that a homomorphism of Kripke structures, as introduced earlier, is the same as a homomorphism of coalgebras when Kripke structures are understood as $\mathbb{P}(-) \times \mathbb{P}(\Phi)$-coalgebras.

Choosing the finite-powerset functor $\mathbb{P}_\omega(-)$ instead of $\mathbb{P}(-)$, coalgebras for the functor $\mathbb{P}_\omega(-) \times \mathbb{P}(\Phi)$ are precisely the image finite Kripke structures.

Saturated Kripke structures, however, lying between image finite and arbitrary Kripke structures, do not seem to allow such a simple modelling by an appropriate *Set*-functor between $\mathbb{P}_\omega(-)$ and $\mathbb{P}(-)$. Instead, we shall have to pass to the category *Top* of topological spaces and continuous mappings and model them as coalgebras over *Top*.

5 Topological Models

Definition 7. *A* topological model *is a Kripke model* $\mathcal{K} = (X, R, v)$ *together with a topology* τ *on* X*, such that*

1. $\forall x \in X.\, R(x)$ *is compact*
2. $\forall O \in \tau.\, \langle R \rangle O \in \tau$
3. $\forall O \in \tau.\, [R] O \in \tau$
4. $\forall p \in \Phi. [\![p]\!] \in \tau$ *and* $(X - [\![p]\!]) \in \tau$.

A *homomorphism* $\varphi : \mathcal{K}_1 \to \mathcal{K}_2$ between topological models $\mathcal{K}_1 = (X_1, R_1, v_1)$ and $\mathcal{K}_2 = (X_2, R_2, v_2)$ is simply a Kripke-homomorphism (see Definition 4) which additionally is continuous with respect to the topologies on X_1 and X_2.

We need two simple technical lemmas:

Lemma 7. *If C is closed, then so are $\langle R\rangle C$ and $[R]C$.*

Proof. Let $C = X - O$ where O is open, then $\langle R\rangle C = \langle R\rangle(X - O) = X - [R]O$ and $[R]C = [R](X - O) = X - \langle R\rangle O$.

Lemma 8. *In every topological model the sets $[\![\phi]\!]$ where $\phi \in \mathcal{L}_\Phi$, are* clopen *(closed and* open*).*

Proof. By induction on the construction of ϕ:

For $p \in \Phi$ the assertion is part of the definition. If the claim is true for ϕ, ϕ_1 and ϕ_2, then it is obviously true for all boolean compositions, in particular for $\neg\phi$ and for $\phi_1 \wedge \phi_2$.

Lemma 7 and Definition 7 ensure that the claim remains true for $\Box\phi$ and $\Diamond\phi$, since $[\![\Box\phi]\!] = [R][\![\phi]\!]$ and $[\![\Diamond\phi]\!] = \langle R\rangle[\![\phi]\!]$.

Topological models with continuous Kripke-Homomorphisms obviously form a category which we shall call $\mathscr{K}_{Top}$.

6 The Compact Vietoris-Functor

Leopold Vietoris, in his 1922 paper [21], defined his *domains of second order* ("Bereiche zweiter Ordnung") as the collection of closed subsets of a compact Hausdorff space. Later several generalizations and modifications of this topology were introduced and studied under the heading of *hypertopology*.

In connection with Kripke structures, Bezhanishvili, Fontaine and Venema [4] consider the Vietoris functor and Vietoris coalgebras over Stone spaces, i.e. compact and totally disconnected Hausdorff spaces.

In compact Hausdorff spaces, all closed subsets are compact. Hence, when extending the Vietoris functor to act on arbitrary topological spaces $\mathcal{X} = (X, \tau)$, one has the choice to take as base set for $\mathbb{V}(\mathcal{X})$ all closed subsets or all compact subsets of X. In [13] the authors show that both choices lead to endofunctors on the category *Top* of topological spaces, the *"lower" Vietoris functor*, and the *compact Vietoris functor*. Here we shall only need to work with the latter, which for us then is "the" Vietoris functor:

Given a topological space $\mathcal{X} = (X, \tau)$, the *Vietoris space* $\mathbb{V}(\mathcal{X})$ takes as base set the collection of all compact subsets $K \subseteq X$.

The *Vietoris topology* on $\mathbb{V}(\mathcal{X})$ is generated by a subbase consisting of all sets

$$\langle O\rangle := \{K \in \mathbb{V}(\mathcal{X}) \mid K \cap O \neq \emptyset\}, \text{ and}$$
$$[O] := \{K \in \mathbb{V}(\mathcal{X}) \mid K \subseteq O\}$$

where $O \in \tau$.

The *Vietoris functor* sends a continuous function $f : \mathcal{X} \to \mathcal{Y}$ to a continuous map $(\mathbb{V}f) : \mathbb{V}(\mathcal{X}) \to \mathbb{V}(\mathcal{Y})$ where $(\mathbb{V}f)(K) := f(K)$. (Recall that the image $f(K)$ of a compact set K by a continuous map f is always compact.) It is easy to calculate that $(\mathbb{V}f)^{-1}(\langle O \rangle) = \langle f^{-1}(O) \rangle$ and $(\mathbb{V}f)^{-1}([O]) = [f^{-1}(O)]$, hence $\mathbb{V}f$ is continuous. In fact, $(\mathbb{V}f)^{-1}$ takes the defining subbase of $\mathbb{V}(\mathcal{Y})$ to the defining subbase of $\mathbb{V}(\mathcal{X})$. Thus $\mathbb{V}$ is indeed an endofunctor on *Top*.

Let now $\mathbb{P}(\Phi)$ be the powerset of Φ, equipped with the topology having as a base the set of all

$$\uparrow p := \{u \subseteq \Phi \mid p \in u\}$$

where $p \in \Phi$, together with their complements $\mathbb{P}(\Phi) - \uparrow p$. This topology, trivially, is Hausdorff, but in general not compact.

Definition 8. *The product $\mathbb{V}(-) \times \mathbb{P}(\Phi)$ of the Vietoris functor $\mathbb{V}$ with the constant functor of value $\mathbb{P}(\Phi)$, carrying the above topology, will be called the Φ-*Vietoris functor, or simply the *Vietoris functor*, when Φ is clear.

The Vietoris functor is an endofunctor on the category *Top* of topological spaces with continuous maps. We can now define:

Vietoris coalgebras are the coalgebras over *Top* for the Φ-Vietoris functor $\mathbb{V}(-) \times \mathbb{P}(\Phi)$. The following result shows that they agree with our topological models:

Theorem 2. *Vietoris coalgebras with coalgebra homomorphisms are the same as topological models with continuous Kripke-homomorphisms.*

Proof. Given a topological model $\mathcal{K} = (X, R, v)$ with underlying space $\mathcal{X} = (X, \tau)$, we can consider it as a Vietoris coalgebra $\mathcal{A} = (\mathcal{X}, \alpha)$ by defining the structure map $\alpha : \mathcal{X} \to \mathbb{V}(\mathcal{X}) \times \mathbb{P}(\Phi)$ as $\alpha(x) := (R(x), v(x))$. To show that α is continuous, we must verify that both components are continuous.

Continuity of (the map) $R : \mathcal{X} \to \mathbb{V}(\mathcal{X})$ needs to be tested only on the subbase for the Vietoris topology on $\mathbb{V}(\mathcal{X})$. Indeed, assume $O \in \tau$, then

$$\begin{aligned} R^{-1}([O]) &= \{x \in X \mid R(x) \in [O]\} \\ &= \{x \in X \mid R(x) \subseteq O\} = [R]O \end{aligned}$$

is open in τ and so is

$$\begin{aligned} R^{-1}(\langle O \rangle) &= \{x \in X \mid R(x) \in \langle O \rangle\} \\ &= \{x \in X \mid R(x) \cap O \neq \emptyset\} = \langle R \rangle O. \end{aligned}$$

To see that v also is continuous, let $\uparrow p \subseteq \mathbb{P}(\Phi)$ be given, then

$$v^{-1}(\uparrow p) = \{x \in X \mid p \in v(x)\} = [\![p]\!] \in \tau$$

as well as

$$v^{-1}(\mathbb{P}(X) - \uparrow p) = \{x \in X \mid p \notin v(x)\} = (X - [\![p]\!]) \in \tau.$$

Conversely, let $(\mathcal{X}, \alpha)$ be a Vietoris coalgebra, with $\mathcal{X} = (X, \tau)$ as base space and $\alpha : \mathcal{X} \to \mathbb{V}(\mathcal{X}) \times \mathbb{P}(\Phi)$ as structure morphism, then $\alpha = (R, v)$ with $R := \pi_1 \circ \alpha : \mathcal{X} \to \mathbb{V}(\mathcal{X})$ and $v := \pi_2 \circ \alpha : \mathcal{X} \to \mathbb{P}(\Phi)$, both of which are continuous. Since $R(x) \in \mathbb{V}(\mathcal{X})$, it is necessarily compact. If O is open in $\mathcal{X} = (X, \tau)$ then $[O]$ is open in $\mathbb{V}(\mathcal{X})$, hence $R^{-1}([O])$ must be open in (X, τ), hence so is

$$[R]O = \{x \in X \mid R(x) \subseteq O\} = \{x \in X \mid R(x) \in [O]\} = R^{-1}([O]).$$

Similarly, for O open in $\mathcal{X} = (X, \tau)$ we have $\langle O \rangle$ open in $\mathbb{V}(\mathcal{X})$, hence $R^{-1}(\langle O \rangle)$ is open in $\mathcal{X}$, which means that

$$\langle R \rangle O = \{x \in X \mid R(x) \cap O \neq \emptyset\} = \{x \in X \mid R(x) \in \langle O \rangle\} = R^{-1}(\langle O \rangle)$$

is open as well.

Finally, for $p \in \Phi$ we have $\uparrow p = \{u \subseteq \Phi \mid p \in u\}$ clopen in the topology on $\mathbb{P}(\Phi)$, so also $[\![p]\!] = \{x \in X \mid p \in v(x)\} = v^{-1}(\{u \subseteq \Phi \mid p \in u\}) = v^{-1}(\uparrow p)$ as well as its complement $X - [\![p]\!]$ are open in τ.

Coalgebra homomorphisms between Vietoris coalgebras, as coalgebras over *Top*, must be continuous and preserve both R and v which means they are the same as continuous Kripke homomorphisms between the corresponding topological models.

7 Characterization Theorem

The following theorem shows that saturated Kripke structures arise precisely as the algebraic reducts of Vietoris coalgebras when forgetting the topology. Put differently, a Kripke structure is saturated if and only if it can be equipped with a topology to turn it into a topological model, resp. into a Vietoris coalgebra.

Bezhanishvili, Fontaine and Venema [4] show one direction of this result for Vietoris coalgebras over Stone spaces. In contrast to their work, we consider the Vietoris functor over arbitrary topological spaces, which allows us to obtain an equivalence:

Theorem 3. *For a Kripke structure $\mathcal{K}$ the following are equivalent:*

1. *$\mathcal{K}$ is saturated,*
2. *$\mathcal{K}$ is the algebraic reduct of a topological model,*
3. *$\mathcal{K}$ is the algebraic reduct of a Vietoris coalgebra.*

Proof. "(1) $\to$ (2)": Assuming that $\mathcal{K} = (X, R, v)$ is saturated, let τ be the topology on X generated by the sets $[\![\phi]\!]$ for $\phi \in \mathcal{L}_\Phi$. It follows that each $[\![\phi]\!]$ is clopen (*clo*sed and *open*), so each open set can be written as $O = \bigcup_{i\in I}[\![\phi_i]\!]$ and each closed set as $C = \bigcap_{i\in I}[\![\phi_i]\!]$.

To show that $\mathcal{K}$ with this topology τ is a topological model, we show first, that $R(x)$ is topologically compact. For that, assume $R(x) \subseteq \bigcup_{i\in I} O_i$, then $R(x) \subseteq \bigcup_{i\in I}\bigcup_{j\in J_i}[\![\phi_j]\!]$, i.e.

$$x \Vdash \Box \bigvee_{i\in I}\bigvee_{j\in J_i} \phi_j.$$

By saturation of $\mathcal{K}$, there are finitely many $j_{i_1} \in J_{i_1}, ..., j_{i_n} \in J_{i_n}$ with

$$x \Vdash \Box(\phi_{j_{i_1}} \vee ... \vee \phi_{j_{i_n}}),$$

so $R(x) \subseteq O_{i_1} \cup ... \cup O_{i_n}$.

Next, to see that $\langle R\rangle O$ is open, we calculate

$$\begin{aligned}
\langle R\rangle O &= \langle R\rangle(\bigcup_{i\in I}[\![\phi_i]\!]) \\
&= \bigcup_{i\in I}\langle R\rangle\,[\![\phi_i]\!] \\
&= \bigcup_{i\in I}\{x \in X \mid x \models \Diamond\phi_i\} \\
&= \bigcup_{i\in I}[\![\Diamond\phi_i]\!],
\end{aligned}$$

which is open, and similarly

$$\begin{aligned}
[R]O &= [R](\bigcup_{i\in I}[\![\phi_i]\!]) \\
&= \{x \in X \mid R(x) \subseteq \bigcup_{i\in I}[\![\phi_i]\!]\} \\
&= \{x \in X \mid R(x) \subseteq \bigcup_{i\in J_x}[\![\phi_i]\!] \text{ for some finite } J_x \subseteq I\} \\
&= \{x \in X \mid x \models \Box\bigvee_{i\in J_x}\phi_i \text{ for some finite } J_x \subseteq I\} \\
&= \bigcup_{J\subseteq I,\, J \text{ finite}} \left[\!\!\left[\Box\bigvee_{i\in J}\phi_i\right]\!\!\right],
\end{aligned}$$

which is open as well.

"(2) $\leftrightarrow$ (3)" is Theorem 2.

"(2) $\to$ (1)": Given a Kripke model $\mathcal{K}$ which is the algebraic reduct of a topological model, assume $x \Vdash \Box\bigvee_{i\in I}\phi_i$, then $R(x) \subseteq \bigcup_{i\in I}[\![\phi_i]\!]$. By Lemma 8, the right hand side is a union of open sets, thus by compactness of $R(x)$ there is a finite subset $I_0 \subseteq I$ with $R(x) \subseteq \bigcup_{i\in I_0} \phi_i$, which means $x \Vdash \Box \bigvee_{i\in I_0} \phi_i$.

Given a saturated Kripke-structure $\mathcal{K} = (X, R, v)$, let $\mathcal{A}(\mathcal{K})$ denote the Vietoris coalgebra, as constructed above, and conversely, given a Vietoris coalgebra $\mathcal{A}$, let $\mathcal{K}(\mathcal{A})$ be the corresponding saturated Kripke structure. On objects, $\mathcal{A}(-)$ and $\mathcal{K}(-)$ are clearly inverses to each other.

On morphisms, this is true as well, since a homomorphism $\varphi : \mathcal{K}_1 \to \mathcal{K}_2$ between saturated Kripke structures preserves (and reflects) modal formulae (see Lemma 2) and the topologies on $\mathcal{A}(\mathcal{K}_1)$ and $\mathcal{A}(\mathcal{K}_2)$ are generated by validity sets of formulae. Conversely, a morphism between Vietoris coalgebras $\mathcal{A}_1$ and $\mathcal{A}_2$ is automatically a Kripke-homomorphism by forgetting continuity.

Corollary 3. *Saturated Kripke structures, topological models, and Vietoris coalgebras are isomorphic as categories.*

8 Closure of Vietoris Structures

In those topological spaces where each point has a countable base for its neighbourhoods, such as, for instance, in metric spaces, continuity can be conveniently dealt with in terms of convergent sequences $(x_n)_{n\in\mathbb{N}}$. For general spaces (X, τ), this intuitive approach is not sufficient, but its spirit and its power can be salvaged if one allows the linearly ordered set $\mathbb{N}$, indexing a sequence, to be replaced by arbitrary directed sets I indexing the elements $(x_i)_{i\in I}$ of a net. Often, a proof based on convergence of sequences can be easily generalized by replacing sequences with nets. Therefore net convergence can be considered more intuitive than the equally powerful notion of filter convergence. The following definitions and results on nets in general topological spaces will be needed. They can be found as a series of exercises in Munkres [18].

8.1 Nets and Subnets

A partially ordered set $(I, \leq)$ is called *directed*, if for each pair $i_1, i_2 \in I$ there is some $i \in I$ such that $i_1 \leq i$ and $i_2 \leq i$, i.e. i is an upper bound for $\{i_1, i_2\}$. It follows that each finite subset $I_0 \subseteq I$ has a common upper bound.

Definition 9. *A subset $J \subseteq I$ is called* cofinal *in I, if for each $i \in I$ there is some $j \in J$ with $i \leq j$. A map $f : J \to I$ between ordered sets $(J, \leq)$ and $(I, \leq)$ is called* cofinal *if its image $f[J]$ is cofinal in I.*

Clearly, if J_1 is cofinal in J_2 and J_2 cofinal in I then J_1 is cofinal in I. Also, compositions of cofinal maps are cofinal.

Let $\mathcal{X} = (X, \tau)$ be a topological space and $x \in X$. By $\mathfrak{U}(x)$ we denote the collection of all open neighborhoods of x. Observe that $\mathfrak{U}(x)$, when ordered by reverse inclusion, is a directed set.

Definition 10. *A* net *in X is a map $\sigma : I \to X$ from a directed set I to X.*

If $\sigma(i) = x_i$, then one often denotes the net σ as $(x_i)_{i\in I}$ and if I is clear from the context one simply writes (x_i).

The net $(x_i)_{i\in I}$ *converges* to $x \in X$ and we shall write $(x_i)_{i\in I} \longrightarrow x$, or when I is understood, simply $(x_i \longrightarrow x)$, provided that

$$\forall U \in \mathfrak{U}(x).\, \exists i_U \in I.\, \forall i \geq i_U.\, x_i \in U. \tag{8.1}$$

In this case, x is called a *limit point* of $(x_i)_{i\in I}$. Colloquially, condition (8.1) can be expressed as "*(x_i) is eventually in every neighborhood of x*".

Limit points need not exist, nor need they be unique, unless X is Hausdorff. In any case though, one has (see [18]):

Proposition 3. *Let (X, τ) and (Y, γ) be arbitrary topological spaces.*

1. A map $\varphi : X \to Y$ is continuous at x if and only if it "preserves convergence", i.e. for all nets $(x_i)_{i\in I}$ in X:

$$(x_i \longrightarrow x) \implies (\varphi(x_i) \longrightarrow \varphi(x))\,.$$

2. Given a subset $A \subseteq X$, then $x \in X$ belongs to the topological closure $\overline{A}$ of A if and only if some net (a_i) in A converges to x. Thus, A is closed iff it contains all limit points of nets in A.

$x \in X$ is called an *accumulation point of the net* $(x_i)_{i\in I}$ if

$$\forall U \in \mathfrak{U}(x).\, \forall i \in I.\, \exists j \geq i.\, x_j \in U. \tag{8.2}$$

Condition (8.2) can be phrased as: "*x_i is frequently in every neighborhood of x*". A characterization of compactness using nets is [18]:

Lemma 9. A subset $A \subseteq X$ is compact if and only if every net in A has an accumulation point in A.

Definition 11. *A net $\lambda : J \to X$ is a* subnet *of $\sigma : I \to X$ if there is a monotonic and cofinal map $f : J \to I$ with $\lambda = \sigma \circ f$:*

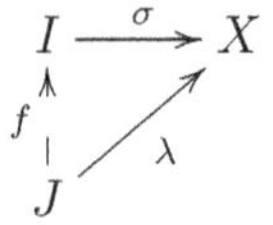

Thus, if $\sigma = (x_i)_{i\in I}$ then $\lambda = (x_{f(j)})_{j\in J}$. One easily checks that the subnet relation is reflexive and transitive, but mainly:

Lemma 10. *If $(x_i)_{i\in I}$ converges to x then so does each subnet $(x_{f(j)})_{j\in J}$.*

Lemma 11. $x \in X$ is an accumulation point of the net $\sigma : I \to X$ if and only if there is a subnet λ of σ converging to x.

Corollary 4. A subset $A \subseteq X$ is compact iff every net in A has a subnet converging to some $a \in A$.

8.2 Convergence in Vietoris Spaces

In this section, we prepare our main result on net convergence in Vietoris spaces. Let $\mathcal{X} = (X, \tau)$ be a topological space. Recall that the *Vietoris space* $\mathbb{V}(\mathcal{X})$ consists of all compact subsets $K \subseteq X$, with a topology generated by a subbase consisting of all sets

$$\langle U \rangle := \{K \in \mathbb{V}(\mathcal{X}) \mid K \cap U \neq \emptyset\}$$
$$[U] := \{K \in \mathbb{V}(\mathcal{X}) \mid K \subseteq U]$$

where U ranges over all open subsets of $\mathcal{X}$. The following result establish the relevant connections between convergence in $\mathbb{V}(\mathcal{X})$ and convergence in $\mathcal{X} = (X, \tau)$.

Lemma 12. *Let $\kappa : I \to \mathbb{V}(\mathcal{X})$ be a net in the Vietoris space $\mathbb{V}(\mathcal{X})$. If $(\kappa_i \longrightarrow K)$ and $K \neq \emptyset$, then κ has a subnet, each member of which is nonempty.*

Proof. Since $K \neq \emptyset$, we have $K \in \langle X \rangle$, so $\langle X \rangle$ is a neighborhood of K in $\mathbb{V}(\mathcal{X})$. As κ converges to K, there must be some $i_0 \in I$ such that $\forall i \geq i_0. \kappa_i \in \langle X \rangle$, i.e. $\forall i \geq i_0. \kappa_i \neq \emptyset$. Put $J = (\{i \in I \mid i \geq i_0\}, \leq)$ and let $f : J \hookrightarrow I$ be the natural inclusion, then f is clearly monotonic and cofinal. Therefore $\tau := \kappa \circ f$ is a subnet of κ and $\tau_j = \kappa_{f(j)} = \kappa_j \neq \emptyset$, owing to $j \in J$.

Lemma 13. *Given a net $\kappa : I \to \mathbb{V}(\mathcal{X})$ converging to $K \in \mathbb{V}(\mathcal{X})$ and $b_i \in \kappa_i$ for each $i \in I$. Then the net $(b_i)_{i \in I}$ has a subnet converging to some $b \in K$.*

Proof. It is enough to show that $(b_i)_{i \in I}$ has an accumulation point $b \in K$. As K is compact, we then obtain a subnet $(b_{f(j)})_{j \in J}$ converging to b. By Lemma 10, the subnet $(\kappa_{f(j)})_{j \in J}$ of κ still converges to K.

For every $x \in K$ which is not an accumulation point of $(b_i)_{i \in I}$, we obtain by negating (8.2) an open neighborhood U_x of x and an $i_x \in I$ such that for all $i \geq i_x$ we have $b_i \notin U_x$.

Assuming that no $x \in K$ is an accumulation point, the family $(U_x)_{x \in K}$ forms an open cover of K. By compactness, there is a finite subcover $U = U_{x_1} \cup ... \cup U_{x_n}$. Choose $i_U \geq i_{x_1}, ..., i_{x_n}$, then for every $i \geq i_U$ we have $b_i \notin U \supseteq K$.

But $[U]$ is also an open neighborhood of K in $\mathbb{V}(\mathcal{X})$ and $(\kappa_i \longrightarrow K)$, so there exists $i_{[U]}$ with $\kappa_i \in [U]$, that is $b_i \in \kappa_i \subseteq U$ for $i \geq i_{[U]}$. For $i \geq \{i_U, i_{[U]}\}$ we enter the contradiction $b_i \in U$ and $b_i \notin U$.

Lemma 14. *Given a net $\kappa : I \to \mathbb{V}(\mathcal{X})$ converging to $K \in \mathbb{V}(\mathcal{X})$ and $a \in K$. Then there is a subnet $(\kappa_j)_{j \in J}$ and elements $a_j \in \kappa_j$ converging to a.*

Proof. By Lemma 12 and Lemma 10, we may assume that $\kappa_i \neq \emptyset$ for all $i \in \mathcal{I}$.

For every open set $U \in \mathfrak{U}(a)$ we have $a \in U \cap K$, so $K \cap U \neq \emptyset$, which means that $K \in \langle U \rangle$, so $\langle U \rangle$ is an open neighborhood of K in $\mathbb{V}(\mathcal{X})$.

Since κ converges to K we have

$$\forall U \in \mathfrak{U}(a).\exists i_U \in I.\forall i \geq i_U.\kappa_i \in \langle U \rangle. \tag{8.3}$$

Consider a partial order on

$$J := \{(i, U) \in I \times \mathfrak{U}(a) \mid \kappa_i \in \langle U \rangle\}$$

by defining:

$$(i_1, U_1) \leq (i_2, U_2) :\iff i_1 \leq i_2 \wedge U_1 \supseteq U_2.$$

To verify that $\mathcal{J} = (J, \leq)$ is directed, let arbitrary $j_1 = (i_1, U_1)$ and $j_2 = (i_2, U_2)$ be given. Pick $U = U_1 \cap U_2$ then by (8.3) there is an $i_U \in I$ with $\kappa_i \in \langle U \rangle$ for all $i \geq i_U$. It suffices to choose $i \geq i_1, i_2, i_U$, then $(i, U) \in J$ and $(i, U) \geq (i_1, U_1), (i_2, U_2)$.

The map $\pi_1 : J \to I$ given as $\pi_1(i, U) := i$ is clearly monotonic. For each $i \in I$ we have $(i, X) \in J$ since $\kappa_i \neq \emptyset$. Hence π_1 is cofinal. Therefore $\kappa \circ \pi_1 : J \to \mathbb{V}(\mathcal{X})$ is a subnet of κ and therefore also converges to K.

For each $(i, U) \in J$ we can pick some $a_{(i,U)} \in \kappa_i \cap U$. This defines a net $(a_j)_{j \in J}$ in X.

To show that $(a_j)_{j \in J}$ converges to a, let U be any open neighborhood of a. By (8.3) there exist some i_U such that in particular $j_U := (i_U, U) \in J$. We therefore have $a_{j_U} := a_{(i_U,U)} \in U$ and for each $j = (i, U') \geq (i_U, U) = j_U$, i.e. for $i \geq i_U$ and $U' \subseteq U$ we have $a_j = a_{(i,U')} \in \kappa_i \cap U' \subseteq U$.

We can combine the previous two lemmas to a theorem relating convergence in Vietoris spaces to convergence in their base spaces:

Theorem 4. *Let $(\kappa_i)_{i \in I}$ converge to K in the Vietoris space $\mathbb{V}(\mathcal{X})$. Then*

1. *for each $a \in K$ there is a subnet $(\kappa_j)_{j \in J}$ and elements $a_j \in \kappa_j$ such that $(a_j \longrightarrow a)$, and*
2. *each net $(b_i)_{i \in I}$ with $b_i \in \kappa_i$ has a subnet $(b_j)_{j \in J}$ converging to some $b \in K$.*

8.3 Closure of Subcoalgebras and Bisimulations

In this section we shall show that in topological Kripke structures, i.e. for Vietoris coalgebras, the topological closure of a substructure is again a substructure and the closure of a bisimulation is a bisimulation. The second of these results has previously been shown for Vietoris coalgebras over Stone spaces in [4], but now we work in the more general context of Vietoris coalgebras over arbitrary topological spaces, so we were forced to prepare our tools in the previous sections.

Theorem 5. *Let $\mathcal{A} = (\mathcal{X}, \alpha)$ be a Vietoris coalgebra. If $U \subseteq X$ is a Kripke substructure of $\mathcal{K}(\mathcal{A})$, then so is its topological closure $\overline{U}$.*

Proof. We may consider $\mathcal{A}$ as a topological model $\mathcal{K} = (X, R, v)$ where $R(x) = (\pi_1 \circ \alpha)(x)$ for each $x \in X$ is the compact set of all successors of x. Assume that

U is a Kripke-substructure, i.e. a subset $U \subseteq X$ such that $R(u) \subseteq U$ for each $u \in U$. We need to show that the same property holds for $\overline{U}$.

Thus let $u \in \overline{U}$ be arbitrary and let w be a successor of u, i.e. $w \in R(u)$. We need to show that $w \in \overline{U}$.

Due to Proposition 3, there is a net $(u_i)_{i \in I}$ converging to u with each $u_i \in U$. By continuity of α, the net $R(u_i)_{i \in I}$ converges to $R(u)$ in the Vietoris topology.

As $w \in R(u)$, we may assume by Lemma 12, that each $R(u_i)$ is nonempty.[2] Next, we may assume by Theorem 4 that we can pick a w_i from each $R(u_i)$ so that the net $(w_i)_{i \in I}$ converges to w in X.

Since U was a subcoalgebra, $R(u_i) \subseteq U$, so each w_i must belong to U. Therefore, we have found a net in U which converges to w, hence $w \in \overline{U}$.

Theorem 6. *If S is a Kripke bisimulation between Vietoris coalgebras $\mathcal{A} = (A, \alpha)$ and $\mathcal{B} = (B, \beta)$, then so is its topological closure $\overline{S}$.*

Proof. Again, we consider $\mathcal{A}$ and $\mathcal{B}$ as topological models with $\alpha = (R_A, v_A)$ and $\beta = (R_B, v_B)$. Given $(a, b) \in \overline{S}$, we need to show that

1. $v_A(a) = v_B(b)$ and
2. whenever $a \twoheadrightarrow u$ then there is some w with $b \twoheadrightarrow w$ and $(u, w) \in \overline{S}$.

The third case of Definition 3 will follow by a symmetric proof.

First note that by Proposition 3 there is a net $(a_i, b_i)_{i \in I}$ converging to (a, b) with each $(a_i, b_i) \in S$. The component nets (a_i), resp. (b_i), converge to a, resp. to b, since the projection maps are continuous.

Also by continuity, $v_A(a_i)$ and $v_B(b_i)$ converge to $v_A(a)$ and $v_B(b) \in \mathbb{P}(\Phi)$. Since $(a_i, b_i) \in S$, we know $v_A(a_i) = v_B(b_i)$ for each $i \in I$. Since the topology on $\mathbb{P}(\Phi)$, the second component of the Vietoris functor, is Hausdorff, we get $v_A(a) = v_B(b)$ as required.

Next, assume $a \twoheadrightarrow u$, i.e. $u \in R_A(a)$, then we need to find some w with $b \twoheadrightarrow w$ and $(u, w) \in \overline{S}$.

By continuity of R_A and R_B, the nets $(R_A(a_i))_{i \in I}$ resp. $(R_B(b_i)_{i \in I})$, converge to $R_A(a)$, resp. to $R_B(b)$ in the Vietoris spaces $\mathbb{V}(\mathcal{A})$, resp. $\mathbb{V}(\mathcal{B})$.

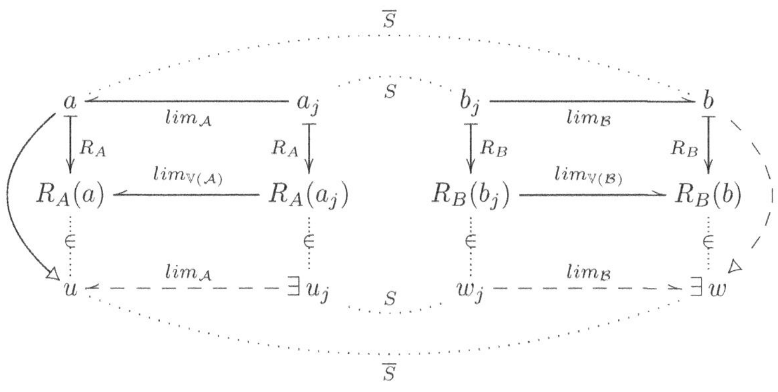

[2] With the phrase "we may assume" we often hide the technicality that we might have to pass to a subnet, such as here to $(\alpha(u_{f(j)}))_{j \in J}$ and retroactively replace $(u_i)_{i \in I}$ by the subnet $(u_{f(j)})_{j \in J}$, which is always justified by Lemma 10.

In the sense mentioned previously, we may assume that the $R_A(a_i)$ are nonempty and further, using part 1 of Theorem 4, and possibly passing to a subnet indexed by some J, we find $u_j \in R_A(a_j)$ with $(u_j \longrightarrow u)$.

Since S is a bisimulation and $a_j \, S \, b_j$ for each j and $u_j \in R_A(a_j)$ it follows that there are $w_j \in R_B(b_j)$ with $(u_j, w_j) \in S$ for each $j \in J$. Since $(b_j \longrightarrow b)$ it follows $(R_B(b_j) \longrightarrow R_B(b))$ by continuity of R_B. Therefore, part 2 of Theorem 4 forces $(w_j)_{j \in J}$ to converge to some $w \in R_B(b)$.

Consequently, $(u_j, w_j) \longrightarrow (u, w)$ where $(u_j, w_j) \in S$ for each $j \in J$, hence by Lemma 3 $(u, w) \in \overline{S}$ as desired.

9 The Terminal Vietoris Coalgebra

To obtain the terminal Vietoris coalgebra, we utilize the equivalence with saturated Kripke structures and look for a terminal saturated Kripke structure instead. This will be found in the "canonical model". The existence of a terminal object in the class of all saturated coalgebras could, in fact, be derived from a very general result by Goldblatt [8], which shows that a terminal coalgebra can actually be found for any *Set*-functor F with a "small logic" having the Hennessy-Milner property.

Here, we describe a construction using the notation introduced above and based on a classical method from [5]. Recall that the *canonical model* for a normal modal logic consists of all *maximally consistent* subsets of $\mathcal{L}_\Phi$. Here $u \subseteq \mathcal{L}_\Phi$ is called

- *consistent*, if no contradiction can be derived from the formulae in u, and
- *maximally consistent*, if additionally for each formula $\phi \in \mathcal{L}_\Phi$, either $\phi \in u$ or $\neg\phi \in u$.

Typical sets of formulas which are maximally consistent arise as

$$[\![x]\!] := \{\phi \mid x \Vdash \phi\},$$

where x is any element of any Kripke structure. Moreover, any consistent set of formulas can be extended to a maximally consistent set.

It is also essential to know that a set u is consistent, if and only if every finite subset $u_0 \subseteq u$ is consistent, see [5].

The *canonical model* is now defined as $\mathcal{M} := (M, \twoheadrightarrow_\mathcal{M}, v_\mathcal{M})$ where M is the collection of all maximally consistent subsets of $\mathcal{L}_\Phi$, and $\twoheadrightarrow_\mathcal{M}$ and $v_\mathcal{M}$ are defined as

$$u \twoheadrightarrow_\mathcal{M} w :\Leftrightarrow \forall\phi.(\Box\phi \in u \implies \phi \in w), \tag{9.1}$$

and

$$v_\mathcal{M}(u) := u \cap \Phi. \tag{9.2}$$

The latter definition extends to the important "truth lemma":

Lemma 15. *For each formula $\phi \in \mathcal{L}_\Phi$ and each $u \in M$ we have:*

$$u \Vdash \phi \iff \phi \in u.$$

As an immediate corollary, we note:

Corollary 5. $\forall u, w \in M.\, u \approx w \implies u = w.$

First, we shall verify, that $\mathcal{M}$ is saturated: Given $u \in M$ and Σ a set of formulas such that for every finite subset $\Sigma_0 \subseteq \Sigma$ there is some w_0 such that $u \twoheadrightarrow w_0$ and $w_0 \Vdash \bigwedge \Sigma_0$. It follows that every finite subset of the set

$$S := \{\phi \mid \Box\phi \in u\} \cup \Sigma$$

is satisfied in some w_0, and hence consistent. Thus, the whole set S itself is consistent. Let w be any maximal consistent set containing S, then $w \in M$ and clearly $u \twoheadrightarrow w$ as well as $w \Vdash \sigma$ for each $\sigma \in \Sigma$. Therefore:

Lemma 16. $\mathcal{M}$ is saturated.

Let us see that moreover:

Theorem 7. $\mathcal{M}$ is the terminal object in the category of all saturated Kripke structures.

Proof. First note that Corollary 5 yields uniqueness: If $\varphi_1, \varphi_2 : \mathcal{K} \to \mathcal{M}$ were different homomorphism, then for some $x \in \mathcal{K}$ we would have $\varphi_1(x) \neq \varphi_2(x)$. However, $x \approx \varphi_1(x)$ as well as $x \approx \varphi_2(x)$ according to (3.1), whence $\varphi_1(x) \approx \varphi_2(x)$, which contradicts Corollary 5.

For any Kripke structure $\mathcal{K}$, we show that the map $[\![-]\!] : \mathcal{K} \to \mathcal{M}$ which sends an element $x \in \mathcal{K}$ to $[\![x]\!] := \{\phi \mid x \Vdash \phi\}$ is a homomorphism, see Definition 4:

First, for each $p \in \Phi$ we have: $x \Vdash p$ in $\mathcal{K}$ implies $p \in [\![x]\!]$, so $[\![x]\!] \Vdash p$ in $\mathcal{M}$, by the *Truth Lemma.*

Next, suppose $x, y \in \mathcal{K}$ and $x \twoheadrightarrow_{\mathcal{K}} y$. Then for each $\phi \in \mathcal{L}_\Phi$ with $x \Vdash \Box\phi$ it follows $y \Vdash \phi$, which by the truth lemma says $\Box\phi \in [\![x]\!] \implies \phi \in [\![y]\!]$, hence $[\![x]\!] \twoheadrightarrow_{\mathcal{M}} [\![y]\!]$ by (9.1).

Finally, let us assume $[\![x]\!] \twoheadrightarrow_{\mathcal{M}} w$ for some maximally consistent set w. We need to find some $y \in \mathcal{K}$ with $x \twoheadrightarrow_{\mathcal{K}} y$ and $[\![y]\!] = w$.

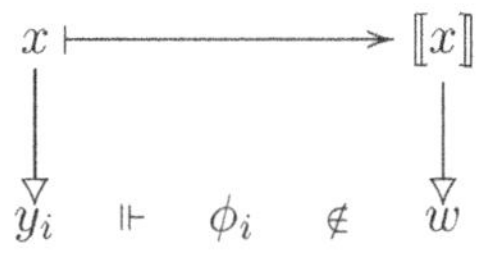

For this we invoke a Hennessy–Milner style argument again: Let $(y_i)_{i \in I}$ be the collection of all successors of x. If $[\![y_i]\!] = w$ for some i, then we are done. Otherwise, assume that $[\![y_i]\!] \neq w$ for each $i \in I$, then there are formulae ϕ_i with $\phi_i \in [\![y_i]\!]$ but $\phi_i \notin w$, or, in other words, $y_i \Vdash \phi_i$, but $w \nVdash \phi_i$.

Hence $x \models \Box \bigvee_{i \in I} \phi_i$. By assumption $\mathcal{K}$ is saturated, so $x \in \mathcal{K}$ is saturated, which means that we can find a finite subset $I_0 \subseteq I$ with $x \Vdash \Box \bigvee_{i \in I_0} \phi_i$. This is now an honest formula, so from $[\![x]\!] \rightarrow w$, and Definition 9.1 we conclude $w \models \bigvee_{i \in I_0} \phi_i$. This means that $w \Vdash \phi_i$ for some $i \in I_0$, contradicting our assumption.

Theorem 3 tells us explicitly, how to obtain the terminal Vietoris coalgebra, so we have:

Theorem 8. *The category of all Vietoris coalgebras has a terminal object. Its base structure is the canonical model, consisting of all maximally consistent sets of $\mathcal{L}_\Phi$-formulas, and its topology is generated by the open sets $[\![\phi]\!]_\mathcal{M} = \{u \in \mathcal{M} \mid \phi \in u\}$ for all $\phi \in \mathcal{L}_\Phi$.*

10 Conclusion

Starting from an arbitrary set Φ of atomic proposition, we have characterized modally saturated Kripke structures as *Top*-coalgebras for $\mathbb{V}(-) \times \mathbb{P}(\Phi)$, which is the compact Vietoris functor on the category Top of topological spaces and continuous mappings, augmented with a constant part, representing sets of atomic propositions.

In fact, the categories of saturated Kripke structures and the category of all Vietoris coalgebras over the category *Top* are isomorphic. We have described the relation of convergence in the Vietoris space $\mathbb{V}(\mathcal{X})$ to convergence in the base space $\mathcal{X}$, from which we could derive that the Kripke-closure of bisimulations and of subcoalgebras are again bisimulations, resp. subcoalgebras. Finally, we have shown that the final Vietoris coalgebra exists, and is derived from the canonical Kripke model.

References

1. Balan, A., Kurz, A.: Finitary functors: from Set to Preord and Poset. In: Corradini, A., Klin, B., Cîrstea, C. (eds.) CALCO 2011. LNCS, vol. 6859, pp. 85–99. Springer, Heidelberg (2011). https://doi.org/10.1007/978-3-642-22944-2_7
2. Barr, M.: Terminal coalgebras in well-founded set theory. Theor. Comput. Sci. **114**(2), 299–315 (1993). https://doi.org/10.1016/0304-3975(93)90076-6
3. Barr, M.: Additions and corrections to 'terminal coalgebras in well-founded set theory'. Theor. Comput. Sci. **124**(1), 189–192 (1994). https://doi.org/10.1016/0304-3975(94)90060-4
4. Bezhanishvili, N., Fontaine, G., Venema, Y.: Vietoris bisimulations. J. Logic Comput. **20**(5), 1017–1040 (2010). https://doi.org/10.1093/logcom/exn091
5. Blackburn, P., Rijke, M.D., Venema, Y.: Modal Logic. Cambridge Tracts in Theoretical Computer Science. Cambridge University Press, Cambridge (2001). https://doi.org/10.1017/CBO9781107050884
6. Doberkat, E.E.: Stochastic Coalgebraic Logic. Monographs in Theoretical Computer Science, Springer, Heidelberg (2009). https://doi.org/10.1007/978-3-642-02995-0

7. Fine, K.: Some connections between elementary and modal logic. In: Blackburn, P., Van Benthem, J., Wolter, F. (eds.) Proceedings of the Third Scandinavian Logic Symposium, vol. 3, pp. 15–31. North-Holland, Amsterdam (1973). https://doi.org/10.1016/S0049-237X(08)70723-7
8. Goldblatt, R.: Final coalgebras and the Hennessy-Milner property. Ann. Pure Appl. Logic **138**, 77–93 (2006)
9. Goranko, V., Otto, M.: 5 model theory of modal logic. In: Blackburn, P., Van Benthem, J., Wolter, F. (eds.) Handbook of Modal Logic, Studies in Logic and Practical Reasoning, vol. 3, pp. 249–329. Elsevier (2007). https://doi.org/10.1016/S1570-2464(07)80008-5
10. Gumm, H.P.: Elements of the general theory of coalgebras. In: LUATCS 1999. Rand Afrikaans University, Johannesburg, South Africa (1999). https://www.mathematik.uni-marburg.de/~gumm/Papers/Luatcs.pdf
11. Gumm, H.P., Schröder, T.: Coalgebras of bounded type. Math. Struct. Comput. Sci. **12**(5), 565–578 (2002). https://doi.org/10.1017/S0960129501003590
12. Hennessy, M., Milner, R.: On observing nondeterminism and concurrency. In: de Bakker, J., van Leeuwen, J. (eds.) ICALP 1980. LNCS, vol. 85, pp. 299–309. Springer, Heidelberg (1980). https://doi.org/10.1007/3-540-10003-2_79
13. Hofmann, D., Neves, R., Nora, P.: Limits in categories of Vietoris coalgebras. Math. Struct. Comput. Sci. **29**(4), 552–587 (2019). https://doi.org/10.1017/S0960129518000269
14. Hollenberg, M.: Hennessy-Milner classes and process algebra. In: Ponse, A., de Rijke M., Venema, Y. (eds.) Modal Logic and Process Algebra. CSLI Lecture Notes, vol. 53, pp. 107–129. CSLI Publications (1995)
15. Jacobs, B.: Introduction to Coalgebra: Towards Mathematics of States and Observation. Cambridge Tracts in Theoretical Computer Science. Cambridge University Press, Cambridge (2016). https://doi.org/10.1017/CBO9781316823187.007
16. Kupke, C., Kurz, A., Venema, Y.: Stone coalgebras. Theor. Comput. Sci. **327**(1), 109–134 (2004). https://doi.org/10.1016/j.tcs.2004.07.023. Selected Papers of CMCS '03
17. Moss, L.S., Viglizzo, I.D.: Final coalgebras for functors on measurable spaces. Inf. Comput. **204**(4), 610–636 (2006). https://doi.org/10.1016/j.ic.2005.04.006. Seventh Workshop on Coalgebraic Methods in Computer Science 2004
18. Munkres, J.: Topology. Prentice Hall, Hoboken (2000)
19. Rutten, J.: Universal coalgebra: a theory of systems. Technical report, R 9652, CWI, Amsterdam (1996). https://ir.cwi.nl/pub/4802
20. Rutten, J.: Universal coalgebra: a theory of systems. Theoret. Comput. Sci. **249**, 3–80 (2000). https://doi.org/10.1016/S0304-3975(00)00056-6
21. Vietoris, L.: Bereiche zweiter Ordnung. Monatshefte für Mathematik und Physik **32**(1), 258–280 (1922). https://doi.org/10.1007/BF01696886

Algebraic Presentation of Semifree Monads

Aloïs Rosset[1(✉)], Helle Hvid Hansen[2], and Jörg Endrullis[1]

[1] Vrije Universiteit Amsterdam, Amsterdam, Netherlands
{alois.rosset,j.endrullis}@vu.nl
[2] University of Groningen, Groningen, Netherlands
h.h.hansen@rug.nl

Abstract. Monads and their composition via distributive laws have many applications in program semantics and functional programming. For many interesting monads, distributive laws fail to exist, and this has motivated investigations into weaker notions. In this line of research, Petrişan and Sarkis recently introduced a construction called the semifree monad in order to study semialgebras for a monad and weak distributive laws. In this paper, we prove that an algebraic presentation of the semifree monad M^{s} on a monad M can be obtained uniformly from an algebraic presentation of M. This result was conjectured by Petrişan and Sarkis. We also show that semifree monads are ideal monads, that the semifree construction is not a monad transformer, and that the semifree construction is a comonad on the category of monads.

Keywords: algebraic theory · monad · algebraic presentation · semifree

1 Introduction

Monads [25,26] are widely used in the coalgebraic semantics of programs with nondeterminism, probabilistic branching and other features, see e.g. [8,14,19,21,28]. In functional programming, monads are used for structuring programs with computational effects, see e.g. [31,33,39]. In order to reason about programs that combine several effects, compositions of monads have been studied such as monad transformers [20,24], coproducts [3,12] and tensors [15,34]. The main approach for studying compositions of two monads with functor parts M and T into a monad with functor part MT is via distributive laws [6]. However, distributive laws do not always exist [22,38,40]. These negative results motivated investigations into weaker notions of distributive laws [9,11,37]. For some of the most prominent examples, including the lack of a distributive law of the probability distributions monad over the powerset monad [38], the failure was located at one of the two unit axioms. To overcome this, Garner [11] defines a notion of *weak distributive law* which drops the problematic axiom. The usefulness of this concept has been demonstrated as several monads have been proven to result

Published by Springer Nature Switzerland AG 2022
H. H. Hansen and F. Zanasi (Eds.): CMCS 2022, LNCS 13225, pp. 110–132, 2022.
https://doi.org/10.1007/978-3-031-10736-8_6

from weak distributive laws: the Vietoris monad [11], the convex powerset monad [13] and the monad of finitely generated convex subsets [7].

Following this line of research, Petrişan & Sarkis [32] defined the *semifree monad* for a monad M on the functor coproduct $M^{\mathrm{s}} := \mathrm{id} + M$. They demonstrated a one-to-one correspondence between weak distributive laws with M and distributive laws with M^{s} satisfying an extra condition. To achieve this, they first proved the existence of an isomorphism of categories between M^{s}-algebras and M-semialgebras, the latter being M-algebras that only satisfy the associativity axiom, but not necessarily the unit axiom. A similar result was proved by Hyland and Tasson [16, Proposition 27] in the context of 2-monads and 2-categories.

The algebraic presentation of finitary monads by algebraic theories allows for equational reasoning about programs with computational effects, and is currently an active area of research, see e.g. [7,8,29,30,40].

The main contribution of this paper is to show that an algebraic presentation of the semifree monad M^{s} can be obtained uniformly from an algebraic presentation of M (Corollary 16). This uniform presentation was conjectured in [32]. We apply the result to obtain algebraic presentations of the semifree monad over the exception monad, the list monad, the multiset monad, the finite powerset monad and the state monad.

We give a brief sketch of the proof. Given an algebraic theory (Σ, E), we first note that it suffices to prove the result for the free monad T of the theory, which is clearly presented by it. Let $(\Sigma^{\mathrm{s}}, E^{\mathrm{s}})$ be the proposed presentation of T^{s}, where the signature Σ^{s} consists of all operations in Σ plus a new unary, idempotent operation denoted a. Due to the isomorphism between T^{s}-algebras and T-semialgebras it suffices to show an isomorphism between the categories of T-semialgebras and $(\Sigma^{\mathrm{s}}, E^{\mathrm{s}})$-algebras.

- Given a T-semialgebra (X, α), we obtain a $(\Sigma^{\mathrm{s}}, E^{\mathrm{s}})$-algebra by interpreting the new unary operation a by embedding elements of X into TX using the unit η_X of T and then applying α. The old operations from Σ are interpreted similarly. We then verify by inductive arguments that this algebra satisfies all equations in E^{s}.
- Given a $(\Sigma^{\mathrm{s}}, E^{\mathrm{s}})$-algebra, we obtain an T-semialgebra (X, α) using that elements of TX are congruence classes of terms. This allows us to show that α is indeed associative by an induction on the term structure of the elements of TTX.

Apart from providing equational reasoning for semifree monads, this uniform presentation result provides a means to study weak distributive laws via the presentations of the monads involved, using a similar approach as Zwart & Marsden [40]. We discuss their work below and in Sect. 6.

The paper is organised as follows. Section 2 introduces preliminary definitions of monads, universal algebra, and semifree monads. Section 3 states and proves the main result, Corollary 16. Section 4 presents examples of the application of Corollary 16. Section 5 investigates the relationship between the semifree construction and the notions of ideal monad and monad transformer. Section 6 concludes and discusses future work. Omitted proofs can be found in an online version on arXiv [36].

Related Work: The work in the present paper concerns the question of how to obtain a monad presentation from existing ones, in a uniform manner.

Zwart & Marsden [40] use presentations of monads to give general results about the non-existence of distributive laws ("no-go theorems"). This approach allowed them to answer several open questions, including the 50 year-old conjecture by Beck [6, Example 4.1] that addition cannot distribute over multiplication. On the positive side, when a distributive law exists, they show how to obtain a presentation of the composite monad from the monads and the distributive law.

Monads of the form $\mathcal{C}+1$ and other modifications of $\mathcal{C}$, the monad of convex subsets of distributions are studied in [29]. Both positive and negative results are obtained in different categories. No uniform presentation result such as Corollary 16 is given.

Recent work on algebraic presentations of specific monads include a presentation of the monad $\mathcal{C}$ [8], and presentations for monads arising via weak distributive laws that combine the monads of semilattices and semimodules [7].

2 Preliminaries

We assume the reader is familiar with basic notions of category theory [4,25,35]. In this section, we recall basic definitions and fix notation concerning monads, algebraic theories and presentations. We also recall basic definitions and results of semifree monads.

Definition 1. *A* **monad** (M,η,μ) *on a category* C *is a triple consisting of an endofunctor* $M:\mathsf{C}\to\mathsf{C}$, *and two natural transformations, the* **unit** $\eta:\mathrm{id}\Rightarrow M$ *and the* **multiplication** $\mu:M^2\Rightarrow M$ *that make (1) and (2) commute. We refer to (2) as the associativity of* μ.

$$\begin{array}{ccccc} M & \xrightarrow{M\eta} & M^2 & \xleftarrow{\eta M} & M \\ & \searrow^{=} & \downarrow{\scriptstyle\mu} & \swarrow_{=} & \\ & & M & & \end{array} \qquad (1) \qquad\qquad \begin{array}{ccc} M^3 & \xrightarrow{\mu M} & M^2 \\ {\scriptstyle M\mu}\downarrow & & \downarrow{\scriptstyle\mu} \\ M^2 & \xrightarrow[\mu]{} & M \end{array} \qquad (2)$$

For convenience, we often refer to a monad (M,η,μ) by its functor part M.

Definition 2. *Given two monads* (M,η^M,μ^M) *and* (T,η^T,μ^T) *on a category* C, *a* **monad morphism** *from* M *to* T *is a natural transformation* $\sigma:M\Rightarrow T$ *that makes (3) and (4) commute, where* $\sigma\sigma:=\sigma_T\circ M\sigma=T\sigma\circ\sigma_M$.

$$\begin{array}{ccc} & \overset{\eta^M}{\nearrow} & M \\ \mathrm{id} & & \downarrow{\scriptstyle\sigma} \\ & \underset{\eta^T}{\searrow} & T \end{array} \qquad (3) \qquad\qquad \begin{array}{ccc} M^2 & \xrightarrow{\sigma\sigma} & T^2 \\ {\scriptstyle\mu^M}\downarrow & & \downarrow{\scriptstyle\mu^T} \\ M & \xrightarrow[\sigma]{} & T \end{array} \qquad (4)$$

If each component of σ is an isomorphism, we say that the two monads are **isomorphic**. *The category of monads on a category* C *and monad morphisms is denoted* $\mathbf{Mon}(\mathsf{C})$.

Definition 3. *Let (M, η, μ) be a monad on category C. An (Eilenberg-Moore) M-***algebra** *is a C-morphism $\alpha : MX \to X$ for some $X \in \mathsf{C}$, denoted (X, α) for short, such that (5) and (6) commute. An M-***semialgebra** *is a C-morphism $\alpha : MX \to X$ that satisfies (6).*

$$\begin{array}{ccc} X & \xrightarrow{\eta_X} & MX \\ & \searrow^{=} & \downarrow \alpha \\ & & X \end{array} \qquad (5) \qquad\qquad \begin{array}{ccc} M^2X & \xrightarrow{\mu_X} & MX \\ M\alpha \downarrow & & \downarrow \alpha \\ MX & \xrightarrow[\alpha]{} & X \end{array} \qquad (6)$$

*An M-***(semi)algebra homomorphism** *$f : (X, \alpha) \to (Y, \beta)$ between two M-(semi)algebras is a function $f : X \to Y$ such that the following diagram commutes:*

$$\begin{array}{ccc} MX & \xrightarrow{Mf} & MY \\ \alpha \downarrow & & \downarrow \beta \\ X & \xrightarrow[f]{} & Y \end{array}$$

The category of M-algebras and M-algebra homomorphisms is denoted $\mathbf{EM}(M)$ *and called the* **Eilenberg-Moore category** *of M. The category* $\mathbf{EM}_s(M)$ *consists of M-semialgebras and M-semialgebra homomorphisms.*

We denote the **coproduct** of two objects X and Y in a category C by $X+Y$, the left and right injections by $\mathsf{inl}^{X+Y} : X \to X+Y$ and $\mathsf{inr}^{X+Y} : Y \to X+Y$, and the arrow given by the universal property of the coproduct for arbitrary $f : X \to Z$ and $g : Y \to Z$ by $[f, g] : X + Y \to Z$. Given $f : X \to X'$ and $g : Y \to Y'$, we denote $f + g := [\mathsf{inl}^{X'+Y'} \circ f, \mathsf{inr}^{X'+Y'} \circ g] : X + Y \to X' + Y'$. We now recall a few basic notions of universal algebra.

Definition 4. – *An* **algebraic signature** *Σ is a set of operation symbols each having its own arity $n \in \mathbb{N}$, denoted $(\mathsf{op} : n)$ for an n-ary $\mathsf{op} \in \Sigma$.*
- *Given an algebraic signature Σ and a set X, the set $\mathcal{T}_\Sigma(X)$ of Σ-terms over X is defined inductively as follows. Elements in X are terms, and they are said to be of depth zero, written $\mathsf{dp}(v) = 0$. If $t_1, \dots, t_n$ are terms, and $(\mathsf{op} : n) \in \Sigma$, then $t := \mathsf{op}(t_1, \dots, t_n)$ is a term of depth $\mathsf{dp}(t) := \max\{\mathsf{dp}(t_1), \dots, \mathsf{dp}(t_n)\}+1$. We define constants (i.e., nullary operations) to have depth 1.*
- *We fix a set $\mathsf{Var} = \{v_1, v_2, v_3, \dots\}$ of variables. To indicate that the variables appearing in $t \in \mathcal{T}_\Sigma(\mathsf{Var})$ are in the set $\{v_1, \dots, v_n\}$, we write $t(v_1, \dots, v_n)$.*
- *A Σ-***algebra** *is a pair $(X, [\![\cdot]\!])$, where X is a set and $[\![\cdot]\!]$ is a collection of interpretations: for each $(\mathsf{op} : n) \in \Sigma$, we have $[\![\mathsf{op}]\!] : X^n \to X$.*
- *Given a Σ-algebra $(X, [\![\cdot]\!])$, any function $f : Y \to X$ extends to a unique homomorphism, $[\![\cdot]\!]_f : \mathcal{T}_\Sigma(Y) \to X$:*

$$[\![y]\!]_f := f(y), \text{ and} \tag{7}$$

$$[\![\mathsf{op}(t_1,\ldots,t_n)]\!]_f := [\![\mathsf{op}]\!]([\![t_1]\!]_f,\ldots,[\![t_n]\!]_f). \tag{8}$$

When f is the identity id_X, the subscript is omitted. The function f is often a variable assignment $\sigma : \mathsf{Var} \to X$ to obtain an interpretation of $\mathcal{T}_\Sigma(\mathsf{Var})$.

- *An* **equation** *over Σ is a pair of terms $(s,t) \in \mathcal{T}_\Sigma(\mathsf{Var}) \times \mathcal{T}_\Sigma(\mathsf{Var})$.*
- *An* **algebraic theory** *is a pair (Σ, E) consisting of an algebraic signature Σ and set E of equations over Σ.*
- *A* **(Σ,E)-algebra** *is a Σ-algebra $(X, [\![\cdot]\!])$ for which the interpretation satisfies all equations in E, meaning that for all $(s,t) \in E$ and all variable assignments σ, $[\![s]\!]_\sigma = [\![t]\!]_\sigma$.*
- *A* **(Σ,E)-algebra homomorphism** *between two (Σ,E)-algebras $(X, [\![\cdot]\!])$ and $(X', [\![\cdot]\!]')$ is a function $f : X \to X'$ such that $f \circ [\![\mathsf{op}]\!] = [\![\mathsf{op}]\!]' \circ f^n$ for all $(\mathsf{op} : n) \in \Sigma$.*
- *The category* $\mathbf{Alg}(\Sigma,E)$ *consists of (Σ,E)-algebras and (Σ,E)-algebra homomorphisms.*
- *Given terms s and t, we write $E \vdash s = t$ to denote that $s = t$ is derivable from E in equational logic, and $E \vDash s = t$ to denote that the equation $s = t$ holds in all (Σ,E)-algebras. Birkhoff's theorem states that $E \vdash s = t \iff E \vDash s = t$.*
- *The* **free (Σ,E)-algebra** *on a set X is the (Σ,E)-algebra $({}^{\mathcal{T}_\Sigma(X)}/_E, [\![\cdot]\!]^X)$ with carrier set consisting of $\mathcal{T}_\Sigma(X)$ modulo the smallest congruence relation containing E. The congruence class of a term t is denoted $\overline{t}$. The interpretation of an operation $(\mathsf{op} : m) \in \Sigma$ is*

$$[\![\mathsf{op}]\!]^X(\overline{t_1},\ldots,\overline{t_n}) := \overline{\mathsf{op}(t_1,\ldots,t_n)}.$$

- *The* **free functor** *$F : \mathsf{Set} \to \mathbf{Alg}(\Sigma,E)$ sends a set X to the free (Σ,E)-algebra $({}^{\mathcal{T}_\Sigma(X)}/_E, [\![\cdot]\!])$. The adjective "free" is also true in the categorical sense, i.e., we have a free-forgetful adjunction*

$$F : \mathsf{Set} \underset{\longleftarrow}{\overset{\perp}{\longrightarrow}} \mathbf{Alg}(\Sigma,E) : U$$

The composite $T_{\Sigma,E} := UF$ is a monad (see e.g. [25, VI.1]). Its unit $\eta_{\Sigma,E}$ sends an element to its equivalence class $x \mapsto \overline{x}$ and its multiplication $\mu_{\Sigma,E}$ flattens terms $\overline{t[\overline{t_i}/v_i]} \mapsto \overline{t[t_i/v_i]}$.

We can finally define the central concept of algebraic presentation.

Definition 5. *An algebraic theory (Σ,E) is an* **algebraic presentation** *of a* Set*-monad (M,η,μ) if $(T_{\Sigma,E}, \eta_{\Sigma,E}, \mu_{\Sigma,E}) \cong (M,\eta,\mu)$.*

Note that a monad can have multiple presentations. From the definition, we immediately have the trivial example that an algebraic theory (Σ,E) is an algebraic presentation of its free monad $T_{\Sigma,E}$.

The class of monads that admit presentations is precisely the class of *finitary monads*; for the definition and also the proof of this correspondence see e.g. [2, Section 3.18, p. 149]. We therefore work only with finitary monads in the article.

Given a Set-monad M and an algebraic theory (Σ, E), the categories of algebras $\mathbf{EM}(M)$ and $\mathbf{Alg}(\Sigma, E)$ are concrete categories, and in this paper, it turns out to be more convenient to work with an equivalent definition of algebraic presentation formulated in terms of concrete isomorphisms.

Definition 6. *A category* C *is* **concrete** *if there is a faithful functor* $U : \mathsf{C} \to \mathsf{Set}$*, usually a forgetful functor. A functor* $F : \mathsf{C} \to \mathsf{D}$ *between concrete categories is itself* **concrete** *if it commutes with the faithful functors* $U_{\mathsf{D}} \circ F = U_{\mathsf{C}}$*. We write* $\mathsf{C} \cong_{\mathsf{conc}} \mathsf{D}$ *to denote that categories* C *and* D *are concretely isomorphic.*

The following lemma is well-known and is a direct consequence of e.g. [5, Theorem III.6.3].

Lemma 7. *For* Set*-monads* $(M, \eta^M, \mu^M), (T, \eta^T, \mu^T)$*, we have*

$$(M, \eta^M, \mu^M) \cong (T, \eta^T, \mu^T) \quad \iff \quad \mathbf{EM}(M) \cong_{\mathsf{conc}} \mathbf{EM}(T).$$

It gives the following alternative formulation of algebraic presentation.

Lemma 8. *An algebraic theory* (Σ, E) *is an algebraic presentation of a (finitary) monad* (M, η, μ) *if and only if* $\mathbf{EM}(M) \cong_{\mathsf{conc}} \mathbf{Alg}(\Sigma, E)$.

Proof. Since $\mathbf{Alg}(\Sigma, E) \cong_{\mathsf{conc}} \mathbf{EM}(T_{\Sigma,E})$ (see e.g. [25, VI.8.1]), the result follows immediately from Lemma 7. □

Remark 9. In the literature, the definition of algebraic presentation is often stated as the condition $\mathbf{EM}(M) \cong \mathbf{Alg}(\Sigma, E)$, i.e., it leaves the "concrete" part implicit, see e.g. [7,8,29,30,32]. The "concrete" part is not necessary in these papers, since they establish algebraic presentations (by indeed establishing concrete isomorphisms), but they do not prove results that assume the existence of an algebraic presentation. In the present paper, we assume that a presentation for M is given, and establish one for M^s, and the proof requires the isomorphism $\mathbf{EM}(M) \cong_{\mathsf{conc}} \mathbf{Alg}(\Sigma, E)$ to be concrete.

The following lemma states two well-known facts that we will need later in proofs.

Lemma 10. *Let* (Σ, E) *be an algebraic theory and* T *denote its free monad* $T_{\Sigma,E}$*. Given a function* $f : Y \to X$*, then the following are* (Σ, E)*-algebra homomorphisms:*

$$\mu_X : (TTX, [\![\cdot]\!]^{TX}) \to (TX, [\![\cdot]\!]^X),$$
$$Tf : (TY, [\![\cdot]\!]^Y) \to (TX, [\![\cdot]\!]^X).$$

The semifree monad M^s for a monad M was introduced in [32] by Petrişan and Sarkis.

Definition 11 ([32]). *Given a monad (M, η, μ) on a category* C *having all finite coproducts, the* **semifree monad** *on M is a monad $(M^{\mathrm{s}}, \eta^{\mathrm{s}}, \mu^{\mathrm{s}})$, where*

$$\begin{aligned} M^{\mathrm{s}} &:= \mathrm{id}_C + M, \\ \eta^{\mathrm{s}} &:= \mathsf{inl}^{\mathrm{id}+M}, \\ \mu^{\mathrm{s}} &:= [\mathrm{id}_{\mathrm{id}+M}, \mathsf{inr}^{\mathrm{id}+M} \circ \mu \circ M[\eta, \mathrm{id}_M]]. \end{aligned}$$

Note that the unit η^{s} injects a set X to its copy on the left in $X + MX$. Petrişan and Sarkis showed in Theorems 3.4 and 4.3 of [32] that:

- There is a (concrete) isomorphism $\mathbf{EM}(M^{\mathrm{s}}) \cong_{\mathsf{conc}} \mathbf{EM}_s(M)$.
- There is a bijection between weak distributive laws $\lambda : MT \Rightarrow TM$ and distributive laws $\delta : M^{\mathrm{s}}T \Rightarrow TM^{\mathrm{s}}$ satisfying an extra condition.

The semifree construction takes a monad as input and outputs another monad. The semifree construction can be made into a functor on $\mathbf{Mon}(\mathsf{C})$ as follows. Given a monad morphism $\sigma : M \Rightarrow T$, and a set X, let

$$\sigma^{\mathrm{s}}_X := \left(X + MX \xrightarrow{\mathrm{id}_X + \sigma_X} X + TX \right). \tag{9}$$

Lemma 12. *The mapping $(-)^{\mathrm{s}} : \mathbf{Mon}(\mathsf{C}) \to \mathbf{Mon}(\mathsf{C})$ is a functor.*

Since functors preserve isomorphism, we have the following consequence.

Corollary 13. *Take two monads $(M, \eta^M, \mu^M), (T, \eta^T, \mu^T)$ on a category* C *that has all finite coproducts. If they are isomorphic $M \cong T$, then their respective semifree monads are also isomorphic $M^{\mathrm{s}} \cong T^{\mathrm{s}}$.*

3 Algebraic Presentation of Semifree Monads

In this section, we state and prove the main result of the paper. We prove that given an algebraic presentation of a (finitary) Set-monad (M, η, μ), we can derive an algebraic presentation of the semifree monad $(M^{\mathrm{s}}, \eta^{\mathrm{s}}, \mu^{\mathrm{s}})$. In particular, if M is a finitary monad, then its semifree monad M^{s} is finitary too.

Before we state the theorem, we give some intuitions for the presentation of $(M^{\mathrm{s}}, \eta^{\mathrm{s}}, \mu^{\mathrm{s}})$. Recall that $M^{\mathrm{s}} = X + MX$ and the unit $\eta^{\mathrm{s}}_X : X \to X + MX$ is the left injection. In terms of presentation, this means that the left copy of X becomes the "new" set of variables. As a consequence, the "old" set of variables $\eta_X(X) \subseteq MX$ is now free in M^{s} of the constraints, such as the unit laws (1) and (5), that it had in M. The inclusion of X via η_X in $M^{\mathrm{s}}X$ corresponds to a new unary operation $(\mathsf{a} : 1)$ in the presentation of M^{s}. On the semantic level, suppose we have an M^{s}-algebra $\gamma : M^{\mathrm{s}}X \to X$. By Theorem 3.4 in [32] and by looking at its proof, we see that γ must be of the form $[\mathrm{id}_X, \alpha]$ where $\alpha : MX \to X$ is an M-semialgebra. Notice the following:

$$\alpha \circ \eta_X \circ \alpha \overset{\eta \text{ nat.}}{=} \alpha \circ M\alpha \circ \eta_{MX} \overset{(6)}{=} \alpha \circ \mu_X \circ \eta_{MX} \overset{(1)}{=} \alpha. \tag{10}$$

Hence, $\alpha \circ \eta_X$ is an idempotent. This map will be our choice for the interpretation of the new symbol $(\mathsf{a} : 1)$.

Definition 14. *Given an algebraic theory* (Σ, E)*, we define a new algebraic theory* $(\Sigma^{\mathrm{s}}, E^{\mathrm{s}})$ *by* $\Sigma^{\mathrm{s}} := \Sigma \uplus \{\mathsf{a} : 1\}$ *and* E^{s} *containing the following:*

$$\mathsf{aa}v_1 = \mathsf{a}v_1, \tag{11}$$

$$\mathsf{a}(\mathsf{op}(v_1, \ldots, v_n)) = \mathsf{op}(v_1, \ldots, v_n), \tag{12}$$

$$\mathsf{op}(\mathsf{a}v_1, \ldots, \mathsf{a}v_n) = \mathsf{op}(v_1, \ldots, v_n), \tag{13}$$

$$t(\mathsf{a}v_1, \ldots, \mathsf{a}v_n) = s(\mathsf{a}v_1, \ldots, \mathsf{a}v_n), \tag{14}$$

for all $(\mathsf{op} : n) \in \Sigma$ *and* $(t(v_1, \ldots, v_n) = s(v_1, \ldots, v_n)) \in E$.

We have the trivial fact that an algebraic theory (Σ, E) is an algebraic presentation of its free monad $T_{\Sigma,E}$. The next theorem states that the algebraic theory $(\Sigma^{\mathrm{s}}, E^{\mathrm{s}})$ of Definition 14 is an algebraic presentation of $T^{\mathrm{s}}_{\Sigma,E}$, the semifree monad on $T_{\Sigma,E}$.

Theorem 15. $(\Sigma^{\mathrm{s}}, E^{\mathrm{s}})$ *is an algebraic presentation of* $(T^{\mathrm{s}}_{\Sigma,E}, \eta^{\mathrm{s}}_{\Sigma,E}, \mu^{\mathrm{s}}_{\Sigma,E})$.

The proof of this theorem is the goal of the rest of Sect. 3. As a direct consequence, we have the following corollary. It was originally formulated as Conjecture 5.4 in [32] by Petrişan and Sarkis.

Corollary 16. *If* (Σ, E) *is an algebraic presentation of a monad* (M, η, μ)*, then* $(\Sigma^{\mathrm{s}}, E^{\mathrm{s}})$ *is an algebraic presentation of* $(M^{\mathrm{s}}, \eta^{\mathrm{s}}, \mu^{\mathrm{s}})$*, the semifree monad on* M.

Proof. Assume that we have a monad isomorphism $T_{\Sigma,E} \cong M$. By Corollary 13, their semifree monads are also isomorphic $T^{\mathrm{s}}_{\Sigma,E} \cong M^{\mathrm{s}}$. By Theorem 15, $T^{\mathrm{s}}_{\Sigma,E} \cong T_{\Sigma^{\mathrm{s}},E^{\mathrm{s}}}$. Hence, $M^{\mathrm{s}} \cong T_{\Sigma^{\mathrm{s}},E^{\mathrm{s}}}$, which means that $(\Sigma^{\mathrm{s}}, E^{\mathrm{s}})$ is an algebraic presentation of M^{s}. □

We will need a few technical lemmas. The next one shows that Eqs. (12) and (13) extend inductively to all terms of depth at least 1.

Lemma 17. *For all* $t \in \mathcal{T}_\Sigma(\mathsf{Var})$ *of depth at least* 1*, an* Σ^{s}*-algebra that satisfies* (11)–(13) *also satisfies*

$$\mathsf{a}t(v_1, \ldots, v_n) = t(v_1, \ldots, v_n), \text{ and} \tag{15}$$

$$t(\mathsf{a}v_1, \ldots, \mathsf{a}v_n) = t(v_1, \ldots, v_n). \tag{16}$$

Proof of Lemma 17. Take $(X, (\!|\cdot|\!))$ that satisfies (11) to (13), and a term $t \in \mathcal{T}_\Sigma(\mathsf{Var})$. The proof goes by induction. When t is of depth 1, these are just Eqs. (12) and (13), which we already know are satisfied. For the induction step, take $t = \mathsf{op}(t_1, \ldots, t_m)$ and suppose that (15) and (16) hold for $t_1, \ldots, t_m$. Among the subterms, say that p of them are of depth at least 1, w.l.o.g. the first p ones $t_1, \ldots, t_p$. Thus, the $q := m - p$ last subterms are variables $w_1, \ldots, w_q \in \{v_1, \ldots, v_n\}$. Then,

$$\begin{aligned}
(\!|\mathsf{a}t(v_1, \ldots, v_n)|\!)_\sigma &= (\!|\mathsf{a}(\mathsf{op}(t_1, \ldots, t_p, w_1, \ldots, w_q)|\!)_\sigma \\
&= (\!|\mathsf{op}(t_1, \ldots, t_p, w_1, \ldots, w_q))|\!)_\sigma && \text{(by (12))} \\
&= (\!|t(v_1, \ldots, v_n)|\!)_\sigma,
\end{aligned}$$

and

$$\begin{aligned}
&\langle\!\langle t(\mathsf{a}v_1,\ldots,\mathsf{a}v_n)\rangle\!\rangle_\sigma \\
&= \langle\!\langle \mathsf{op}\rangle\!\rangle\left(\langle\!\langle t_1(\mathsf{a}v_1,\ldots,\mathsf{a}v_n)\rangle\!\rangle_\sigma,\ldots,\langle\!\langle t_p(\mathsf{a}v_1,\ldots,\mathsf{a}v_n)\rangle\!\rangle_\sigma,\langle\!\langle \mathsf{a}w_1\rangle\!\rangle_\sigma,\ldots,\langle\!\langle \mathsf{a}w_q\rangle\!\rangle_\sigma\right) \\
&= \langle\!\langle \mathsf{op}\rangle\!\rangle\left(\langle\!\langle t_1(v_1,\ldots,v_n)\rangle\!\rangle_\sigma,\ldots,\langle\!\langle t_p(v_1,\ldots,v_n)\rangle\!\rangle_\sigma,\langle\!\langle \mathsf{a}w_1\rangle\!\rangle_\sigma,\ldots,\langle\!\langle \mathsf{a}w_q\rangle\!\rangle_\sigma\right) && \text{(I.H.)} \\
&= \langle\!\langle \mathsf{op}(t_1,\ldots,t_p,\mathsf{a}w_1,\ldots,\mathsf{a}w_q)\rangle\!\rangle_\sigma \\
&= \langle\!\langle \mathsf{op}(\mathsf{a}t_1,\ldots,\mathsf{a}t_p,\mathsf{aa}w_1,\ldots,\mathsf{aa}w_q)\rangle\!\rangle_\sigma && \text{(by (13))} \\
&= \langle\!\langle \mathsf{op}(\mathsf{a}t_1,\ldots,\mathsf{a}t_p,\mathsf{a}w_1,\ldots,\mathsf{a}w_q)\rangle\!\rangle_\sigma && \text{(idempotency (11))} \\
&= \langle\!\langle \mathsf{op}(t_1,\ldots,t_p,w_1,\ldots,w_q)\rangle\!\rangle_\sigma && \text{(by (13))} \\
&= \langle\!\langle t(v_1,\ldots,v_n)\rangle\!\rangle_\sigma .
\end{aligned}$$

□

In Definition 14, Eq. (14) tells us that equations from E give rise to equations in E^{s} by substituting $v \mapsto \mathsf{a}v$, for all variables v. The next lemma indicates that the same procedure can be done for theorems, i.e., that a theorem deducible from E becomes, after the substitution $v \mapsto \mathsf{a}v$, a theorem deducible from E^{s}.

Lemma 18. *For all terms $t(v_1,\ldots,v_n), s(v_1,\ldots,v_n) \in \mathcal{T}_\Sigma(\mathsf{Var})$,*

$$E \vdash t(v_1,\ldots,v_n) = s(v_1,\ldots,v_n) \implies E^{\mathrm{s}} \vdash t(\mathsf{a}v_1,\ldots,\mathsf{a}v_n) = s(\mathsf{a}v_1,\ldots,\mathsf{a}v_n).$$

The last lemma is purely technical. Its reasoning appears multiple times in different proofs. Stating it here allows us to prove it once for all.

Lemma 19. *Let (Σ, E) be an algebraic theory with free monad $(T_{\Sigma,E}, \eta_{\Sigma,E}, \mu_{\Sigma,E})$. For every $T_{\Sigma,E}$-semialgebra $\alpha : T_{\Sigma,E}X \to X$ and operation symbol $(\mathsf{op} : n) \in \Sigma$, we have*

$$\alpha \circ [\![\mathsf{op}]\!]^X = \alpha \circ [\![\mathsf{op}]\!]^X \circ \eta_X^n \circ \alpha^n. \tag{17}$$

We now tackle the proof of Theorem 15. For simplicity, we will denote in the rest of this section the free monad $(T_{\Sigma,E}, \eta_{\Sigma,E}, \mu_{\Sigma,E})$ simply by (T, η, μ). To prove that $(\Sigma^{\mathrm{s}}, E^{\mathrm{s}})$ is an algebraic presentation of T^{s}, it suffices by Lemma 8 to prove that $\mathbf{Alg}(\Sigma^{\mathrm{s}}, E^{\mathrm{s}}) \cong_{\mathsf{conc}} \mathbf{EM}(T^{\mathrm{s}})$. Recall that $\mathbf{EM}(T^{\mathrm{s}}) \cong_{\mathsf{conc}} \mathbf{EM}_s(T)$, by Theorem 3.4 in [32], i.e., T^{s}-algebras are concretely isomorphic to T-semialgebras. Therefore, it suffices to prove that $\mathbf{EM}_s(T) \cong_{\mathsf{conc}} \mathbf{Alg}(\Sigma^{\mathrm{s}}, E^{\mathrm{s}})$.

3.1 From T-semialgebras to $(\Sigma^{\mathrm{s}}, E^{\mathrm{s}})$-algebras

In the forward direction, suppose we have a semialgebra $\alpha : TX \to X$. We want to obtain an $(\Sigma^{\mathrm{s}}, E^{\mathrm{s}})$-algebra. It will be constructed with carrier X, since we are aiming for a concrete isomorphism.

Definition 20. *We define the mapping*

$$G : \mathbf{EM}_s(T) \to \mathbf{Alg}(\Sigma^{\mathrm{s}}, E^{\mathrm{s}})$$

by $G(X, \alpha) := (X, \langle\!\langle \cdot \rangle\!\rangle)$ *on objects, where the interpretation* $\langle\!\langle \cdot \rangle\!\rangle$ *is defined on the operation symbols* $\mathsf{a} : 1$ *and* $(\mathsf{op} : n) \in \Sigma$ *as*

$$\langle\!\langle \mathsf{a} \rangle\!\rangle := \left(X \xrightarrow{\eta_X} TX \xrightarrow{\alpha} X \right), \tag{18}$$

$$\langle\!\langle \mathsf{op} \rangle\!\rangle := \left(X^n \xrightarrow{(\eta_X)^n} (TX)^n \xrightarrow{[\![\mathsf{op}]\!]^X} TX \xrightarrow{\alpha} X \right), \tag{19}$$

and $G(f) := f$ *on morphisms.*

The goal now is to demonstrate that G is well-defined on objects and arrows. It then follows immediately that G is a functor due to being essentially identity on arrows. To this end, we first establish in Lemma 21, a property that generalises both (18) and (19) into one formula. Then, we show in Lemma 22 that G indeed outputs $(\Sigma^{\mathrm{s}}, E^{\mathrm{s}})$-algebras. Lastly, we show in Lemma 23 that G outputs $(\Sigma^{\mathrm{s}}, E^{\mathrm{s}})$-algebra homomorphisms, and hence is also well-defined on arrows.

Lemma 21. *For all terms* $t(v_1, \ldots, v_n) \in \mathcal{T}_\Sigma(\mathsf{Var})$ *of depth at least* 1*, and all variable assignments* $\sigma : \mathsf{Var} \to X$*, we have*

$$\langle\!\langle t \rangle\!\rangle_\sigma = \alpha \circ [\![t]\!]^X_{\eta_X \circ \sigma}. \tag{20}$$

Lemma 22. *For all T-semialgebras* $\alpha : TX \to X, G(X, \alpha)$ *is a* $(\Sigma^{\mathrm{s}}, E^{\mathrm{s}})$*-algebra.*

Proof. We check that $(X, \langle\!\langle \cdot \rangle\!\rangle) := G(X, \alpha)$ satisfies the equations in E^{s}. Let $\sigma \colon \mathsf{Var} \to X$ be a variable assignment.

(i) For E^{s}-equations arising from (11), we have:

$$\begin{aligned}
\langle\!\langle \mathsf{aa}v_1 \rangle\!\rangle_\sigma &= \langle\!\langle \mathsf{a} \rangle\!\rangle \langle\!\langle \mathsf{a} \rangle\!\rangle (\sigma v_1) \\
&= (\alpha \circ \eta_X) \circ (\alpha \circ \eta_X)(\sigma v_1) && \text{(by (18))} \\
&= \alpha \circ \eta_X(\sigma v_1) && \text{(by (10))} \\
&= \langle\!\langle \mathsf{a}v_1 \rangle\!\rangle_\sigma.
\end{aligned}$$

(ii) For E^{s}-equations arising from (12).

$$\begin{aligned}
&\langle\!\langle \mathsf{a}(\mathsf{op}(v_1, \ldots, v_n)) \rangle\!\rangle_\sigma \\
&= \langle\!\langle \mathsf{a} \rangle\!\rangle \circ \langle\!\langle \mathsf{op} \rangle\!\rangle (\sigma v_1, \ldots, \sigma v_n) \\
&= \alpha \circ \eta_X \circ \alpha \circ [\![\mathsf{op}]\!]^X \circ (\eta_X)^n (\sigma v_1, \ldots, \sigma v_n) && \text{(by (18), (19))} \\
&= \alpha \circ [\![\mathsf{op}]\!]^X \circ (\eta_X)^n (\sigma v_1, \ldots, \sigma v_n) && \text{(by (10))} \\
&= \langle\!\langle \mathsf{op}(v_1, \ldots, v_n) \rangle\!\rangle_\sigma && \text{(by (19))}
\end{aligned}$$

(iii) For E^{s}-equations arising from (13), we have:

$$\begin{aligned}
&\llparenthesis \mathsf{op}(\mathsf{a}v_1,\ldots,\mathsf{a}v_n)\rrparenthesis_\sigma \\
&= \llparenthesis \mathsf{op}\rrparenthesis(\llparenthesis \mathsf{a}v_1\rrparenthesis_\sigma,\ldots,\llparenthesis \mathsf{a}v_n\rrparenthesis_\sigma) \\
&= \alpha \circ \llbracket \mathsf{op}\rrbracket^X \circ \eta^n_X(\alpha\circ\eta_X(\sigma v_1),\ldots,\alpha\circ\eta_X(\sigma v_n)) && \text{(by (18), (19))} \\
&= \alpha \circ \llbracket \mathsf{op}\rrbracket^X \circ \eta^n_X \circ \alpha^n \circ \eta^n_X(\sigma v_1,\ldots,\sigma v_n) \\
&= \alpha \circ \llbracket \mathsf{op}\rrbracket^X \circ \eta^n_X(\sigma v_1,\ldots,\sigma v_n) && \text{(by (17))} \\
&= \llparenthesis \mathsf{op}(v_1,\ldots,v_n)\rrparenthesis_\sigma. && \text{(by (19))}
\end{aligned}$$

(iv) For E^{s}-equations arising from (14), let $(t(v_1,\ldots,v_n) = s(v_1,\ldots,v_n)) \in E$. We have so far verified (11)–(13) and we can thus invoke Lemma 17. Since $(TX, \llbracket\cdot\rrbracket^X)$ is a (Σ, E)-algebra, and $\eta_X \circ \sigma : \mathsf{Var} \to TX$ is a variable assignement, we have $\llbracket t\rrbracket^X_{\eta_X\circ\sigma} = \llbracket s\rrbracket^X_{\eta_X\circ\sigma}$. We distinguish cases:

- Suppose t and s are variables, v_1 and v_2 respectively:

$$\begin{aligned}
\llbracket v_1\rrbracket^X_{\eta_X\circ\sigma} = \llbracket v_2\rrbracket^X_{\eta_X\circ\sigma} &\Leftrightarrow \eta_X\circ\sigma(v_1) = \eta_X\circ\sigma(v_2) && \text{(def (7))} \\
&\Rightarrow \alpha\circ\eta_X\circ\sigma(v_1) = \alpha\circ\eta_X\circ\sigma(v_2) \\
&\Rightarrow \llparenthesis \mathsf{a}v_1\rrparenthesis_\sigma = \llparenthesis \mathsf{a}v_2\rrparenthesis_\sigma. && \text{(def (18))}
\end{aligned}$$

- Suppose one is a variable and the other is not, say w.l.o.g. that t is v_1:

$$\begin{aligned}
\llbracket v_1\rrbracket^X_{\eta_X\circ\sigma} = \llbracket s\rrbracket^X_{\eta_X\circ\sigma} &\Leftrightarrow \eta_X\circ\sigma(v_1) = \llbracket s\rrbracket^X_{\eta_X\circ\sigma} && \text{(def (7))} \\
&\Rightarrow \alpha\circ\eta_X\circ\sigma(v_1) = \alpha\circ\llbracket s\rrbracket^X_{\eta_X\circ\sigma} \\
&\Rightarrow \llparenthesis \mathsf{a}v_1\rrparenthesis_\sigma = \llparenthesis s(v_1,\ldots,v_n)\rrparenthesis_\sigma && \text{(def (18); (20))} \\
&\Rightarrow \llparenthesis \mathsf{a}v_1\rrparenthesis_\sigma = \llparenthesis s(\mathsf{a}v_1,\ldots,\mathsf{a}v_n)\rrparenthesis_\sigma. && \text{(by (16))}
\end{aligned}$$

- Suppose neither of t and s is a variable:

$$\begin{aligned}
\llbracket t\rrbracket^X_{\eta_X\circ\sigma} = \llbracket s\rrbracket^X_{\eta_X\circ\sigma} &\Rightarrow \alpha\circ\llbracket t\rrbracket^X_{\eta_X\circ\sigma} = \alpha\circ\llbracket s\rrbracket^X_{\eta_X\circ\sigma} \\
&\Rightarrow \llparenthesis t(v_1,\ldots,v_n)\rrparenthesis_\sigma = \llparenthesis s(v_1,\ldots,v_n)\rrparenthesis_\sigma && \text{(by (20))} \\
&\Rightarrow \llparenthesis t(\mathsf{a}v_1,\ldots,\mathsf{a}v_n)\rrparenthesis_\sigma = \llparenthesis s(\mathsf{a}v_1,\ldots,\mathsf{a}v_n)\rrparenthesis_\sigma. && \text{(by (16))}
\end{aligned}$$

□

Lemma 23. *G maps T-semialgebra homomorphisms to $(\Sigma^{\mathrm{s}}, E^{\mathrm{s}})$-algebra homomorphisms.*

Proof. Suppose $f : (X,\alpha) \to (Y,\beta)$ is a T-semialgebra homomorphism. Let $(X, \llparenthesis\cdot\rrparenthesis^X) := G(X,\alpha)$ and $(Y, \llparenthesis\cdot\rrparenthesis^Y) := G(Y,\beta)$. We check that $G(f)$, which is defined as f in Definition 20, is an $(\Sigma^{\mathrm{s}}, E^{\mathrm{s}})$-algebra homomorphism, or in other words, that it commutes with the interpretations of the operations a and each $(\mathsf{op} : n) \in \Sigma$.

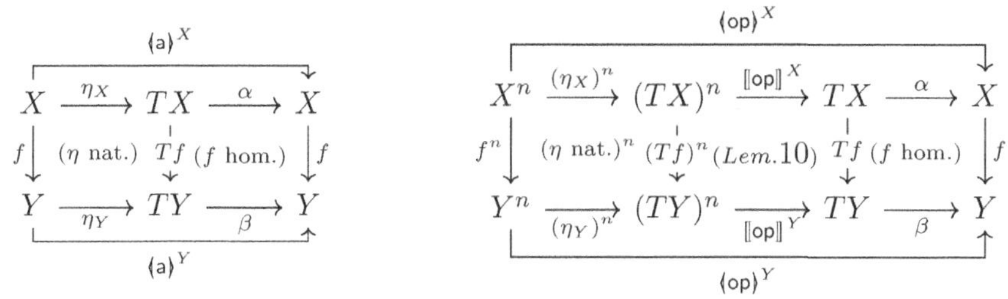

□

3.2 From (Σ^{s}, E^{s})-algebras to T-semialgebras

For the backward direction, given a (Σ^{s}, E^{s})-algebra $(X, (\!|\cdot|\!))$, we want to define a T-semialgebra $\alpha : TX \to X$. Since the elements of TX are equivalence classes of terms, we can construct the desired semialgebra and the backward functor H as follows.

Definition 24. *We define the mapping*

$$H : \mathbf{Alg}(\Sigma^{s}, E^{s}) \to \mathbf{EM}_s(T)$$

by $H(X, (\!|\cdot|\!)) := (X, \alpha)$ *on objects, where* α *is defined as follows:*

$$\alpha : TX \to X : \bar{t} \mapsto (\!|t|\!)_{(\!|\mathsf{a}|\!)}, \tag{21}$$

and $H(f) := f$ *on morphisms.*

We now show that H is well-defined on objects and morphisms. It then follows immediately that H is a functor due to being essentially identity on morphisms. Since the definition of α relies on equivalence classes, we first show in Lemma 25 that α is well-defined, i.e., that changing representatives does not matter. Then, we prove in Lemma 27 that α is indeed a T-semialgebra. Lastly, we show in Lemma 28 that H outputs T-semialgebra homomorphisms.

Lemma 25. *Given a* (Σ^{s}, E^{s})*-algebra* $(X, (\!|\cdot|\!))$*, and* $(X, \alpha) := H(X, (\!|\cdot|\!))$*, then* α *from* (21) *is a well-defined function.*

The next lemma states two identities that give more insight into how α works on distinct elements. It makes future manipulations of α easier.

Lemma 26. *Given a* (Σ^{s}, E^{s})*-algebra* $(X, (\!|\cdot|\!))$*, and* $(X, \alpha) := H(X, (\!|\cdot|\!))$*, then for all* $x \in X$*,* $(\mathsf{op} : n \in \Sigma)$*, and* $c_1, \ldots, c_n \in \mathcal{T}_\Sigma(X)/E$*, it holds that*

$$\alpha \circ \eta_X(x) = (\!|\mathsf{a}|\!)(x), \ \ \textit{and} \tag{22}$$

$$\alpha \circ [\![\mathsf{op}]\!]^X(c_1, \ldots, c_n) = (\!|\mathsf{op}|\!)(\alpha c_1, \ldots, \alpha c_n), \tag{23}$$

The proof of the associativity of α highlights the use of term representatives.

Lemma 27. *Given a* (Σ^{s}, E^{s})*-algebra* $(X, (\!|\cdot|\!))$*, and* $(X, \alpha) := H(X, (\!|\cdot|\!))$*, then* α *is a* T*-semialgebra, i.e., it satisfies the associativity axiom (6).*

Proof of Lemma 27. We have to show that $\alpha \circ T\alpha(\overline{t}) = \alpha \circ \mu_X(\overline{t})$ for all $\overline{t} \in TTX = {}^{T_\Sigma(TX)}/_E$. We do an induction on t:

- For the base case suppose t is some $\overline{s} \in TX = {}^{T_\Sigma(X)}/_E$. The goal can be reformulated as $\alpha \circ T\alpha \circ \eta_{TX}(\overline{s}) = \alpha \circ \mu_X \circ \eta_{TX}(\overline{s})$. By the unit law (1), it is the same as proving $\alpha \circ T\alpha \circ \eta_{TX}(\overline{s}) = \alpha(\overline{s})$. We distinguish cases for s:
  - If s is some $x \in X$, i.e., $\overline{s} = \eta_X(x)$:

$$\begin{aligned} \alpha \circ T\alpha \circ \eta_{TX} \circ \eta_X(x) &= \alpha \circ \eta_X \circ \alpha \circ \eta_X(x) && (\eta \text{ nat.}) \\ &= \langle\mathsf{a}\rangle \circ \langle\mathsf{a}\rangle(x) && (\text{by } (22)) \\ &= \langle\mathsf{a}\rangle(x) && (\text{idempotence } (11)) \\ &= \alpha \circ \eta_X(x) && (\text{by } (22)) \end{aligned}$$

  - If $s = \mathsf{op}(s_1, \ldots, s_m)$ for $s_1, \ldots, s_n \in T_\Sigma(X)$ and $(\mathsf{op} : m) \in \Sigma$:

$$\begin{aligned} \alpha \circ T\alpha \circ \eta_{TX}(\overline{s}) &= \alpha \circ T\alpha \circ \eta_{TX}\left(\overline{\mathsf{op}(s_1, \ldots, s_m)}\right) \\ &= \alpha \circ T\alpha \circ \eta_{TX}\left([\![\mathsf{op}]\!]^X(\overline{s_1}, \ldots, \overline{s_m})\right) && (\text{def. } [\![\cdot]\!]^X) \\ &= \alpha \circ \eta_X \circ \alpha \circ [\![\mathsf{op}]\!]^X(\overline{s_1}, \ldots, \overline{s_m}) && (\eta \text{ nat.}) \\ &= \langle\mathsf{a}\rangle \circ \langle\mathsf{op}\rangle(\alpha(\overline{s_1}), \ldots, \alpha(\overline{s_m})) && (\text{by } (22), (23)) \\ &= \langle\mathsf{op}\rangle(\alpha(\overline{s_1}), \ldots, \alpha(\overline{s_m})) && (\text{equation } (12)) \\ &= \alpha \circ [\![\mathsf{op}]\!]^X(\overline{s_1}, \ldots, \overline{s_m}) && (\text{by } (23)) \\ &= \alpha\left(\overline{\mathsf{op}(s_1, \ldots, s_m)}\right) && (\text{def. } [\![\cdot]\!]) \\ &= \alpha(\overline{s}). \end{aligned}$$

- For the induction step of t, suppose $\alpha \circ T\alpha(\overline{t_i}) = \alpha \circ \mu_X(\overline{t_i})$ holds for $t_1, \ldots, t_n \in T_\Sigma(TX)$ and let us prove it for $t = \mathsf{op}(t_1, \ldots, t_n)$.

$$\begin{aligned} \alpha \circ T\alpha(\overline{t}) &= \alpha \circ T\alpha\left(\overline{\mathsf{op}(t_1, \ldots, t_n)}\right) \\ &= \alpha \circ T\alpha \circ [\![\mathsf{op}]\!]^{TX}(\overline{t_1}, \ldots, \overline{t_n}) && (\text{def. } [\![\cdot]\!]^{TX}) \\ &= \alpha \circ [\![\mathsf{op}]\!]^X \circ (T\alpha)^n(\overline{t_1}, \ldots, \overline{t_n}) && (\text{Lemma } 10) \\ &= \langle\mathsf{op}\rangle \circ (\alpha \circ T\alpha)^n(\overline{t_1}, \ldots, \overline{t_n}) && (\text{by } (23)) \\ &= \langle\mathsf{op}\rangle \circ (\alpha \circ \mu_X)^n(\overline{t_1}, \ldots, \overline{t_n}) && (\text{by I.H.}) \\ &= \alpha \circ [\![\mathsf{op}]\!]^X \circ \mu_X^n(\overline{t_1}, \ldots, \overline{t_n}) && (\text{by } (23)) \\ &= \alpha \circ \mu_X \circ [\![\mathsf{op}]\!]^{TX}(\overline{t_1}, \ldots, \overline{t_n}) && (\text{Lemma } 10) \\ &= \alpha \circ \mu_X\left(\overline{\mathsf{op}(t_1, \ldots, t_m)}\right) && (\text{def. } [\![\cdot]\!]^{TX}) \\ &= \alpha \circ \mu_X(\overline{t}) \end{aligned}$$

This concludes the proof that α is associative. □

Lemma 28. *H maps $(\Sigma^{\mathrm{s}}, E^{\mathrm{s}})$-algebra homomorphisms to T-semialgebra homomorphisms.*

Proof. Suppose $f : (X, \langle\cdot\rangle^X) \to (Y, \langle\cdot\rangle^Y)$ is a $(\Sigma^{\mathrm{s}}, E^{\mathrm{s}})$-algebra homomorphism. Let $(X, \alpha) := H(X, \langle\cdot\rangle^X)$ and $(Y, \beta) := H(Y, \langle\cdot\rangle^Y)$. We want to check that $H(f)$, which is equal to f, is a T-semialgebra homomorphism, i.e. the following commute:

$$\begin{array}{ccc} TX & \xrightarrow{Tf} & TY \\ {\scriptstyle\alpha}\downarrow & & \downarrow{\scriptstyle\beta} \\ X & \xrightarrow[f]{} & Y \end{array}$$

We prove $f \circ \alpha(\overline{t}) = \beta \circ Tf(\overline{t})$ for all $\overline{t} \in TX = {}^{\mathcal{T}_\Sigma(X)}/_E$ by an induction on t:

- Suppose t is some $x \in X$:

$$\begin{aligned} f \circ \alpha(\overline{x}) &= f \circ \alpha \circ \eta_X(x) && (\text{def. } \eta_X) \\ &= f \circ \langle \mathsf{a} \rangle^X(x) && (\text{by (22)}) \\ &= \langle \mathsf{a} \rangle^Y \circ f(x) && (f \text{ homom.}) \\ &= \beta \circ \eta_Y \circ f(x) && (\text{by (22)}) \\ &= \beta \circ Mf \circ \eta_X(x) && (\eta \text{ nat.}) \\ &= \beta \circ Mf(\overline{x}). && (\text{def. } \eta_X) \end{aligned}$$

- Suppose that it holds for $t_1, \ldots, t_n$ and let us prove it for $t = \mathsf{op}(t_1, \ldots, t_n)$:

$$\begin{aligned} f \circ \alpha(\overline{\mathsf{op}(t_1, \ldots, t_n)}) &= f \circ \alpha \circ [\![\mathsf{op}]\!]^X(\overline{t_1}, \ldots, \overline{t_n}) && (\text{def. } [\![\cdot]\!]^X) \\ &= f \circ \langle \mathsf{op} \rangle^X(\alpha(\overline{t_1}), \ldots, \alpha(\overline{t_n})) && (\text{by (23)}) \\ &= \langle \mathsf{op} \rangle^Y (f \circ \alpha)^n(\overline{t_1}, \ldots, \overline{t_n}) && (f \text{ homom.}) \\ &= \langle \mathsf{op} \rangle^Y (\beta \circ Tf)^n(\overline{t_1}, \ldots, \overline{t_n}) && (\text{by I.H.}) \\ &= \beta \circ [\![\mathsf{op}]\!]^Y \circ (Tf)^n(\overline{t_1}, \ldots, \overline{t_n}) && (\text{by (23)}) \\ &= \beta \circ Tf \circ [\![\mathsf{op}]\!]^X(\overline{t_1}, \ldots, \overline{t_n}) && (\text{Lemma 10}) \\ &= \beta \circ Tf(\overline{\mathsf{op}(t_1, \ldots, t_m)}). && (\text{def. } [\![\cdot]\!]^X) \end{aligned}$$

□

3.3 Joining both Constructions

We have shown in the two previous sections that we have functors G and H as shown here:

$$G : \mathbf{EM}_s(T) \rightleftarrows \mathbf{Alg}(\Sigma^{\mathrm{s}}, E^{\mathrm{s}}) : H.$$

It remains to show that they are inverses.

Lemma 29. *The functors G and H are inverses.*

Proof. – Suppose we start with a T-semialgebra $\alpha : TX \to X$. Let $(X, \llparenthesis\cdot\rrparenthesis) := G(X, \alpha)$ and $(X, \alpha') := H(X, \llparenthesis\cdot\rrparenthesis)$. We prove $\alpha'(\overline{t}) = \alpha(\overline{t})$ for all $\overline{t} \in TX = \mathcal{T}_\Sigma(X)/E$ by induction on t:

- Suppose t is some $x \in X$:

$$\begin{aligned} \alpha'(\overline{x}) &= \alpha' \circ \eta_X(x) && (\text{def.}\, \eta_X) \\ &= \llparenthesis \mathsf{a} \rrparenthesis(x) && (\text{by (22)}) \\ &= \alpha \circ \eta_X(x) && (\text{def.of}\, \llparenthesis \mathsf{a} \rrparenthesis\, \text{in}(18)) \\ &= \alpha \circ (\overline{x}). && (\text{def.}\, \eta_X) \end{aligned}$$

- Suppose it holds for $t_1, \ldots, t_n$ and let us prove it for $t = \mathsf{op}(t_1, \ldots, t_n)$:

$$\begin{aligned} \alpha'(\overline{t}) &= \alpha'(\overline{\mathsf{op}(t_1, \ldots, t_n)}) \\ &= \alpha' \circ [\![\mathsf{op}]\!]^X(\overline{t_1}, \ldots, \overline{t_m}) && (\text{def.}\, [\![\cdot]\!]^X) \\ &= \llparenthesis \mathsf{op} \rrparenthesis(\alpha'(\overline{t_1}), \ldots, \alpha'(\overline{t_n})) && (\text{by (23)}) \\ &= \llparenthesis \mathsf{op} \rrparenthesis(\alpha(\overline{t_1}), \ldots, \alpha(\overline{t_n})) && (\text{by I.H.}) \\ &= \alpha \circ [\![\mathsf{op}]\!]^X \circ \eta_X^n \circ \alpha^n(\overline{t_1}, \ldots, \overline{t_n}) && (\text{def.}\, \llparenthesis \mathsf{op} \rrparenthesis\, \text{in}(19)) \\ &= \alpha \circ [\![\mathsf{op}]\!]^X(\overline{t_1}, \ldots, \overline{t_n}) && (\text{by (17)}) \\ &= \alpha(\overline{\mathsf{op}(t_1, \ldots, t_n)}) && (\text{def.}\, [\![\cdot]\!]^X) \\ &= \alpha(\overline{t}) \end{aligned}$$

– Suppose we start with an $(\Sigma^{\mathrm{s}}, E^{\mathrm{s}})$-algebra $(X, \llparenthesis\cdot\rrparenthesis)$. Let $(X, \alpha) := H(X, \llparenthesis\cdot\rrparenthesis)$ and $(X, \llparenthesis\cdot\rrparenthesis') := G(X, \alpha)$. We want to prove that $\llparenthesis\cdot\rrparenthesis' = \llparenthesis\cdot\rrparenthesis$. Let us check first for $(\mathsf{a} : 1) \in \Sigma^{\mathrm{s}}$:

$$\begin{aligned} \llparenthesis \mathsf{a} \rrparenthesis' &= \alpha \circ \eta_X && (\text{def.}\, \llparenthesis\cdot\rrparenthesis'\, \text{in}(18)) \\ &= \llparenthesis \mathsf{a} \rrparenthesis, && (\text{by (22)}) \end{aligned}$$

and then for $(\mathsf{op} : n) \in \Sigma$:

$$\begin{aligned} \llparenthesis \mathsf{op} \rrparenthesis' &= \alpha \circ [\![\mathsf{op}]\!]^X \circ (\eta_X)^n && (\text{def.}\, \llparenthesis\cdot\rrparenthesis'\, \text{in}(19)) \\ &= \llparenthesis \mathsf{op} \rrparenthesis \circ (\alpha \circ \eta_X)^n && (\text{by (23)}) \\ &= \llparenthesis \mathsf{op} \rrparenthesis \circ \llparenthesis \mathsf{a} \rrparenthesis^n && (\text{by (22)}) \\ &= \llparenthesis \mathsf{op} \rrparenthesis. && (\text{equation (13) in}\, E^{\mathrm{s}}) \end{aligned}$$

□

The proof of Theorem 15 is now complete and contained in Lemmas 22, 23, 25, 27, 28, and 29.

4 Examples

We now give multiple examples to illustrate the applicability of Corollary 16. First, notice that some equations in E^{s} can be simplified to equations in E. More precisely, equations in E^{s} that arise via (14) from an equation $t = s$ in E where t and s are terms of depth at least 1, reduces to $t = s$ due to Lemma 17. Similarly, if t is variable v and s is not a variable, then the equation in E^{s} arising from (14) becomes $\mathsf{a}v = s$; the case when s is a variable and t is not is analogous.

The presentations of the semifree monads on the maybe monad $(-) + 1$, the semigroup monad $(-)^+$ and the distribution monad $\mathcal{D}$ were proven each individually by hand in [32]. Those three examples gave a strong intuition that a uniform presentation was possible, which lead the authors of [32] to conjecture Corollary 16 that we proved in this article.

Example 30. The *exception* monad $X \mapsto X + K$, where K is a fixed set (meant to contain a list of possible exception states) is presented by the theory of K-pointed sets $\Sigma = \{c_k : 0 \mid k \in K\}$, $E = \varnothing$. Its semifree monad has functor $X \mapsto X + (X + K)$, and presentation

$$\Sigma^{\mathsf{s}} = \{c_k : 0 \mid k \in K\} \cup \{\mathsf{a} : 1\},$$
$$E^{\mathsf{s}} = \{\mathsf{a}\mathsf{a}v = \mathsf{a}v\} \cup \{\mathsf{a}c_k = c_k \mid k \in K\}.$$

Its algebras $(X, [\![\cdot]\!])$ are sets X with a retract $Y = \mathrm{im}([\![\mathsf{a}]\!]) \subseteq X$, the retraction being $[\![\mathsf{a}]\!] : X \to X$, and with a set of distinguished elements $\{y_k \mid k \in K\} \subseteq Y$ in the retract.

Example 31. The *list* monad $X \mapsto X^* = \bigsqcup_{n \geqslant 0} X^n$ is presented by the theory of monoids $\Sigma_* = \{e : 0, \cdot : 2\}$, $E_* = \{(u \cdot v) \cdot w = u \cdot (v \cdot w),\ \ e \cdot v = v,\ \ v \cdot e = v\}$, see e.g. [4, Example 10.7]. Its semifree monad has functor $X \mapsto X + X^*$. Notice that its presentation can be simplified. By (14) and Lemma 17, we obtain the equations $e \cdot v = \mathsf{a}v$ and $v \cdot e = \mathsf{a}v$. These equations show that the new symbol $(\mathsf{a} : 1)$ is not needed since it can be expressed using the operations $(e : 0)$ and $(\cdot : 2)$. However, we then need to add the equation $e \cdot v = v \cdot e$. If we continue this simplification, we end up with

$$E_*^{\mathsf{s}} = \left\{ \begin{array}{cc} e \cdot v = v \cdot e, & e \cdot e = e, \\ e \cdot (u \cdot v) = u \cdot v, & (u \cdot v) \cdot w = u \cdot (v \cdot w) \end{array} \right\}.$$

This corresponds to the theory of semigroups $(S, \cdot)$ that admit a retract $R = S \cdot S$ which is a monoid $(R, \cdot, e)$. The retraction is $e \cdot (-) = (-) \cdot e : S \to S$.

Example 32. The *multiset* monad $\mathcal{M}(X) = \{\phi : X \to \mathbb{N} \mid \mathsf{supp}(\phi) \text{ finite}\}$, where $\mathsf{supp}(\phi)$ means the support of ϕ, is presented by the theory of commutative monoids $\Sigma_{\mathcal{M}} = \{e : 0, \cdot : 2\}$, $E_{\mathcal{M}} = E_* \uplus \{u \cdot v = v \cdot u\}$ [18, Section 2]. Its semifree monad has functor $\mathcal{M}^{\mathsf{s}}(X) = X + \mathcal{M}(X)$. Its presentation can be built direct on the simplified presentation of the semifree monoid monad of Example 31. Furthermore, $u \cdot v = v \cdot u$ renders the equation $e \cdot v = v \cdot e$ redundant. Therefore, this presentation ends up being $\Sigma^{\mathsf{s}}_{\mathcal{M}} = \{e : 0, \cdot : 2\}$ and

$$E^{\mathsf{s}}_{*} = \left\{ \begin{matrix} u \cdot v = v \cdot u, & e \cdot e = e, \\ e \cdot (u \cdot v) = u \cdot v, & (u \cdot v) \cdot w = u \cdot (v \cdot w) \end{matrix} \right\}.$$

This corresponds to the theory of commutative semigroups $(S, \cdot)$ that admits a retract $R = S \cdot S$, which is a commutative monoid $(R, \cdot, e)$. The retraction is $e \cdot (-) : S \to S$.

Example 33. The *finite powerset* monad $\mathcal{P}(X) = \{Y \subseteq X \mid Y \text{ finite}\}$ is presented by the theory of join-semilattices with bottom $\Sigma_{\mathcal{P}} = \{e : 0, \cdot : 2\}$, $E_{\mathcal{P}} = E_{\mathcal{M}} \uplus \{v \cdot v = v\}$ [17, p. 81]. Its semifree monad has functor $\mathcal{P}^{\mathsf{s}}(x) = X + \mathcal{P}(X)$. The equation $v \cdot v = v$ becomes $v \cdot v = \mathsf{a}v$, and as in Example 31, the symbol a can be replaced by $e \cdot -$. We end up with

$$\Sigma^{\mathsf{s}}_{\mathcal{P}} = \{e : 0, \cdot : 2\}, \qquad E^{\mathsf{s}}_{\mathcal{P}} = E^{\mathsf{s}}_{\mathcal{M}} \uplus \{v \cdot v = e \cdot v\}.$$

Example 34. The *state* monad $\mathsf{State}(X) = (S \times X)^S$, where S is a fixed finite set (generally of states), is presented by the theory of global states (see [27,33]) $\Sigma_{\mathsf{State}} = \{f : n\} \cup \{g_i : 1 \mid 1 \leqslant i \leqslant n\}$, where $n := |S|$, and

$$E_{\mathsf{State}} = \left\{ g_i g_j v = g_j v, \quad g_i f(v_1, \ldots, v_n) = g_i v_i, \quad f(g_1 v, \ldots, g_n v) = v \right\}.$$

The semifree monad on State has functor $\mathsf{State}^{\mathsf{s}}(x) = X + (S \times X)^S$. Its presentation can be simplified. The equation $f(g_1 v, \ldots, g_n v) = \mathsf{a}v$ implies that a can be expressed as $f(g_1(-), \ldots, g_n(-))$. Then, some equations turn out to be redundant, like $g_i \mathsf{a} v = g_i v$ or $\mathsf{aa}v = \mathsf{a}v$. After simplifications, we end up with $\Sigma^{\mathsf{s}}_{\mathsf{State}} = \{f : n\} \cup \{g_i : 1 \mid 0 \leqslant i \leqslant n\}$, and

$$E^{\mathsf{s}}_{\mathsf{State}} = \left\{ \begin{matrix} f(g_1 v_1, \ldots, g_n v_n) = f(v_1, \ldots, v_n), & g_i g_j v = g_j v, \\ f(g_i v, \ldots, g_i v) = g_i v, & g_i f(v_1, \ldots, v_n) = g_i v_i \end{matrix} \right\}.$$

Example 35. Consider the repeated semifree construction on the identity monad id on Set.

- id is presented by $\Sigma = E = \varnothing$. Note that id-algebras are identity morphisms because of the unit axiom (1), and (Σ, E)-algebras are just sets with no operations. The presentation sends an id-algebra (X, id_X) to the (Σ, E)-algebra $(X, \varnothing)$.
- $\mathsf{id}^{\mathsf{s}}(X) = X + X$ is presented by $\Sigma^{\mathsf{s}} = \{\mathsf{a} : 1\}$ and $E^{\mathsf{s}} = \{\mathsf{aa}v = \mathsf{a}v\}$.
- $\mathsf{id}^{\mathsf{s}^2}(X) = X + (X + X)$ is presented by $\Sigma^{\mathsf{s}^2} = \{\mathsf{a}, \mathsf{b} : 1\}$ and

$$E^{\mathsf{s}^2} = \{\mathsf{bb}v = \mathsf{b}v, \quad \mathsf{ba}v = \mathsf{a}v, \quad \mathsf{ab}v = \mathsf{a}v, \quad \mathsf{aa}v = \mathsf{a}v\}.$$

The equation directly given by (14) is $\mathsf{aab}v = \mathsf{ab}v$, which simplifies to $\mathsf{aa}v = \mathsf{a}v$ using Lemma 17. We can summarize E^{s^2} as saying that a and b are idempotent, and that a absorbs b on the left and on the right.

- Repeating the procedure n times, we inductively obtain a monad on $n+1$ disjoint copies of X, $\mathrm{id}^{\mathrm{s}^n}(X) = X+\ldots+X \cong n_\infty \times X$, where $n_\infty := n \uplus \{\infty\} = \{0, \ldots, n-1, \infty\}$. The set n_∞ is a linear order, as a subset of the extended natural numbers $\mathbb{N} \uplus \{\infty\}$. It is also a monoid, with unit ∞ and operation min. Hence, we have a writer monad. Its presentation is given by n unary idempotents $\Sigma^{\mathrm{s}^n} = \{\mathsf{a}_0, \ldots, \mathsf{a}_{n-1} : 1\}$. An idempotent a_i absorbs another one a_j on both sides whenever $i < j$:

$$E^{\mathrm{s}^n} = \{\mathsf{a}_i \mathsf{a}_j v = \mathsf{a}_{\mathsf{min}(i,j)} v \mid 0 \leqslant i, j \leqslant n-1\}.$$

An $\mathrm{id}^{\mathrm{s}^n}$-algebra with carrier set X corresponds then to having a left-action of n_∞ on X. This is not a surprise, as algebras of writer monads are always actions. Being a linear order, n_∞ is also a meet-semilattice. Such an $\mathrm{id}^{\mathrm{s}^n}$-algebra can thus be viewed as a semi-lattice automaton, similar to the lattice automata of [23]. The carrier X is the set of states, the a_i are the letters of the alphabet, and a transition on a_i is simply associated with the value of a_i in the semilattice n_∞.

5 Relation Between Semifree Monads and Other Monad Constructions

Ideal monads were introduced by Aczel et al. [1] to study solutions to guarded recursive equations using coalgebraic methods. The authors investigated completely iterative monads, following earlier work from Elgot et al. [10] on iterative algebraic theories. Ideal monads are an abstraction of the core properties of completely iterative monads, and, in particular, completely iterative monads are ideal. Moreover, Ghani & Uustalu [12] showed that ideal monads have the right mathematical structure to avoid the problems encountered in the construction of monad coproducts, and they gave a simple construction of coproducts of ideal monads by investigating coalgebraic fixed points. This same construction was later proved to be applicable to the class of consistent Set-monads [3].

We now define ideal monads and show that semifree monads are ideal.

Definition 36. *In a category with finite coproducts, an* **ideal monad** *is a quintuple (T, η, μ, T_0, m_0) where (T, η, μ) is a monad with functor $T = \mathrm{id} + T_0$, unit $\eta = \mathsf{inl}^{\mathrm{id}+T} : \mathrm{id} \to \mathrm{id} + T_0$, and where $\mu : TT \to T$ "restricts" to the natural transformation $m_0 : T_0 T \to T_0$, meaning that the following square commutes.*

$$\begin{array}{ccc} T_0 T & \xrightarrow{\mathsf{inr}^{\mathrm{id}+T} \circ T} & TT \\ {\scriptstyle m_0}\downarrow & & \downarrow{\scriptstyle \mu} \\ T_0 & \xrightarrow[\mathsf{inr}^{\mathrm{id}+T}]{} & T \end{array} \tag{24}$$

Notice that having (24) commute is equivalent to having

$$\mu = [\mathrm{id}_{\mathrm{id}+T_0}, \mathsf{inr}^{\mathrm{id}+T_0} \circ m_0]. \tag{25}$$

Examples of ideal monads include (free) completely iterative monads [1, Example 4.4], algebraically free monads, exception monads, and interactive output monads [12].

Example 37. Semifree monads are ideal. Take a semifree monad (M^s, η^s, μ^s). It is enough to show that its multiplication μ^s satisfies (25). For that, let

$$m_0 := (MM^s \xrightarrow{M[\eta, \mathrm{id}_M]} MM \xrightarrow{\mu} M)$$

and notice that the equation on the right-hand side of (25) becomes the exact definition of μ^s in Definition 11. Hence, $(M^s, \eta^s, \mu^s, M, m_0)$ is ideal.

Semifree monads thus enjoy all the nice properties of ideal monads such as having simple coproducts [12].

The concrete isomorphism between M^s-algebras and M-semialgebras can be phrased as, any semialgebra structure for M on an object X can be replaced by an Eilenberg-Moore algebra structure for M^s. We will show that this property has an analogue for ideal monads. However, since T_0 is only assumed to be an endofunctor, there is no notion of T_0-semialgebra, but we can use the ideal monad structure to define an analogue of the associativity diagram of Eilenberg-Moore algebras (6). This leads us to consider functor T_0-algebras (i.e., morphisms of the type $a\colon T_0X \to X$), such that a square, analogous to the associativity square of (6), commutes (diagram (26) below). We denote the category of functor T_0-algebras and T_0-algebra morphisms by $\mathbf{Alg}(T_0)$.

Lemma 38. *Let (T, η, μ, T_0, m_0) be an ideal monad. There is an isomorphism between $\mathbf{EM}(T)$ and the full subcategory of $\mathbf{Alg}(T_0)$ of all $a : T_0X \to X$ that makes (26) commute.*

$$\begin{array}{ccc} T_0TX & \xrightarrow{T_0[\mathrm{id}_X, a]} & T_0X \\ {\scriptstyle m_{0,X}}\downarrow & & \downarrow{\scriptstyle a} \\ T_0X & \xrightarrow[a]{} & X \end{array} \tag{26}$$

As observed in Lemma 12, the semifree construction is an endofunctor on the category of monads $\mathbf{Mon}(\mathsf{C})$. We therefore ask whether $(-)^s$ is a *monad transformer*, i.e., a pointed endofunctor on $\mathbf{Mon}(\mathsf{C})$ [24]. Here, *pointed* means to admit a natural transformation $\mathrm{id}_{\mathbf{Mon}(\mathsf{C})} \Rightarrow (-)^s$. It turns out that this functor is not pointed, but it is *co-pointed*, meaning that it admits a natural transformation $\epsilon\colon (-)^s \Rightarrow \mathrm{id}_{\mathbf{Mon}(\mathsf{C})}$. We have, in fact, a comonad $((-)^s, \epsilon, \delta)$ on $\mathbf{Mon}(\mathsf{C})$.

We collect the above observations in the following lemmas.

Lemma 39. *Given a category C with finite coproducts, the semifree endofunctor $(-)^s : \mathbf{Mon}(\mathsf{C}) \to \mathbf{Mon}(\mathsf{C})$ is not pointed.*

Lemma 40. *Given a category* C *with finite coproducts, the triple* $((-)^{\mathrm{s}}, \epsilon, \delta)$ *is a comonad on* $\mathbf{Mon}(\mathsf{C})$ *by defining for a monad* (M, η, μ) *and an object* X*:*

$$\epsilon_{M,X} : X + MX \xrightarrow{[\eta_X, \mathrm{id}_{MX}]} MX, \text{ and}$$

$$\delta_{M,X} : X + MX \xrightarrow{\mathrm{id}_X + \mathrm{inr}^{X+MX}} X + (X + MX).$$

As consequence of the above, the semifree construction is not a monad transformer. In general, a monad transformer T_L is defined on a "base" monad L such that $L = T_L(\mathrm{id}_{\mathsf{C}})$. In our case, we have $(\mathrm{id}_{\mathsf{C}})^{\mathrm{s}} = \mathrm{id}_{\mathsf{C}} + \mathrm{id}_{\mathsf{C}}$. This is an interesting monad in itself, even though it was not the starting point of the semifree construction.

6 Conclusion

In this paper, we proved a uniform algebraic presentation of semifree monads $M^{\mathrm{s}}(X) = X + M(X)$ when an algebraic presentation of the monad M is known. We also showed that semifree monads are instances of ideal monads, and that the semifree construction is not a monad transformer, but it is a comonad on the category of monads.

There are several directions for future work. Given that the functor part of the semifree monad M^{s} is a functor coproduct $\mathrm{id} + M$, it would be interesting to understand better the relationship to coproducts of monads, and whether Corollary 16 could be generalised to give presentations of (certain) monad coproducts [12]. Similarly, we would like to investigate if the observations we made in Example 35 on algebras for the iterated semifree monad on identity can be generalised to other monads.

The presentation of semifree monads provides a means to study no-go theorems for weak distributive laws, using the correspondence between weak distributive laws for M and certain distributive laws for M^{s} [32], and a similar approach as in [40]. In [40], no-go theorems for distributive laws are proved using criteria on presentations of the monads. However, their no-go theorems are not directly applicable for the semifree monad, as they require equations where one side is a single variable, e.g., unitality $e * v = v$ or idempotence $v * v = v$. Such equations never occur in the presentations of M^{s}, as the simplest terms that those equations contain are of the form $\mathsf{a}v$. Hence new problematic equations in presentations of monads must be identified. Furthermore, weak distributive laws for M correspond to distributive laws for M^{s} satisfying an extra condition, and hence to a subclass of the composite theories for M^{s}. This subclass may satisfy more equations, which could be helpful in establishing no-go theorems for weak distributive laws for M.

A more fundamental question related to the definition of algebraic presentation is whether, over the category Set, an isomorphism between Eilenberg-Moore categories of algebras implies an isomorphism of Set-monads. This fails over general categories, but an argument may exist for the specific case of Set.

Finally, since Set-monads sometimes have interesting liftings to other categories, one could consider the following question. Suppose we have a uniform presentation for a construction on a certain Set-monad which can be lifted to a category C, does this presentation also lift to C? This was investigated in [29] for the non-empty convex distribution monad and its Hausdorff-Kantorovich lifting to metric spaces.

Acknowledgment. We thank Ralph Sarkis, and Roy Overbeek for useful discussion, suggestions and corrections. We also thank all anonymous reviewers for their valuable feedback and suggestions. Aloïs Rosset and Jörg Endrullis received funding from the Netherlands Organization for Scientific Research (NWO) under the Innovational Research Incentives Scheme Vidi (project. No. VI.Vidi.192.004).

References

1. Aczel, P., Adámek, J., Milius, S., Velebil, J.: Infinite trees and completely iterative theories: a coalgebraic view. Theoret. Comput. Sci. **300**(1–3), 1–45 (2003). https://doi.org/10.1016/S0304-3975(02)00728-4
2. Adamek, J., Rosicky, J.: Locally Presentable and Accessible Categories. London Mathematical Society Lecture Note Series. Cambridge University Press, Cambridge (1994). https://doi.org/10.1017/CBO9780511600579
3. Adámek, J., Milius, S., Bowler, N.J., Levy, P.B.: Coproducts of monads on set. In: Proceedings of the 27th Annual IEEE Symposium on Logic in Computer Science, LICS 2012, Dubrovnik, Croatia, 25–28 June 2012, pp. 45–54. IEEE Computer Society (2012). https://doi.org/10.1109/LICS.2012.16
4. Awodey, S.: Category Theory. Oxford Logic Guides. Ebsco Publishing (2006)
5. Barr, M., Wells, C.: Toposes, Triples and Theories. Comprehensive Studies in Mathematics. Springer, New York (1985)
6. Beck, J.: Distributive laws. In: Eckmann, B. (ed.) Seminar on Triples and Categorical Homology Theory. LNM, vol. 80, pp. 119–140. Springer, Heidelberg (1969). https://doi.org/10.1007/BFb0083084
7. Bonchi, F., Santamaria, A.: Combining semilattices and semimodules. In: FOSSACS 2021. LNCS, vol. 12650, pp. 102–123. Springer, Cham (2021). https://doi.org/10.1007/978-3-030-71995-1_6
8. Bonchi, F., Sokolova, A., Vignudelli, V.: The theory of traces for systems with nondeterminism and probability. In: 34th Annual ACM/IEEE Symposium on Logic in Computer Science, LICS 2019, pp. 1–14. IEEE (2019). https://doi.org/10.1109/LICS.2019.8785673
9. Böhm, G.: The weak theory of monads. Adv. Math. **225**, 1–32 (2010). https://doi.org/10.1016/j.aim.2010.02.015
10. Elgot, C.C., Bloom, S.L., Tindell, R.: On the algebraic structure of rooted trees. J. Comput. Syst. Sci. **16**, 361–399 (1978). https://doi.org/10.1007/978-1-4613-8177-8_7
11. Garner, R.: The vietoris monad and weak distributive laws. Appl. Categ. Struct. **28**(2), 339–354 (2019). https://doi.org/10.1007/s10485-019-09582-w
12. Ghani, N., Uustalu, T.: Coproducts of ideal monads. RAIRO Theor. Inform. Appl. **38**(4), 321–342 (2004). https://doi.org/10.1051/ita:2004016

13. Goy, A., Petrisan, D.: Combining probabilistic and non-deterministic choice via weak distributive laws. In: Hermanns, H., Zhang, L., Kobayashi, N., Miller, D. (eds.) LICS 2020: 35th Annual ACM/IEEE Symposium on Logic in Computer Science, pp. 454–464. ACM (2020). https://doi.org/10.1145/3373718.3394795
14. van Heerdt, G., Sammartino, M., Silva, A.: Learning automata with side-effects. In: Petrişan, D., Rot, J. (eds.) CMCS 2020. LNCS, vol. 12094, pp. 68–89. Springer, Cham (2020). https://doi.org/10.1007/978-3-030-57201-3_5
15. Hyland, M., Plotkin, G.D., Power, J.: Combining effects: sum and tensor. Theoret. Comput. Sci. **357**(1–3), 70–99 (2006). https://doi.org/10.1016/j.tcs.2006.03.013
16. Hyland, M., Tasson, C.: The linear-non-linear substitution 2-monad. In: Spivak, D.I., Vicary, J. (eds.) Proceedings of the 3rd Annual International Applied Category Theory Conference 2020, ACT 2020. EPTCS, vol. 333, pp. 215–229 (2020). https://doi.org/10.4204/EPTCS.333.15
17. Jacobs, B.: Semantics of weakening and contraction. Ann. Pure Appl. Logic **69**(1), 73–106 (1994). https://doi.org/10.1016/0168-0072(94)90020-5
18. Jacobs, B.: Convexity, duality and effects. In: Calude, C.S., Sassone, V. (eds.) TCS 2010. IAICT, vol. 323, pp. 1–19. Springer, Heidelberg (2010). https://doi.org/10.1007/978-3-642-15240-5_1
19. Jacobs, B., Silva, A., Sokolova, A.: Trace semantics via determinization. J. Comput. Syst. Sci. **81**(5), 859–879 (2015). https://doi.org/10.1016/j.jcss.2014.12.005
20. Jaskelioff, M.: Modular monad transformers. In: Castagna, G. (ed.) ESOP 2009. LNCS, vol. 5502, pp. 64–79. Springer, Heidelberg (2009). https://doi.org/10.1007/978-3-642-00590-9_6
21. Katsumata, S., Rivas, E., Uustalu, T.: Interaction laws of monads and comonads. In: Hermanns, H., Zhang, L., Kobayashi, N., Miller, D. (eds.) LICS 2020: 35th Annual ACM/IEEE Symposium on Logic in Computer Science, pp. 604–618. ACM (2020). https://doi.org/10.1145/3373718.3394808
22. Klin, B., Salamanca, J.: Iterated covariant powerset is not a monad. In: Staton, S. (ed.) Proceedings of the Thirty-Fourth Conference on the Mathematical Foundations of Programming Semantics, MFPS 2018. Electronic Notes in Theoretical Computer Science, vol. 341, pp. 261–276. Elsevier (2018). https://doi.org/10.1016/j.entcs.2018.11.013
23. Kupferman, O., Lustig, Y.: Lattice automata. In: Cook, B., Podelski, A. (eds.) VMCAI 2007. LNCS, vol. 4349, pp. 199–213. Springer, Heidelberg (2007). https://doi.org/10.1007/978-3-540-69738-1_14
24. Liang, S., Hudak, P., Jones, M.P.: Monad transformers and modular interpreters. In: Cytron, R.K., Lee, P. (eds.) Conference Record of POPL 1995: 22nd ACM SIGPLAN-SIGACT Symposium on Principles of Programming Languages, San Francisco, California, USA, 23–25 January 1995, pp. 333–343. ACM Press (1995). https://doi.org/10.1145/199448.199528
25. MacLane, S.: Categories for the Working Mathematician. Graduate Texts in Mathematics, vol. 5. Springer, New York (1971)
26. Manes, E.: Algebraic Theories. Graduate Texts in Mathematics, vol. 26. Springer, New York (1976). https://doi.org/10.1007/978-1-4612-9860-1
27. Métayer, F.: State monads and their algebras. arXiv:math/0407251, Category Theory (2004). https://doi.org/10.48550/arXiv.math/0407251
28. Milius, S., Pattinson, D., Schröder, L.: Generic trace semantics and graded monads. In: Moss, L.S., Sobocinski, P. (eds.) 6th Conference on Algebra and Coalgebra in Computer Science, CALCO 2015. LIPIcs, vol. 35, pp. 253–269. Schloss Dagstuhl - Leibniz-Zentrum für Informatik (2015). https://doi.org/10.4230/LIPIcs.CALCO.2015.253

29. Mio, M., Sarkis, R., Vignudelli, V.: Combining nondeterminism, probability, and termination: equational and metric reasoning. In: 36th Annual ACM/IEEE Symposium on Logic in Computer Science, LICS 2021, pp. 1–14. IEEE (2021). https://doi.org/10.1109/LICS52264.2021.9470717
30. Mio, M., Vignudelli, V.: Monads and quantitative equational theories for nondeterminism and probability. In: Konnov, I., Kovács, L. (eds.) 31st International Conference on Concurrency Theory, CONCUR 2020. LIPIcs, vol. 171, pp. 28:1–28:18. Schloss Dagstuhl - Leibniz-Zentrum für Informatik (2020). https://doi.org/10.4230/LIPIcs.CONCUR.2020.28
31. Moggi, E.: Notions of computation and monads. Inf. Comput. **93**(1), 55–92 (1991). https://doi.org/10.1016/0890-5401(91)90052-4. Selections from 1989 IEEE Symposium on Logic in Computer Science
32. Petrisan, D., Sarkis, R.: Semialgebras and weak distributive laws. In: Sokolova, A. (ed.) Proceedings 37th Conference on Mathematical Foundations of Programming Semantics, MFPS 2021. EPTCS, vol. 351, pp. 218–241 (2021). https://doi.org/10.4204/EPTCS.351.14
33. Plotkin, G.D., Power, J.: Notions of computation determine monads. In: Nielsen, M., Engberg, U. (eds.) Foundations of Software Science and Computation Structures, vol. 2303, pp. 342–356. Springer, Heidelberg (2002). https://doi.org/10.1007/3-540-45931-6_247
34. Power, J.: Discrete Lawvere theories. In: Fiadeiro, J.L., Harman, N., Roggenbach, M., Rutten, J. (eds.) CALCO 2005. LNCS, vol. 3629, pp. 348–363. Springer, Heidelberg (2005). https://doi.org/10.1007/11548133_22
35. Riehl, E.: Category Theory in Context. Dover Modern Math Originals, Dover Publications, Aurora (2017)
36. Rosset, A., Hansen, H.H., Endrullis, J.: Algebraic presentation of semifree monads. arXiv:cs.LO (2022). https://doi.org/10.48550/ARXIV.2205.05392
37. Street, R.: Weak distributive laws. Theory Appl. Categories **22**, 313–320 (2009)
38. Varacca, D., Winskel, G.: Distributing probabililty over nondeterminism. Math. Struct. Comput. Sci. **16**, 87–113 (2006). https://doi.org/10.1017/S0960129505005074
39. Wadler, P.: The essence of functional programming. In: Sethi, R. (ed.) Conference Record of the Nineteenth Annual ACM SIGPLAN-SIGACT Symposium on Principles of Programming Languages, pp. 1–14. ACM Press (1992). https://doi.org/10.1145/143165.143169
40. Zwart, M., Marsden, D.: No-go theorems for distributive laws. In: 34th Annual ACM/IEEE Symposium on Logic in Computer Science, LICS 2019, pp. 1–13. IEEE (2019). https://doi.org/10.1109/LICS.2019.8785707

Corecursion Up-to via Causal Transformations

Damien Pous[1], Jurriaan Rot[2(✉)], and Ruben Turkenburg[2]

[1] Plume, LIP, CNRS, ENS de Lyon, Université de Lyon, UMR 5668, Lyon, France
damien.pous@ens-lyon.fr

[2] Radboud University, Nijmegen, Netherlands
{jurriaan.rot,ruben.turkenburg2}@ru.nl

Abstract. Up-to techniques are a widely used family of enhancements of corecursion and coinduction. The soundness of these techniques can be shown systematically through the use of distributive laws. In this paper we propose instead to present up-to techniques as causal transformations, which are a certain type of natural transformations over the final sequence of a functor. These generalise the approach to proving soundness via distributive laws, and inherit their good compositionality properties. We show how causal transformations give rise to valid up-to techniques both for corecursive definitions and coinductive proofs.

Keywords: Coalgebra · Corecursion · Coinduction · Final sequence

1 Introduction

We assume familiarity with the most basic concepts of universal coalgebra [21] in this introduction; we formally define them in Sect. 2.

Let us recall the corecursion up-to principle from [31,32], which encompasses (and is implicit in) various results from the literature [8,11,20,24,26,39].

Let B be a functor with a final coalgebra (Z, ζ), and let F be a functor with an algebra $a\colon FZ \to Z$ on the final coalgebra. *Corecursion up to the algebra* a *is valid* if for every BF-coalgebra (X, f), there exists a unique morphism $f^a\colon X \to Z$ making the following diagram commute.

$$\begin{array}{ccc} X & \xrightarrow{\qquad f^a \qquad} & Z \\ {\scriptstyle f}\downarrow & & \downarrow{\scriptstyle \zeta} \\ BFX & \xrightarrow{BFf^a} BFZ \xrightarrow{Ba} & BZ \end{array} \tag{1}$$

When F is the identity functor (and a the identity morphism), this is just plain corecursion. Plain corecursion makes it possible, for instance, to define pointwise

This work was supported by the LABEX MILYON (ANR-10-LABX-0070) of Université de Lyon, within the program "Investissements d'Avenir" (ANR-11-IDEX-0007) operated by the French National Research Agency (ANR).

Published by Springer Nature Switzerland AG 2022
H. H. Hansen and F. Zanasi (Eds.): CMCS 2022, LNCS 13225, pp. 133–154, 2022.
https://doi.org/10.1007/978-3-031-10736-8_7

addition on streams. Indeed, streams (of real numbers, $\mathbb{R}^\omega$) form the final coalgebra for the functor $BX = \mathbb{R} \times X$ on sets. Let us write x_0 for the first element of a stream x, and x' for its tail. If f is the following B-coalgebra structure on $(\mathbb{R}^\omega)^2$,

$$\begin{aligned}(\mathbb{R}^\omega)^2 &\to B((\mathbb{R}^\omega)^2)\\ (x, y) &\mapsto (x_0 + y_0,\ (x', y'))\end{aligned}$$

then $f^{\mathsf{id}} \colon (\mathbb{R}^\omega)^2 \to \mathbb{R}^\omega$ is nothing but pointwise addition on streams: the only binary operation $\oplus$ satisfying the following equations.

$$\begin{aligned}(x \oplus y)_0 &= x_0 + y_0\\ (x \oplus y)' &= x' \oplus y'\end{aligned}$$

(Where x_0 and x' respectively denote the head and tail of a stream x.)

Corecursion up-to proves useful to define more complex operations like shuffle product ($\otimes$), satisfying the following equations:

$$\begin{aligned}(x \otimes y)_0 &= x_0 \times y_0\\ (x \otimes y)' &= (x \otimes y') \oplus (x' \otimes y)\end{aligned}$$

Indeed, in such a situation we need to call a function (pointwise addition) on objects which are not fully defined yet (the two corecursive calls $x \otimes y'$ and $x' \otimes y$). Using the functor $FX = X^2$ and seeing $\oplus$ as an F-algebra on $\mathbb{R}^\omega$, we can define shuffle product using corecursion up-to (1) and the following BF-coalgebra:

$$\begin{aligned}(\mathbb{R}^\omega)^2 &\to BF((\mathbb{R}^\omega)^2)\\ (x, y) &\mapsto (x_0 \times y_0,\ ((x, y'),\ (x', y)))\end{aligned}$$

Here, the inner pairs ((x, y') and (x', y)) correspond to the corecursive calls to $\otimes$, while the intermediate pair $((x, y'),\ (x', y))$ corresponds to a call to the F-algebra, i.e., in this case, pointwise addition.

Of course, not every algebra on a final coalgebra yields a valid corecursion up-to principle. Here are two sufficient conditions:

1. [6,7] a is induced by a distributive law $\lambda \colon FB \Rightarrow BF$ and the base category has countable coproducts (or F is a monad and λ a distributive law of this monad over the functor B);
2. [31] B is a polynomial set functor, and a is a *causal* F-algebra.

While the first condition is nice and well-known, it requires the machinery of distributive laws, and it is not always easy to show that a given algebra arises from a distributive law. The second condition does not suffer from this: causality is rather simple to check in practice (for instance, an algebra on streams is causal if and only if the n-th element of its output only depends on the n first elements of its inputs), but the condition that the starting functor is a polynomial set functor is often too strong (e.g., the finite powerset functor is not polynomial).

Shuffle product on streams is simple enough, so that the two approaches can both be used: the algebra we want to use, $\oplus$, arises from a simple distributive law, and it is obviously causal. In fact, the algebra induced by a distributive law on the final coalgebra of an ω-continuous[1] set functor is always causal.

Causal Transformations

Here we propose yet another condition based on the final sequence $\overline{B}$ of the functor B. Assuming that B is an endofunctor on a complete category $\mathcal{C}$, recall that this final sequence is a sequence of objects indexed by ordinals, which, if it stabilises, yields a final coalgebra for B (cf. [5, Theorem 1.3], or earlier for the dual case of algebras [2]). Here we shall present this sequence as a functor $\overline{B}\colon \mathsf{Ord}^{\mathsf{op}} \to \mathcal{C}$ from the category of ordinals to the base category. We call a natural transformation of type $F\overline{B} \Rightarrow \overline{B}$ a *causal transformation for B*.

Assuming that the final sequence stabilises at ordinal κ, so that for all causal transformations α, α_κ is an algebra on the final coalgebra, our new condition is the following:

3. a is the κ-th element α_κ of a causal transformation α.

Intuitively, looking at the final sequence as a sequence of approximations of the final coalgebra, an algebra satisfying the above condition must be defined not only on the final coalgebra, but also on all its approximations.

Example 1.1. Let $\mathcal{P}_f$ be the finite powerset functor, whose final coalgebra $(\mathcal{T}_f, c)$ consists of all finitely branching trees quotiented by bisimilarity, and the function c mapping a tree to its finite set of children. Consider the "delay" function $d\colon \mathcal{T}_f \to \mathcal{T}_f$ that adds a unary node at the root of the given tree (formally, $d(t) = c^{-1}(\{t\})$), and suppose we want to define the function e that corecursively delays all inner nodes of a given tree. For all trees t, this function should satisfy

$$c(e(t)) = \mathcal{P}_f(d \circ e)(c(t)) .$$

Graphically, on two examples, we have:

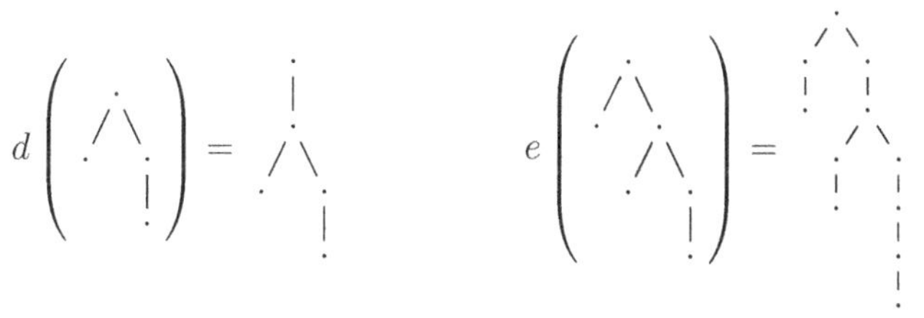

[1] I.e., preserving limits of ω^{op}-chains.

We can obtain e by corecursion up-to, using the identity functor for F, d for the algebra a, and c as $\mathcal{P}_f\mathsf{Id}$-coalgebra: in this case, diagram (1) precisely yields the above equation for the solution $e = c^d$.

In this example, condition 2/ is not satisfied: $\mathcal{P}_f$ is not polynomial. Moreover, it is not obvious how to construct a distributive law yielding the algebra d, in order to fulfil condition 1/. In contrast, one can easily extend the algebra d into a causal transformation $\delta\colon \overline{\mathcal{P}_f} \Rightarrow \overline{\mathcal{P}_f}$. To this end, recall that the final sequence of $\mathcal{P}_f$ stabilises at $\omega+\omega$ [41], and that it consists of all finite trees of depth at most n at all finite ordinals n, and of all *compactly branching* trees which are finitely branching up-to depth n at all ordinals $\omega+n$ for n finite. Intuitively, we can thus define a counterpart to the function d at all stages of this sequence: take a tree, add a unary node at the root, and, for the finite stages, truncate the resulting tree at the given depth. This is even easier formally: just set $\delta_k(t) = \overline{\mathcal{P}_f}(k+1,k)(\{t\})$, where $\overline{\mathcal{P}_f}(k+1,k)$ is the morphism from $\overline{\mathcal{P}_f}(k+1)$ to $\overline{\mathcal{P}_f}(k)$ in the final sequence. $\square$

For polynomial set functors, the three conditions turn out to be equivalent (a consequence of [32, Theorem 8.6 and Corollary 9.6]) In the general case, condition 3/ is implied by condition 1/: every distributive law yields a causal transformation whose κ-th element is its induced algebra on the final coalgebra [32, Lemma 6.2]. The converse is not true, cf. end of Sect. 5, but it is if the functor B has a *companion* [32]. Condition 3/ also generalises condition 2/: on sets, when B is ω-continuous, there is a one-to-one correspondence between causal algebras and causal transformations [32, Theorem 8.6].

That correspondence is non-trivial to establish, like the fact that condition 2/ provides corecursion up-to (1). For the latter, the approach in [32] goes via the construction of a distributive law starting from a causal algebra.

We use a much simpler path here, and we prove directly that condition 3/ implies validity of corecursion up-to (1), without mentioning any distributive law. And we actually get more: we obtain a corecursion up-to principle even in those cases where the final sequence does not stabilise.

Recall the notion of *corecursive algebra* [12] (dual to recursive coalgebras [14]): a B-algebra (A, a) is said to be *corecursive* if for all B-coalgebras (X, f), there is a unique morphism $f'\colon X \to A$ such that the following diagram commutes

$$\begin{array}{ccc} X & \xrightarrow{f'} & A \\ {\scriptstyle f}\downarrow & & \uparrow{\scriptstyle a} \\ BX & \xrightarrow{Bf'} & BA \end{array}$$

Paul Levy observed that all elements of the final sequence (which are B-algebras by definition), are corecursive [25, p. 5, footnote 2].

When we have a causal transformation $\alpha\colon F\overline{B} \Rightarrow \overline{B}$, we prove here that for each stage k of the final sequence and for every BF-coalgebra (X, f), there is a unique morphism $f_k^\alpha\colon X \to \overline{B}(k)$ such that

$$\begin{array}{ccccc} X & & \xrightarrow{\quad f_k^\alpha \quad} & & \overline{B}(k) \\ {\scriptstyle f}\downarrow & & & & \uparrow{\scriptstyle \overline{B}(k+1,k)} \\ BFX & \xrightarrow{BFf_k^\alpha} & BF\overline{B}(k) & \xrightarrow{B\alpha_k} & B\overline{B}(k) \end{array}$$

In other words, the BF-algebra $\overline{B}(k, k+1) \circ B\alpha_k$ is corecursive. This generalises Levy's observation when F is the identity functor, and we recover that condition 3/ guarantees validity of corecursion up-to α_κ when the final sequence stabilises at κ (so that $\overline{B}(\kappa, \kappa+1)$ is an isomorphism, with inverse the final coalgebra).

This result shows that we can use up-to techniques to define operations on all approximations of the final coalgebra, even when this coalgebra does not exist!

Example 1.2. Consider the full powerset functor ($\mathcal{P}$), which does not admit a final coalgebra. At ordinal ω, its final sequence yields *compactly branching trees* $\mathcal{T}_\omega \triangleq \overline{\mathcal{P}}(\omega)$ [41]. Therefore, in order to define an operation on compactly branching trees, we can use $\mathcal{P}F$-coalgebras for any causal transformation $F\overline{\mathcal{P}} \Rightarrow \overline{\mathcal{P}}$. For instance, we can define a similar operation as in Example 1.1. Call $c\colon \mathcal{T}_\omega \to \mathcal{P}\mathcal{T}_\omega$ the function mapping a (compactly branching) tree to its set of children, and define $\delta\colon \overline{\mathcal{P}} \Rightarrow \overline{\mathcal{P}}$ as before: $\delta_k(t) \triangleq \overline{\mathcal{P}}(k+1,k)(\{t\})$, so that $d \triangleq \delta_\omega\colon \mathcal{T}_\omega \to \mathcal{T}_\omega$ is a "delay" function on compactly branching trees. At ordinal ω, we get a unique function e such that

$$\begin{array}{ccccc} \mathcal{T}_\omega & & \xrightarrow{\quad e \quad} & & \mathcal{T}_\omega \\ {\scriptstyle c}\downarrow & & & & \uparrow{\scriptstyle \overline{\mathcal{P}}(\omega+1,\omega)} \\ \mathcal{P}\mathcal{T}_\omega & \xrightarrow{\mathcal{P}e} & \mathcal{P}\mathcal{T}_\omega & \xrightarrow{\mathcal{P}d} & \mathcal{P}\mathcal{T}_\omega \end{array}$$

(Note however that here, unlike in Example 1.1, c is not an inverse of the morphism on the right: we only have $\overline{\mathcal{P}}(\omega+1,\omega) \circ c = \mathsf{id}$. Therefore, this diagram is weaker than the one with a c going down on the right. □

We prove the aforementioned results in Sect. 3. Then we discuss compositionality of causal transformations (Sect. 4), and their lifting to coinductive predicates (Sect. 5).

Compositionality

A delicate point about up-to techniques for corecursion is compositionality. Indeed, given two algebras $a\colon FZ \to Z$ and $b\colon GZ \to Z$ on a final coalgebra, which are both valid for corecursion up-to (1), nothing guarantees that their composition ($a \circ Fb\colon FGZ \to Z$) or their coproduct ($[a;b]\colon (F+G)Z \to Z$) remains valid for corecursion up-to. For instance, knowing that for streams, both corecursion up-to $\oplus$ and corecursion up-to $\otimes$ are valid is not enough to deduce that corecursion up-to both $\oplus$ and $\otimes$ is valid too.

In the context of bisimilarity and coinductive predicates, such questions have been studied extensively by Davide Sangiorgi [36] early after the introduction of up-to techniques by Robin Milner [27]. This resulted in the concepts of *respectful functions* and then *compatible functions* [29], subclasses of valid up-to techniques enjoying nice compositionality properties.

In the context of categorical corecursion, distributive laws (condition 1/) and causal algebras (condition 2/) also enjoy such compositionality properties [11,32]. We show in Sect. 4 that the situation is similar with causal transformations (condition 3/): these can be organised as a category with arbitrary products, in which some generic and useful basic transformations always exist. As a consequence, simple causal transformations can be assembled into more complex ones, achieving the expected modularity for corecursion up-to.

Liftings and Coinductive Predicates

Corecursion (up-to) makes it possible to define operations on final coalgebras (e.g., $\oplus$ and $\otimes$ on streams). Once such operations have been defined, one often needs to reason about them, to establish some of their properties (e.g., both $\oplus$ and $\otimes$ are associative and commutative). This is why we also need to develop the theory of coinductive predicates, and to provide up-to proof techniques for those. Typically, reasoning about an operation defined by corecursion up-to requires related coinduction up-to techniques.

We exploit the fibrational approach to coinductive predicates [10,15,16] in Sect. 5, where we show how to get coinductive up-to techniques from causal transformations, provided that these causal transformation lift.

2 Preliminaries

We recall the basic categorical concepts we use in the paper. The reader familiar with universal coalgebra [21] may safely skip this section.

Coalgebras, Algebras. For a category $\mathcal{C}$ and a functor $F\colon \mathcal{C} \to \mathcal{C}$, an *F-coalgebra* is a pair (X, f) with X an object in $\mathcal{C}$ and $f\colon X \to FX$. A *homomorphism* of F-coalgebras $h\colon (X, f) \to (Y, g)$ is a map $h\colon X \to Y$ such that $g \circ h = Fh \circ f$. Coalgebras for F form a category and an F-coalgebra is *final* if it is final in that category. An *F-algebra* is a pair (X, a) with X an object of $\mathcal{C}$ and $a\colon FX \to X$.

Ordinals. We write ω for the first infinite ordinal. The category Ord of ordinals has as objects the ordinals themselves, and there is a unique arrow $j \to k$ iff $j \leq k$. This is similar to the usual view of a poset as a category, except that the ordinals do not form a set.

Final Sequence. The *final sequence* of a functor $B\colon \mathcal{C} \to \mathcal{C}$ in a complete category $\mathcal{C}$ is an ordinal indexed sequence of objects B_i with connecting morphisms $B_{j,i}\colon B_j \to B_i, i \leq j$ constructed in the following way. The first object is $B_0 \triangleq 1$, the final object of $\mathcal{C}$ and for a successor ordinal $j+1$ we have $B_{j+1} \triangleq BB_j$. Further, $B_{i,i} \triangleq \mathsf{id}$, $B_{i,0} \triangleq !_i\colon B_i \to 1$ and $B_{j+1,i+1} \triangleq BB_{j,i}$. For a limit ordinal λ we have $B_\lambda \triangleq \lim_{i<\lambda} B_i$ and $(B_{\lambda,i})_{i<\lambda}$ forms a limiting cone. We write $\overline{B}$ for the final sequence of B, seen as a functor $\mathsf{Ord}^{\mathsf{op}} \to \mathcal{C}$. Accordingly, we write $\overline{B}(i)$ for B_i and $\overline{B}(j,i)$ for $B_{j,i}$.

We say that the final sequence $\overline{B}$ *stabilises at ordinal* κ if $\overline{B}(\kappa+1,\kappa)$ is an isomorphism. In this case, $\overline{B}(\kappa)$ is a final coalgebra [5, Theorem 1.3] (shown for the dual case of algebras in [2]). For ω-continuous functors (e.g., polynomial set functors), the final sequence stabilises at ω.

Fact 2.1. *Given a coalgebra $f\colon X \to BX$, we can construct a cone $f_i\colon X \to B_i$ over the final sequence, inductively. We start with the unique map $f_0 \triangleq !_X\colon X \to 1$ and for a successor we define $f_{i+1} = Bf_i \circ f$. For a limit ordinal k, the map $f_k\colon X \to B_k$ is the unique map obtained from the induction hypothesis and universality of B_k.*

Distributive Laws. For functors $F, B\colon \mathcal{C} \to \mathcal{C}$, a *distributive law* of F over B is a natural transformation $\lambda\colon FB \Rightarrow BF$. When F comes with a monad structure (F, η, μ), we call $\lambda\colon FB \Rightarrow BF$ a *distributive law of a monad* if it satisfies $B\eta = \lambda \circ \eta B$ and $\lambda \circ \mu B = B\mu \circ \lambda F \circ F\lambda$.

Every distributive law $\lambda\colon FB \Rightarrow BF$ over a functor B admitting a final coalgebra (Z, ζ) *induces an algebra* on Z by considering the coalgebra $\lambda_Z \circ F\zeta\colon FZ \to BFZ$ and using finality.

3 Corecursion Up-to Causal Transformations

Our main result does not explicitly mention corecursion. Remember that we call a natural transformation of the form $F\overline{B} \Rightarrow \overline{B}$ a *causal transformation*. Given such a causal transformation, we inductively construct morphisms from any BF-coalgebra into each object of the final sequence $\overline{B}$. We get validity of corecursion up-to as a special case, when we have a final B-coalgebra at ordinal κ in the final sequence (Corollary 3.4 below).

Theorem 3.1. *Let $B, F\colon \mathcal{C} \to \mathcal{C}$ be endofunctors on a complete category $\mathcal{C}$ and let $\alpha\colon F\overline{B} \Rightarrow \overline{B}$ be a causal transformation. For every BF-coalgebra $g\colon X \to BFX$ and every ordinal k, there is a unique map $g_k^\dagger$ making the following diagram commute:*

$$
\begin{array}{ccccc}
X & & \xrightarrow{\quad g_k^\dagger \quad} & & \overline{B}(k) \\
{\scriptstyle g}\downarrow & & & & \uparrow{\scriptstyle \overline{B}(k+1,k)} \\
BFX & \xrightarrow[BFg_k^\dagger]{} & BF\overline{B}(k) & \xrightarrow[B\alpha_k]{} & B\overline{B}(k)
\end{array}
$$

Proof. We proceed by transfinite induction on the ordinal k, additionally proving that for all ordinals $i < k$, we have

$$\overline{B}(k, i) \circ g_k^\dagger = g_i^\dagger \tag{2}$$

When $k = 0$, we have

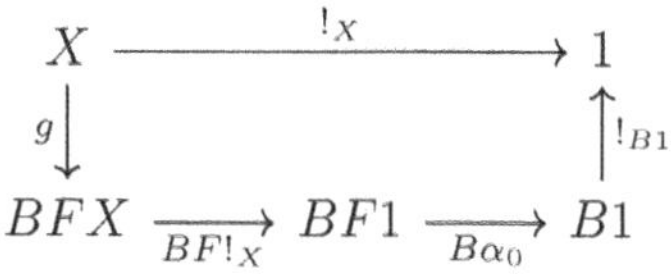

Uniqueness and commutativity both follow from the uniqueness of the arrow $!_X$ from X into the final object. The property (2) holds trivially.

For the case $k = j + 1$ for some ordinal j, we assume the following commutative diagram:

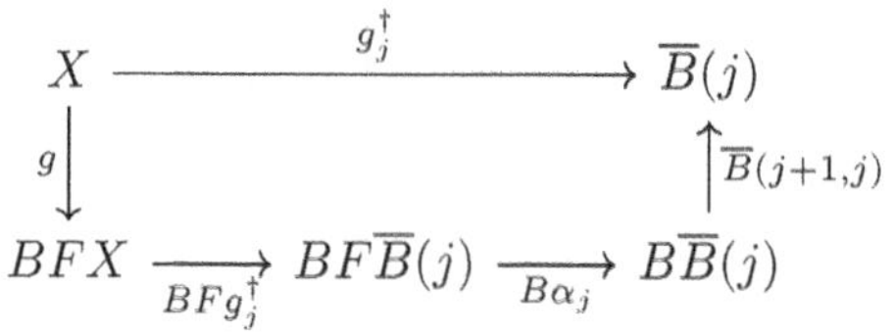

(3)

Now consider the following diagram, where $g_{j+1}^\dagger$ is the map we wish to define:

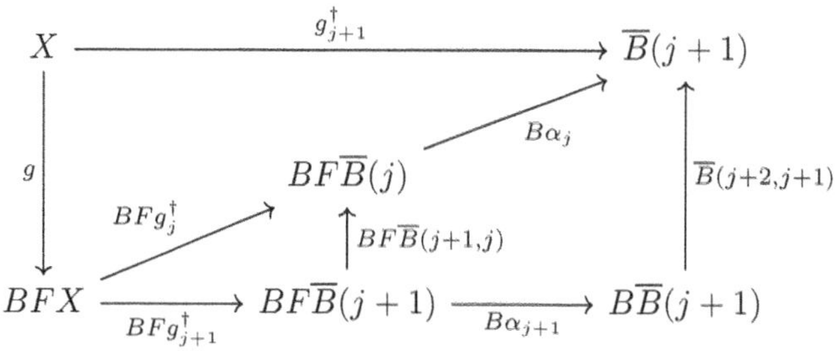

(4)

The lower left triangle commutes as we have $\overline{B}(j+1,j) \circ g_{j+1}^\dagger = \overline{B}(j+1,j) \circ B\alpha_j \circ BFg_j^\dagger \circ g = g_j^\dagger$ by Eq. (3), meaning also $BF\overline{B}(j+1,j) \circ BFg_{j+1}^\dagger = BFg_j^\dagger$. The lower right trapezium commutes by naturality of α and functoriality of B:

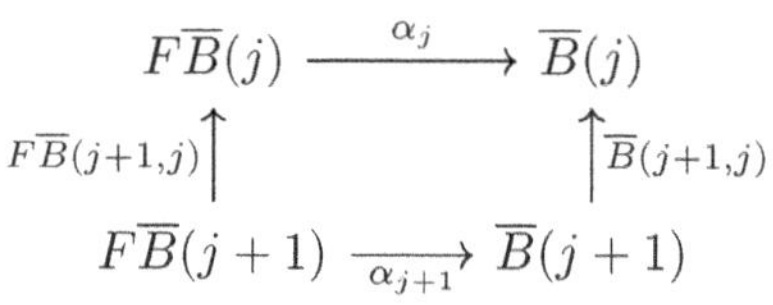

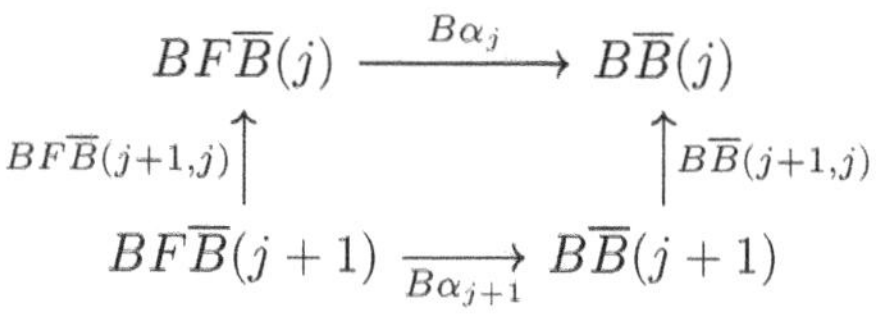

Taking $g_{j+1}^\dagger \triangleq B\alpha_j \circ BFg_j^\dagger \circ g$, we see that this is the unique map we require, as it makes the diagram of Eq. (4) commute, and any such map making the successor case of (3) commute, must satisfy the above equation.

To show the property (2), let $i < k$ and consider the following diagram

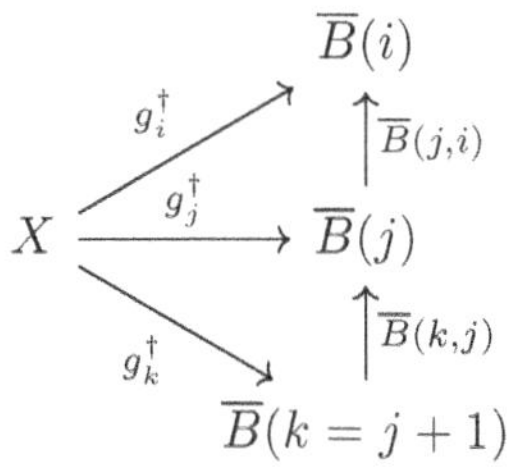

Then (2) follows by the induction hypothesis and definition of $g_k^\dagger$.

Finally, we have the case of a limit ordinal k. In this case, we assume we have the diagram as in (3) commuting for all $j < k$ and note that, by definition of the final sequence, the maps $\overline{B}(k, j)$ for $j < k$ form a limiting cone. We would like to use the universal property of such a limiting cone to construct a map into its apex, the object $\overline{B}(k)$. To do this, we require that the maps $g_j^\dagger \colon X \to \overline{B}(j)$ form a cone over the final sequence. This holds by the induction hypothesis, specifically the property of (2) for all $j < k$, and so by the universal property, we have a map $g_k^\dagger \colon X \to \overline{B}(k)$. This also immediately establishes the limit case of property (2).

Now consider the following diagram:

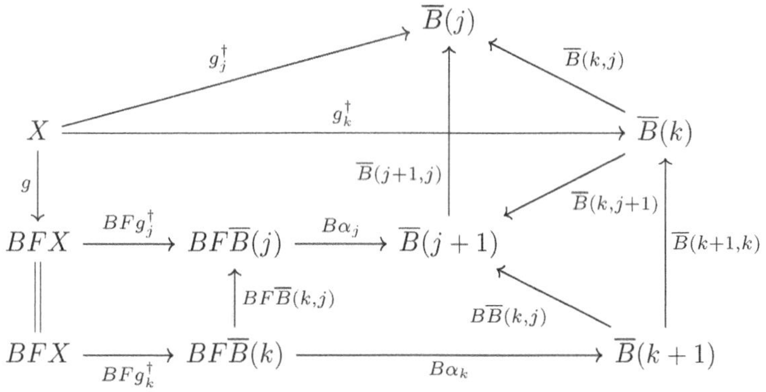

(5)

We would like, for all $j < k$, the outer route from X to $\overline{B}(j)$ to be equal to $g_j^\dagger$ (in an equation: $g_j^\dagger = \overline{B}(k, j) \circ \overline{B}(k+1, k) \circ B\alpha_k \circ BFg_k^\dagger \circ g$). Then, by definition of $g_k^\dagger$ as the unique map such that $\overline{B}(k, j) \circ g_k^\dagger = g_j^\dagger$ we will have commutativity of diagram (3) for the ordinal k, i.e., $g_k^\dagger = \overline{B}(k + 1, k) \circ B\alpha_k \circ BFg_k^\dagger \circ g$.

Equationally, the proof goes as follows:

$$\overline{B}(k,j) \circ \overline{B}(k+1,k) \circ B\alpha_k \circ BFg_k^\dagger \circ g \tag{6}$$
$$= \overline{B}(j+1,j) \circ \overline{B}(k,j+1) \circ \overline{B}(k+1,k) \circ B\alpha_k \circ BFg_k^\dagger \circ g \tag{7}$$
$$= \overline{B}(j+1,j) \circ B\overline{B}(k,j) \circ B\alpha_k \circ BFg_k^\dagger \circ g \tag{8}$$
$$= \overline{B}(j+1,j) \circ B\alpha_j \circ BF\overline{B}(k,j) \circ BFg_k^\dagger \circ g \tag{9}$$
$$= \overline{B}(j+1,j) \circ B\alpha_j \circ BFg_j^\dagger \circ g \tag{10}$$
$$\stackrel{I.H.}{=} g_j^\dagger \tag{11}$$

To show this diagrammatically, we have included the inner part in Eq. (5). Then the upper right and right-hand triangles commute by definition of the final sequence (7), (8). The lower right trapezium commutes by naturality of α and functoriality of B (9). Commutativity of the lower left square follows by property (2) as well as functoriality of BF (10). The final equality (11) holds by the induction hypothesis. Together, this gives the required commutativity, and uniqueness of $g_k^\dagger$. □

As announced in the introduction, the above theorem actually generalises some well-known facts:

Corollary 3.2 ([25, p. 5, footnote 2]). *Every element $\overline{B}(k)$ of the final sequence $\overline{B}$, seen as a B-algebra with structure map $\overline{B}(k+1,k)$, is corecursive.*

Proof. This is the special case of Theorem 3.1 where we take the identity functor for F, and the identity causal transformation for α. □

Corollary 3.3. *[2,5] If the final sequence $\overline{B}$ stabilises at ordinal κ, then $\overline{B}(\kappa)$ is a final coalgebra (with structure map $\overline{B}(\kappa+1,\kappa)^{-1}$).*

Proof. This is the special case of the previous corollary where we select the κ-th element of the final sequence. □

More importantly for the present paper, Theorem 3.1 justifies condition 3/ from the introduction, for corecursion up to an algebra.

Corollary 3.4 (Corecursion up-to from causal transformations). *If the final sequence $\overline{B}$ stabilises at ordinal κ, and if α is a causal transformation for B, then corecursion up-to the algebra α_κ is valid for B.*

The above corollary does not require any distributive law to start with, only a causal transformation. This is the point we want to emphasise in the present work, and we shall study causal transformations in the following sections. Nevertheless, this result also gives a new way to get validity of corecursion up-to from a distributive law, which we discuss in the remainder of this section:

Corollary 3.5 (Corecursion up-to from distributive laws). *Let $\lambda\colon FB \Rightarrow BF$ be a distributive law. If the final sequence $\overline{B}$ yields a final coalgebra at ordinal κ, then corecursion up-to the algebra induced by λ on this final coalgebra is valid.*

Proof. It is shown in [32, Lemma 6.2] that such a distributive law induces a unique $\alpha\colon F\overline{B} \Rightarrow \overline{B}$ with the property that $\alpha_\kappa\colon F\overline{B}(\kappa) \to \overline{B}(\kappa)$ is the algebra induced by the distributive law on $\overline{B}(\kappa)$. These properties allow us to apply the above corollary, giving the required result. □

The statement of this last corollary is very close to the work of Falk Bartels on generalised coinduction [6,7]: in our terminology, Theorems 3.8 and 3.9 of [6] can be summarised as follows:

Theorem 3.6. *Let $\lambda\colon FB \Rightarrow BF$ be a distributive law. If B has a final coalgebra and either of the following two conditions holds:*

- *the category $\mathcal{C}$ has countable coproducts, or*
- *F is a monad and λ is a distributive law of a monad,*

then corecursion up-to the algebra induced by λ on the final coalgebra is valid.

In [32, Theorem 9.2] we showed a result analogous to Corollary 3.4 for the case of polynomial functors on Set (with $\kappa = \omega$). The proof uses the *companion* of B (the final distributive law in a suitable sense) and uses Theorem 3.6 to conclude. The above direct proof from causal transformations is much simpler.

Let us highlight the differences between Theorem 3.6 and Corollary 3.5, first concerning the statements: 1/ we require a complete category where Bartels only needs countable coproducts (if any); and 2/ we require that the final sequence stabilises where he only needs the existence of a final coalgebra. Those differences disappear in the category of sets, which is complete and where the mere existence of a final coalgebra ensures that the final sequence stabilises [3].

Now let us compare the two proofs.

Under the first assumption of Theorem 3.6, Bartels uses the given distributive law to construct a B-coalgebra with the countable coproduct $\sum_{i=0}^{\infty} F^i X$ as carrier. There is a unique map from that coalgebra into the final B-coalgebra, which is used to obtain the unique solution for the BF-coalgebra structure. Under the second assumption, the monad structure on F can be used to construct more directly a B-coalgebra with carrier FX. This idea also underlies the generalised powerset construction developed in [37]: one determinises the given BF-coalgebra into a B-coalgebra with a larger carrier.

Interestingly, we never construct such a B-coalgebra in our proof, instead giving a direct construction of the required map into the final coalgebra, by transfinite induction. The downside is that we need ordinals and transfinite induction, where Falk Bartels does not. Therefore, in a sense his argument is more constructive (e.g., it can be formalised in type theory).

4 Compositionality

In this section, we show that the natural transformations of the form $F\overline{A} \Rightarrow \overline{B}$, which we will call *causal transformations from A to B*, enjoy good compositionality properties. This generalises the transformations $F\overline{B} \Rightarrow \overline{B}$ of earlier sections and in fact, we define a category in which such causal transformations are the morphisms. This category has all products, and we obtain compositionality properties which are similar to those established in [11, Proposition 3.3] for compatible functors.

Definition 4.1. *For functors* $F\colon \mathcal{C} \to \mathcal{D}$, $A\colon \mathcal{C} \to \mathcal{C}$ *and* $B\colon \mathcal{D} \to \mathcal{D}$, *a* causal transformation from A to B *is a natural transformation* $\alpha\colon F\overline{A} \Rightarrow \overline{B}$.

Theorem 4.2. *We have a category* $\mathcal{K}$ *with the following data:*

- *objects are pairs* $(A, \mathcal{C})$ *with* $A\colon \mathcal{C} \to \mathcal{C}$ *a functor on a complete category* $\mathcal{C}$.
- *morphisms from* $(A, \mathcal{C})$ *to* $(B, \mathcal{D})$ *are pairs* (F, α) *with* $\alpha\colon F\overline{A} \Rightarrow \overline{B}$.

Proof. The identity on an object $(A, \mathcal{C})$ is the pair $(\mathsf{Id}, \mathsf{id})$. The composition of two morphisms $(F, \alpha)\colon (X, \mathcal{C}) \to (Y, \mathcal{D})$ and $(G, \beta)\colon (Y, \mathcal{D}) \to (Z, \mathcal{E})$ is given by

$$(G \circ F)\overline{X} \xRightarrow{G\alpha} G\overline{Y} \xRightarrow{\beta} \overline{Z}$$

□

When the categories $\mathcal{C}$ and $\mathcal{D}$ are clear from the context, we write $A \dashrightarrow B$ for the homset from $(A, \mathcal{C})$ to $(B, \mathcal{D})$ in $\mathcal{K}$.

Theorem 4.3. *The category* $\mathcal{K}$ *has all products.*

Proof. We only deal with binary products to ease notation, leaving the general case to the reader.

For objects $(A_1, \mathcal{C}_1)(A_2, \mathcal{C}_2)$ in $\mathcal{K}$ we construct the pair $(A_1 \times A_2, \mathcal{C}_1 \times \mathcal{C}_2)$, taking products in the category Cat. The projections must be causal transformations $p_i\colon F_i\overline{A_1 \times A_2} \Rightarrow \overline{A_i}$. To obtain these we take F_i to be exactly the projection π_i in Cat, together with families of maps $p_i(k)\colon \pi_i \circ \overline{A_1 \times A_2}(k) \to \overline{A_i}(k)$ which necessarily consist of identity maps, as we show now.

First, we claim that $\overline{A_1 \times A_2} = \langle \overline{A_1}, \overline{A_2} \rangle$. To prove this we use transfinite induction. For the successor case, assume that $\overline{A_1 \times A_2}(i) = \langle \overline{A_1}, \overline{A_2} \rangle(i) = (\overline{A_1}(i), \overline{A_2}(i))$ for some i. Then, we have

$$\begin{aligned}
\overline{A_1 \times A_2}(i+1) &= (A_1 \times A_2)(\overline{A_1 \times A_2}(i)) && (12)\\
&= (A_1 \times A_2)(\overline{A_1}(i), \overline{A_2}(i)) && (13)\\
&= (A_1(\overline{A_1}(i)), A_2(\overline{A_2}(i))) && (14)\\
&= (\overline{A_1}(i+1), \overline{A_2}(i+1)) && (15)\\
&= \langle \overline{A_1}, \overline{A_2} \rangle(i+1) && (16)
\end{aligned}$$

where Eq. (13) follows from the induction hypothesis.

For the limit case, let k be some limit ordinal and assume that $\overline{A_1 \times A_2}(l) = \langle \overline{A_1}, \overline{A_2} \rangle(l)$ holds for all $l < k$. Now, we have

$$\begin{aligned}
\overline{A_1 \times A_2}(k) &= \lim_{l<k} \overline{A_1 \times A_2}(l) && (17)\\
&= \lim_{l<k} \langle \overline{A_1}, \overline{A_2} \rangle(l) && (18)\\
&= \langle \overline{A_1}, \overline{A_2} \rangle(k) && (19)
\end{aligned}$$

where Eq. (18) follows from the induction hypothesis. The last step, Eq. (19), follows from how limits are computed in a product category; the important point being that cones over a pairing of functors are the same as pairs of cones over the component functors.

From the established equality $\overline{A_1 \times A_2} = \langle \overline{A_1}, \overline{A_2} \rangle$, we conclude that $\pi_i \circ \overline{A_1 \times A_2}(k) = \overline{A_i}(k)$ so that we can take each component $p_i(k)$ of our projections to be the identity. Naturality then holds trivially. It remains to show that the above construction is universal.

Suppose we have an object $(Q \colon \mathcal{D} \to \mathcal{D}, \mathcal{D}) \in \mathcal{K}$ with morphisms $(F_i, \alpha_i) \colon Q \to (A_i, \mathcal{C}_i)$ for $i = 1, 2$. Then we can construct the pair $(\langle F_1, F_2 \rangle, (\alpha_1, \alpha_2))$ of the functors and causal transformations so that $(\alpha_1, \alpha_2) \colon \langle F_1, F_2 \rangle \overline{Q} \to \overline{A_1 \times A_2}$ is a map in $\mathcal{C}_1 \times \mathcal{C}_2$. As this is a product in Cat, we have the required property that for all $j = 1, 2$

$$(\pi_j, \mathsf{id}) \circ (\langle F_1, F_2 \rangle, (\alpha_1, \alpha_2)) = (F_j, \alpha_j)$$

More concretely, the components of the causal transformations are maps of type

$$\langle F_1, F_2 \rangle \overline{Q}(l) \to (\overline{A_1}(l), \overline{A_2}(l)) = (F_1 \overline{Q}(l), F_2 \overline{Q}(l)) \to (\overline{A_1}(l), \overline{A_2}(l))$$

which, by definition of the product category, consist of maps in $\mathcal{C}_1$ and $\mathcal{C}_2$ which we consider in parallel. Further, uniqueness follows from the definition of the pairing $\langle F_1, F_2 \rangle$ and of maps in the product category. Finally, naturality follows by assumption on the α_i, thus, we indeed have a categorical product. □

Finally, we have the following basic morphisms in $\mathcal{K}$, giving access to up-to constant techniques and to coproduct of up-to techniques.

Proposition 4.4. *For every endofunctor $B \colon \mathcal{C} \to \mathcal{C}$, we have the following morphisms in $\mathcal{K}$:*

1. *$(\Delta_X, \delta_f) : B \dot{\to} B$ where X is the carrier of a coalgebra $f \colon X \to BX$ and Δ_X is the constant functor associated to X.*
2. *$(\coprod \colon \mathcal{C}^I \to \mathcal{C}, \gamma) : B^I \dot{\to} B$, assuming that coproducts in $\mathcal{C}$ exist.*

Proof. For Item 1, we define δ_f as the cone given by Fact 2.1. Item 2 follows from the universal property of coproduct. □

Together with Theorem 4.2 and Theorem 4.3, the above proposition makes it possible to define complex causal transformations out of basic ones, thus enabling compositional proofs of validity for complex corecursion up-to schemes.

Example 4.5. Suppose we work with streams, and we want to define a unary operation f satisfying the following equations

$$f(x)_0 = x_0$$
$$f(x)' = (x \oplus f(x')) \oplus f(x'')$$

In order to use corecursion up-to, we need an algebra combining the ability to call $\oplus$ twice in a row, with arguments which are either an existing stream (x), or corecursive calls to f on some existing streams ($f(x')$ and $f(x'')$).

We can use for that the functor $F(F(\mathbb{R}^\omega + \mathsf{Id}) + \mathsf{Id})$ where $FX = X^2$, and the associated algebra $\oplus \circ [\oplus \circ [\mathsf{id}; \mathsf{id}]; \mathsf{id}]$ on $\mathbb{R}^\omega$.

Thanks to the above results, showing that such an algebra arises from a causal transformation amounts to showing that $\oplus$ arises from a causal transformation, which is straightforward. This is sufficient because we know that we have identity morphisms in $\mathcal{K}$ and we can cope with the constant $\mathbb{R}^\omega$ functor via Item 1 of Proposition 4.4 since $\mathbb{R}^\omega$ is a coalgebra (the final one). Since we can take coproducts in Set, we can finally apply Item 2 of Proposition 4.4 and compose these constructions in $\mathcal{K}$ to obtain the required causal transformation. □

As is the case for distributive laws [22,24,32,34,40], we can also define maps between causal transformations.

Definition 4.6. *Given two causal transformations $(F, \alpha), (G, \beta) \colon A \dot{\to} B$, an arrow from (F, α) to (G, β) is a natural transformation $\kappa \colon F \Rightarrow G$ such that $\beta \circ \kappa\overline{A} = \alpha$. These arrows turn $\mathcal{K}$ into a 2-category.*

We claim that the arrows of the above definition turn $\mathcal{K}$ into a 2-category, however we have not checked the details and the consequences of this are currently unclear. Of most interest is a possible correspondence with an analogous category DL with endofunctors as objects and distributive laws as maps [32, Definition 6.1]. The definition of maps between distributive laws again yields a 2-category, from which there may be a 2-functor whose image on distributive laws are exactly the causal transformations obtained via the construction in [32, Lemma 6.2]. There, it is already shown that the construction extends to a functor from distributive laws of A over B to causal transformations from A to B, so giving a 2-functor should generalise this result to the setting where A and B are not fixed. It is further known that, under certain conditions (e.g. the existence of the companion), we can also go back from causal transformations to distributive laws. We would like to further investigate this correspondence in the 2-categorical context and its relation to up-to techniques.

5 Up-to Techniques for Coinductive Proofs

As explained in the previous sections, a causal transformation $\alpha \colon F\overline{B} \Rightarrow \overline{B}$ gives rise to a valid corecursion-up-to principle. In this section we show how it also induces up-to techniques for *coinductive proofs*. We take a fibrational view on coinductive predicates [15,16], where coalgebras in the base category are viewed as state-based systems, and coinductive proofs arise as coalgebras in the *fibre* above the state space. These fibres are often assumed to be a complete lattices in this setting. The key technical result in this section is that any causal transformation in the base category gives rise to a causal transformation in the fibre above the final coalgebra (assumed to be a complete lattice), which enables its use as an up-to technique for coinductive proofs.

5.1 Coinduction Up-to in a Lattice

Let us first briefly recall the basic notions of coinduction and up-to techniques in complete lattices [33]. This can be viewed as a special case of the theory of coalgebras and corecursion (up-to), by instantiating the base category with a complete lattice viewed as a posetal category.

Let $b\colon L \to L$ be a monotone map on a complete lattice L. By the Knaster-Tarski theorem, b has a greatest fixed point νb, which is also the greatest post-fixed point. This gives a coinduction principle: if $x \leq b(x)$, then $x \leq \nu b$.

The definition of corecursion up-to given in the Introduction (1) instantiates to the following coinduction up-to principle. Given a function $f\colon L \to L$ such that $f(\nu b) \leq \nu b$ (this is the algebra structure), coinduction up to f is valid if $x \leq bf(x)$ implies $x \leq \nu b$. This amounts to the established notion of soundness in this setting, with the additional requirement that f preserves νb.

Conditions similar to 1/ from the Introduction have been developed independently in this setting [29,36]: a function f is called *b-compatible* if $fb \leq bf$, that is, there is a distributive law of f over b. Compatibility implies validity in the above sense, and enjoys good compositionality properties (which the class of valid or sound functions does not).

Condition 2/ has no clear counterpart in this setting. In contrast, we considered condition 3/ in previous work [31,32]. In this case, a causal transformation $f\overline{b} \Rightarrow \overline{b}$ is just a function which preserves the final sequence at any point, i.e., $fb^i(\top) \leq b^i(\top)$ for every ordinal i. In *op. cit.* we show that f satisfies this property if and only if $f \leq t$, where t is the *companion* of b, that is, the greatest compatible function [30]. Such a property is very close to Parrow and Weber's characterisation of the greatest respectful function [28] (which happens to coincide with the greatest compatible function—the companion). A constructive version was also used later in the context of Agda [13].

In the lattice-theoretic setting, these causal transformations form a class of valid enhancements which can be more convenient to work with than the stricter requirement of being compatible. In the remainder of this section, we show how to move from "categorical" causal transformations, in a base category where coalgebras are state-based systems, to these lattice-theoretic causal transformations. This is enabled by the use of fibrations, which provide us precisely with the infrastructure to move from coalgebras (as state-based systems) to coinductive proofs thereon.

5.2 Background on Coinduction in a Fibrational Setting

We recall the basics of coinductive predicates in a fibration, but only briefly; see, for instance, [15] for a detailed introduction. First, let $B\colon \mathsf{Set} \to \mathsf{Set}$ and recall that *relation lifting* assigns to every relation $R \subseteq X \times X$ a relation $\mathsf{Rel}(B)(R)$ on BX, defined by

$$\mathsf{Rel}(B)(R) = \{(B\pi_1(t), B\pi_2(t)) \mid t \in BR\}.$$

A *bisimulation* on a B-coalgebra (X, f) is then a relation $R \subseteq X \times X$ such that $R \subseteq (f \times f)^{-1}(\mathsf{Rel}(B)(R))$, and bisimilarity is the greatest fixed point of the monotone map $(f \times f)^{-1}(\mathsf{Rel}(B)(-))\colon \mathsf{Rel}_X \to \mathsf{Rel}_X$, where Rel_X is the lattice of relations on X. It arises as the limit of the final sequence of $(f \times f)^{-1}(\mathsf{Rel}(B)(-))$.

The situation can be massively generalised by moving from relations on sets to a *fibration* $p\colon \mathcal{E} \to \mathcal{C}$, and from the relation lifting to arbitrary liftings of endofunctors on $\mathcal{C}$. We omit the definition of fibration here (see [19]); we however recall the key notions associated to them.

For such a fibration p, we say an object R in $\mathcal{E}$ is *above* an object X in $\mathcal{C}$ if $p(R) = X$, and similarly for morphisms. Further, the *fibre* $\mathcal{E}_X$ above an object X in $\mathcal{C}$ is the subcategory of $\mathcal{E}$ consisting of all objects above X, and all morphisms above id_X. For every arrow $f\colon X \to Y$ there is a *reindexing functor* $f^*\colon \mathcal{E}_Y \to \mathcal{E}_X$.

Throughout this section, we assume that $p\colon \mathcal{E} \to \mathcal{C}$ is a $\mathsf{CLat}_\wedge$*-fibration* (e.g., [38]), which means that each fibre $\mathcal{E}_X$ is a complete lattice, and reindexing preserves arbitrary meets. Below, we shall often refer explicitly to this poset structure, by writing $R \leq S$ if there exists an arrow from R to S in $\mathcal{E}_X$. These are instances of topological functors [17]. Every $\mathsf{CLat}_\wedge$-fibration is a bifibration, which means every reindexing functor f^* has a left adjoint $\coprod_f$.

A *lifting* of a functor $B\colon \mathcal{C} \to \mathcal{C}$ is a functor $\mathbb{B}\colon \mathcal{E} \to \mathcal{E}$ such that $p \circ \mathbb{B} = B \circ p$. For such a lifting and an object X in $\mathcal{C}$, the functor $\mathbb{B}$ restricts to a functor between fibres $\mathbb{B}_X\colon \mathcal{E}_X \to \mathcal{E}_{BX}$. A lifting $(B, \mathbb{B})$ is a *fibration map* if

$$(Bf)^* \circ \mathbb{B}_Y = \mathbb{B}_X \circ f^*$$

for any arrow $f\colon X \to Y$ in $\mathcal{C}$ (the inequality from right to left holds for any lifting).

Given a B-coalgebra (X, g) and a lifting $(B, \mathbb{B})$, we define the functor (that is, monotone map)

$$g^* \circ \mathbb{B}_X\colon \mathcal{E}_X \to \mathcal{E}_X\,.$$

Its final coalgebra (greatest fixed point) $\nu(g^* \circ \mathbb{B}_X)$ exists by the assumption that each fibre is a complete lattice, and is referred to as the *coinductive predicate* defined by $\mathbb{B}_X$. It is the greatest post-fixed point (coalgebra) of $g^* \circ \mathbb{B}_X$; such post-fixed points are called *invariants* in [15]. This gives rise to the lattice-theoretic coinductive proof technique: to prove that an object R in $\mathcal{E}$ is below the greatest fixed point, it suffices to show it is a post-fixed point of $g^* \circ \mathbb{B}_X$.

Example 5.1. Consider the category Rel where an object is a pair (R, X) of sets with $R \subseteq X \times X$, and an arrow from (R, X) to (S, Y) is a map $f\colon X \to Y$ such that $f(R) \subseteq S$. Reindexing is given by inverse image. The forgetful functor $p\colon \mathsf{Rel} \to \mathsf{Set}$ mapping (R, X) to X is a $\mathsf{CLat}_\wedge$-fibration. The relation lifting $\mathsf{Rel}(B)$ is a lifting of B, often referred to as the *canonical* lifting in this general setting. For a coalgebra (X, g), $g^* \circ \mathsf{Rel}(B)_X$ is precisely the monotone map described at the beginning of this subsection, whose greatest fixed point is bisimilarity on (X, g). □

Other liftings of B give rise to other coinductive predicates. For instance, for the powerset functor $\mathcal{P}\colon \mathsf{Set} \to \mathsf{Set}$, consider the lifting

$$\mathsf{Rel}_{\leq}(\mathcal{P})(R) = \{(S, U) \mid \forall x \in S.\, \exists y \in U.\, (x, y) \in R\}\,.$$

Coalgebras for $\mathcal{P}$ are transition systems, post-fixed points of $g^* \circ \mathsf{Rel}_{\leq}(\mathcal{P})$ are simulations, and its greatest fixed point is similarity. This is an instance of a much more general fibrational characterisation of similarity [18]. Other examples of coinductive predicates that have been explored in a fibrational setting are behavioural distances [4,9,38] and various unary predicates and invariants [10, 15] in the fibration of predicates over sets.

In the abstract setting of coinductive predicates via liftings, we can consider *up-to techniques* in the fibre as well, basically by instantiating the setting in Sect. 3. A systematic construction of such up-to techniques in a fibration is in [11]. Of particular interest is the *contextual closure*: given a lifting $\mathbb{F}$ of F and an algebra $a\colon FX \to X$, it is defined as the map $\coprod_a \circ \mathbb{F}_X \colon \mathcal{E}_X \to \mathcal{E}_X$.

Example 5.2. On streams, the algebra $\oplus\colon F\mathbb{R}^\omega \to \mathbb{R}^\omega$ for the squaring functor $FX = X^2$, together with the canonical lifting of F, gives rise to the following monotone function on relations on streams:

$$\begin{aligned} \lfloor \oplus \rfloor \colon \mathcal{P}(\mathbb{R}^\omega \times \mathbb{R}^\omega) &\to \mathcal{P}(\mathbb{R}^\omega \times \mathbb{R}^\omega) \\ R &\mapsto \{(x \oplus y, z \oplus t) \mid x\, R\, y \text{ and } z\, R\, t\} \end{aligned}$$

Such a function often proves useful as an up-to technique in bisimulation proofs on streams: it makes it possible to use the coinductive hypothesis under calls to pointwise addition, and to get rid of common sub-expressions. This is typically convenient to reason about operations defined by corecursion up to $\oplus$, like shuffle product (see, e.g., the example in [30, Section 5]). □

In [10, Theorem 6.7], it is shown that $\coprod_a \circ \mathbb{F}_X$ is valid (even compatible) if there is a distributive law $\lambda\colon FB \Rightarrow BF$ such that (X, a, g) is a λ-bialgebra, and λ lifts to a distributive law $\mathbb{FB} \Rightarrow \mathbb{BF}$.

5.3 Causal Transformations in the Fibre

We now show how to move from a causal transformation in the base category of a fibration to one in the fibre above the state space of the final coalgebra.

Assumption 5.3. *Throughout this subsection, we assume:*

- *a* $\mathsf{CLat}_\wedge$*-fibration* $p\colon \mathcal{E} \to \mathcal{C}$ *into a complete category* $\mathcal{C}$*;*
- *endofunctors* $B, F\colon \mathcal{C} \to \mathcal{C}$ *such that the final sequence of* B *stabilises at some ordinal* κ*; thus the final coalgebra is given by* $(\overline{B}(\kappa), \zeta)$*;*
- *liftings* $\mathbb{B}, \mathbb{F}$ *of* B *and* F *respectively;*
- *a causal transformation* $\alpha\colon F\overline{B} \Rightarrow \overline{B}$.

Now, consider the final B-coalgebra $(\overline{B}(\kappa), \zeta)$. The key idea is to extract from the above data a causal transformation

$$\coprod_{\alpha_\kappa} \circ \mathbb{F}_{\overline{B}(\kappa)} \circ \overline{(\zeta^* \circ \mathbb{B}_{\overline{B}(\kappa)})} \leq \overline{(\zeta^* \circ \mathbb{B}_{\overline{B}(\kappa)})} \,.$$

Here, $f \triangleq \coprod_{\alpha_\kappa} \circ \mathbb{F}_{\overline{B}(\kappa)} \colon \mathcal{E}_{\overline{B}(\kappa)} \to \mathcal{E}_{\overline{B}(\kappa)}$ is the up-to technique induced by the lifting $\mathbb{F}$ and the algebra α_κ, whereas $b \triangleq (\zeta^* \circ \mathbb{B}_{\overline{B}(\kappa)}) \colon \mathcal{E}_{\overline{B}(\kappa)} \to \mathcal{E}_{\overline{B}(\kappa)}$ is the monotone map in the fibre above $\overline{B}(\kappa)$, whose greatest fixed point (final coalgebra) is the coinductive predicate defined by $\mathbb{B}$. Having such a causal transformation means that the up-to technique f is a valid enhancement for b. This is the contents of Theorem 5.5.

In the proof of Theorem 5.5, we use the following lemma, which relates the final sequence of $\zeta^* \circ \mathbb{B}_{\overline{B}(\kappa)}$ in the fibre to the final sequence of the lifting $\mathbb{B}$. Recall from Fact 2.1 that every coalgebra (X, g) induces a cone $g_i \colon X \to \overline{B}(i)$ over the final sequence.

Lemma 5.4. *Suppose $(B, \mathbb{B})$ is a fibration map, and let $g \colon X \to BX$ be a coalgebra. For any ordinal i, we have $\overline{(g^* \circ \mathbb{B}_X)}(i) = g_i^* \circ \overline{\mathbb{B}}(i)$.*

Proof. By transfinite induction on i. The base and successor case are shown in [23, Lemma 5.4]. For k a limit ordinal, we compute:

$$\begin{aligned}
g_k^*(\lim_{i<k} \overline{\mathbb{B}}(i)) &= g_k^*(\bigwedge_{i<k} B_{k,i}^*(\overline{\mathbb{B}}(i))) \\
&= \bigwedge_{i<k} g_k^* \circ \overline{B}(k,i)^*(\overline{\mathbb{B}}(i)) \\
&= \bigwedge_{i<k} (\overline{B}(k,i) \circ g_k)^*(\overline{\mathbb{B}}(i)) \\
&= \bigwedge_{i<k} g_i^*(\overline{\mathbb{B}}(i)) \\
&= \bigwedge_{i<k} \overline{(g^* \circ \mathbb{B}_X)}(i) \,.
\end{aligned}$$

The first step follows from the computation of limits in $\mathsf{CLat}_\wedge$-fibrations, see, e.g., [38]; this is a consequence of [19, Prop. 9.2.1]. The second step follows from the definition of $\mathsf{CLat}_\wedge$-fibrations. The third since $\mathsf{CLat}_\wedge$-fibrations are split. The fourth since the g_i's form a cone over the final sequence. And the last by the induction hypothesis. □

This brings us to the main result of this section:

Theorem 5.5. *Suppose $(B, \mathbb{B})$ is a fibration map, and suppose that for every ordinal i, we have $\coprod_{\alpha_i}(\mathbb{F}\overline{\mathbb{B}}(i)) \leq \overline{\mathbb{B}}(i)$ Then for every ordinal i:*

$$\coprod_{\alpha_\kappa} \circ \mathbb{F}_{\overline{B}(\kappa)} \circ \overline{(\zeta^* \circ \mathbb{B}_{\overline{B}(\kappa)})}(i) \leq \overline{(\zeta^* \circ \mathbb{B}_{\overline{B}(\kappa)})}(i) \,.$$

Proof. For any i, we have

$$\begin{aligned}
&\coprod_{\alpha_\kappa} \circ \mathbb{F}_{\overline{B}(\kappa)} \circ \overline{(\zeta^* \circ \mathbb{B}_{\overline{B}(\kappa)})}(i) \\
&= \coprod_{\alpha_\kappa} \circ \mathbb{F}_{\overline{B}(\kappa)} \circ \zeta_i^*(\overline{\mathbb{B}}(i)) && \text{(Lemma 5.4)} \\
&\leq \coprod_{\alpha_\kappa} \circ (F\zeta_i)^* \circ \mathbb{F}_{\overline{B}(i)}(\overline{\mathbb{B}}(i)) && \text{(basic property liftings)} \\
&\leq \zeta_i^* \circ \coprod_{\alpha_i} \circ \mathbb{F}_{\overline{B}(i)}(\overline{\mathbb{B}}(i)) && \text{(follows from naturality } \alpha\text{)} \\
&\leq \zeta_i^*(\overline{\mathbb{B}}(i)) && \text{(by assumption)}
\end{aligned}$$

For the one-but-last step, note that $\zeta_i = \overline{B}(\kappa, i)$, and thus naturality of α implies $\alpha_i \circ F\zeta_i = \zeta_i \circ \alpha_\kappa$, and hence $(F\zeta_i)^* \circ \alpha_i^* = \alpha_\kappa^* \circ \zeta_i^*$. The desired inequality is obtained as the mate:

$$\coprod_{\alpha_\kappa} \circ (F\zeta_i)^* \leq \coprod_{\alpha_\kappa} \circ (F\zeta_i)^* \circ \alpha_i^* \circ \coprod_{\alpha_i} = \coprod_{\alpha_\kappa} \circ \alpha_\kappa^* \circ \zeta_i^* \circ \coprod_{\alpha_i} \leq \zeta_i^* \circ \coprod_{\alpha_i},$$

using the unit of the adjunction $\coprod_{\alpha_i} \dashv \alpha_i^*$ in the first step, and the counit of $\coprod_{\alpha_\kappa} \dashv \alpha_\kappa^*$ in the last step. □

The condition that $\coprod_{\alpha_i}(\mathbb{F}\overline{\mathbb{B}}(i)) \leq \overline{\mathbb{B}}(i)$ holds for all i is equivalent to the requirement that α lifts to a natural transformation $\mathbb{F}\overline{\mathbb{B}} \Rightarrow \overline{\mathbb{B}}$. In our setting of $\mathsf{CLat}_\wedge$-fibrations, it basically says that $\mathbb{F}$ and α need to preserve all approximations of the coinductive predicate of interest. For instance, for the lifting for similarity of transition systems defined below Example 5.1, the i-th component of the final sequence of $\mathsf{Rel}_\leq(\mathcal{P})$ consists of the i-steps similarity relation $\leq_i$; and the condition that α lifts means that the direct image of $\mathbb{F}(\leq_i)$ under $\alpha \times \alpha$ is contained in $\leq_i$. This is the case, for instance, if $\mathbb{F}$ takes the transitive closure (and F is the identity functor, α the identity map); in that case, the requirement simply amounts to the fact that each element in the final sequence of $\mathsf{Rel}_\leq(\mathcal{P})$ is transitive. Indeed, up-to transitive closure is valid for similarity.

If $\mathbb{F} = \mathsf{Rel}(F)$ and $\mathbb{B} = \mathsf{Rel}(B)$ in the relation fibration $\mathsf{Rel} \to \mathsf{Set}$, the condition that α lifts vacuously holds. This follows, for instance, by the following lemma, which is closely related to the coinduction principle in [16]. It makes use of the equality functor $\mathsf{Eq}\colon \mathsf{Set} \to \mathsf{Rel}$, which maps a set X to the diagonal $\{(x,x) \mid x \in X\}$.

Lemma 5.6. *For any functor* $B\colon \mathsf{Set} \to \mathsf{Set}$, *we have* $\overline{\mathsf{Rel}(B)} = \mathsf{Eq} \circ \overline{B}$.

Proof. By transfinite induction. For a limit ordinal k, use that Eq has a left adjoint (quotients, see [16]) so that it preserves limits. Then

$$\overline{\mathsf{Rel}(B)}(k) = \lim_{i<k} \overline{\mathsf{Rel}(B)}(i) = \lim_{i<k} \mathsf{Eq}(\overline{B}(i)) = \mathsf{Eq}(\lim_{i<k} \overline{B}(i)) = \mathsf{Eq} \circ \overline{B}(k)\,.$$

For a successor ordinal, we use that relation lifting preserves equality [21], that is, $\mathsf{Rel}(B) \circ \mathsf{Eq} = \mathsf{Eq} \circ B$:

$$\begin{aligned}\overline{\mathsf{Rel}(B)}(i+1) &= \mathsf{Rel}(B)(\overline{\mathsf{Rel}(B)}(i)) \\ &= \mathsf{Rel}(B)(\mathsf{Eq}(\overline{B}(i))) \\ &= \mathsf{Eq}(B\overline{B}(i)) \\ &= \mathsf{Eq}(\overline{B}(i+1)) \,.\end{aligned}$$

□

Using the above lemma we get $\mathsf{Rel}(F) \circ \overline{\mathsf{Rel}(B)} = \mathsf{Rel}(F) \circ \mathsf{Eq} \circ \overline{B} = \mathsf{Eq} \circ F \circ \overline{B}$, again using that relation lifting preserves equality, and thus we can define the lifting of α simply as $\mathsf{Eq}(\alpha)$. Alternatively, we expect the fact that any causal transformation lifts in this way also follows similarly to the fact that any distributive law lifts in this case [10], using that Rel is a 2-functor [21].

Curiously, in Theorem 5.5 we assumed that $(B, \mathbb{B})$ is a fibration map. For the canonical relation lifting $\mathsf{Rel}(B)$, this means that B needs to preserve weak pullbacks for the above theorem to apply. This is in contrast to the result in [10], which does not make this requirement.

However, Theorem 5.5 cannot be easily generalised to functors that do not preserve weak pullbacks. Consider, for instance, the case where $F = \mathsf{Id}$, $\alpha = \mathsf{id}$, and $\mathbb{F}(R)$ is the least equivalence relation containing R. The requirement that α lifts then says that each element of the final sequence of $\mathbb{B}$ is an equivalence relation.

A classical simple example of a functor $B\colon \mathsf{Set} \to \mathsf{Set}$ which does not preserve pullbacks is the one defined on objects as $B(X) = \{(x, y, z) \mid |\{x, y, z\}| \leq 2\}$, see [1]. The final sequence of B stabilises immediately. The final sequence of its canonical lifting $\mathsf{Rel}(B)$ consists simply of the equality relation on a singleton (cf. Lemma 5.6). This is clearly an equivalence relation., and therefore the condition that α lifts holds in this case (thus taking $\mathbb{B}$ to be $\mathsf{Rel}(B)$). But up-to-equivalence is not sound for this functor (a counterexample for up-to-bisimilarity is given in [35], this can be adapted).

Indeed, the condition from [10] that the distributive law between F and B lifts, is much stronger: it says that $\mathsf{Rel}(B)$ should commute with the equivalence closure functor $\mathbb{F}$. This is not the case in general, if B does not preserve weak pullbacks. In fact, this example shows that not all causal transformations are definable by a distributive law.

References

1. Aczel, P., Mendler, N.: A final coalgebra theorem. In: Pitt, D.H., Rydeheard, D.E., Dybjer, P., Pitts, A.M., Poigné, A. (eds.) Category Theory and Computer Science. LNCS, vol. 389, pp. 357–365. Springer, Heidelberg (1989). https://doi.org/10.1007/BFb0018361

2. Adámek, J.: Free algebras and automata realizations in the language of categories. Commentationes Mathematicae Universitatis Carolinae **15**(4), 589–602 (1974). http://eudml.org/doc/16649
3. Adámek, J., Koubek, V.: On the greatest fixed point of a set functor. Theor. Comput. Sci. **150**(1), 57–75 (1995). https://doi.org/10.1016/0304-3975(95)00011-K
4. Baldan, P., Bonchi, F., Kerstan, H., König, B.: Coalgebraic behavioral metrics. Log. Methods Comput. Sci. **14**(3) (2018)
5. Barr, M.: Algebraically compact functors. J. Pure Appl. Algebr. **82**(3), 211–231 (1992). https://doi.org/10.1016/0022-4049(92)90169-G
6. Bartels, F.: Generalised coinduction. In: Proceedings of the CMCS, vol. 44 of Electronic Notes in Theoretical Computer Science, pp. 67–87. Elsevier (2001). https://doi.org/10.1016/S1571-0661(04)80903-4
7. Bartels, F.: Generalised coinduction. Math. Struct. Comput. Sci. **13**(2), 321–348 (2003). https://doi.org/10.1017/S0960129502003900
8. Bartels, F.: On generalised coinduction and probabilistic specification formats. PhD thesis, CWI, Amsterdam, April 2004
9. Bonchi, F., König, B., Petrişan, D.: Up-to techniques for behavioural metrics via fibrations. In: Proceedings of the CONCUR, vol.118 of LIPIcs, pp. 17:1–17:17. Schloss Dagstuhl - Leibniz-Zentrum für Informatik (2018)
10. Bonchi, F., Petrişan, D., Pous, D., Rot, J.: Coinduction up-to in a fibrational setting. In: Proceedings of the CSL-LICS, pp. 20:1–20:9. ACM (2014). https://doi.org/10.1145/2603088.2603149
11. Bonchi, F., Petrişan, D., Pous, D., Rot, J.: A general account of coinduction up-to. Acta Inform. **54**(2), 127–190 (2017). https://doi.org/10.1007/s00236-016-0271-4
12. Capretta, V., Uustalu, T., Vene, V.: Corecursive algebras: a study of general structured corecursion. In: Oliveira, M.V.M., Woodcock, J. (eds.) SBMF 2009. LNCS, vol. 5902, pp. 84–100. Springer, Heidelberg (2009). https://doi.org/10.1007/978-3-642-10452-7_7
13. Danielsson, N.A.: Up-to techniques using sized types. Proc. ACM Program. Lang. **2**(POPL), 43:1–43:28 (2018)
14. Girard, J.Y., Lafont, Y., Taylor, P.: Proofs and types. Cambridge Tracts in Theoretical Computer Science 7. Cambridge University Press (1988)
15. Hasuo, I., Kataoka, T., Cho, K.: Coinductive predicates and final sequences in a fibration. Math. Struct. Comput. Sci. **28**(4), 562–611 (2018)
16. Hermida, C., Jacobs, B.: Structural induction and coinduction in a fibrational setting. Inf. Comput. **145**(2), 107–152 (1998). https://doi.org/10.1006/inco.1998.2725
17. Herrlich, H.: Topological functors. Gen. Topol. Appl. **4**(2), 125–142 (1974). https://doi.org/10.1016/0016-660X(74)90016-6
18. Hughes, J., Jacobs, B.: Simulations in coalgebra. Theor. Comput. Sci. **327**(1–2), 71–108 (2004)
19. Jacobs, B.: Categorical Logic and Type Theory. Elsevier, Amsterdam (1999)
20. Jacobs, B.: Distributive laws for the coinductive solution of recursive equations. Inf. Comput. **204**(4), 561–587 (2006). https://doi.org/10.1016/j.ic.2005.03.006
21. Jacobs, B.: Introduction to coalgebra: towards mathematics of states and observation, vol. 59 of Cambridge Tracts in Theoretical Computer Science. Cambridge University Press (2016). https://doi.org/10.1017/CBO9781316823187
22. Klin, B., Nachyła, B.: Presenting morphisms of distributive laws. In: Proceedings of the CALCO, vol. 35 of LIPIcs, pp. 190–204. Schloss Dagstuhl - Leibniz-Zentrum für Informatik (2015). https://doi.org/10.4230/LIPIcs.CALCO.2015.190

23. Kupke, C., Rot, J.: Expressive logics for coinductive predicates. Log. Methods Comput. Sci. **17**(4) (2021). https://doi.org/10.46298/lmcs-17(4:19)2021
24. Lenisa, M., Power, J., Watanabe, H.: Distributivity for endofunctors, pointed and co-pointed endofunctors, monads and comonads. Electr. Notes Theor. Comput. Sci. **33**, 230–260 (2000). https://doi.org/10.1016/S1571-0661(05)80350-0
25. Levy, P.B.: A ghost at ω_1. Log. Methods Comput. Sci. **14**(3) (2018). https://doi.org/10.23638/LMCS-14(3:4)2018
26. Milius, S., Moss, L.S., Schwencke, D.: Abstract GSOS rules and a modular treatment of recursive definitions. Log. Methods Comput. Sci. **9**(3) (2013). https://doi.org/10.2168/LMCS-9(3:28)2013
27. Milner, R.: Communication and Concurrency. Prentice Hall, Hoboken (1989)
28. Parrow, J., Weber, T.: The largest respectful function. Log. Methods Comput. Sci. **12**(2) (2016). https://doi.org/10.2168/LMCS-12(2:11)2016
29. Pous, D.: Complete lattices and up-to techniques. In: Shao, Z. (ed.) APLAS 2007. LNCS, vol. 4807, pp. 351–366. Springer, Heidelberg (2007). https://doi.org/10.1007/978-3-540-76637-7_24
30. Pous, D.: Coinduction all the way up. In: Proceedings of the LICS, pp. 307–316. ACM (2016). https://doi.org/10.1145/2933575.2934564
31. Pous, D., Rot, J.: Companions, codensity and causality. In: Proceedings of the FOSSACS 2017, pp. 106–123 (2017). https://doi.org/10.1007/978-3-662-54458-7_7
32. Pous, D., Rot, J.: Companions, causality and codensity. Log. Methods Comput. Sci. **15**(3) (2019). https://doi.org/10.23638/LMCS-15(3:14)2019
33. Pous, D., Sangiorgi, D.: Advanced topics in bisimulation and coinduction, chapter about Enhancements of the coinductive proof method. Cambridge University Press (2011). http://www.cambridge.org/gb/knowledge/isbn/item6542021
34. Power, J., Watanabe, H.: Combining a monad and a comonad. Theor. Comput. Sci. **280**(1–2), 137–162 (2002). https://doi.org/10.1016/S0304-3975(01)00024-X
35. Rot, J., Bonchi, F., Bonsangue, M., Pous, D., Rutten, J., Silva, A.: Enhanced coalgebraic bisimulation. Math. Struct. Comput. Sci. **27**(7), 1236–1264 (2017). https://doi.org/10.1017/S0960129515000523
36. Sangiorgi, D.: On the bisimulation proof method. Math. Struct. Comput. Sci. **8**, 447–479 (1998). https://doi.org/10.1017/S0960129598002527
37. Silva, A., Bonchi, F., Bonsangue, M., Rutten, J.: Generalizing the powerset construction, coalgebraically. In: Proceedings of the FSTTCS, vol. 8 of LIPIcs, pp. 272–283. Schloss Dagstuhl - Leibniz-Zentrum für Informatik (2010). https://doi.org/10.4230/LIPIcs.FSTTCS.2010.272
38. Sprunger, D., Katsumata, S., Dubut, J., Hasuo, I.: Fibrational bisimulations and quantitative reasoning: extended version. J. Log. Comput. **31**(6), 1526–1559 (2021). https://doi.org/10.1093/logcom/exab051
39. Uustalu, T., Vene, V., Pardo, A.: Recursion schemes from comonads. Nord. J. Comput. **8**(3), 366–390 (2001). http://www.cs.helsinki.fi/njc/References/uustaluvp2001:366.html
40. Watanabe, H.: Well-behaved translations between structural operational semantics. Electr. Notes Theor. Comput. Sci. **65**(1), 337–357 (2002). https://doi.org/10.1016/S1571-0661(04)80372-4
41. Worrell, J.: On the final sequence of a finitary set functor. Theor. Comput. Sci. **338**(1–3), 184–199 (2005). https://doi.org/10.1016/j.tcs.2004.12.009

Corecursive Algebras in Nature

Lawrence S. Moss(✉) and Victoria Noquez

Department of Mathematics, Indiana University, Bloomington, IN, USA
lmoss@indiana.edu, vnoquez@iu.edu

Abstract. Pavlović and Pratt obtained several presentations of sets such as the half-open interval $[0, 1)$ as the final coalgebra of a very basic functor on the category of sets, namely product with the natural numbers. We re-prove and extend some of their results, and we establish some new presentations as well. More importantly in this paper, we exhibit several corecursive algebra structures on sets of real numbers, and we connect to continued fractions and to linear fractional transformations. We present a general result which, under hypotheses, shows that a corecursive algebra has as a subalgebra the final coalgebra with the inverse structure.

Keywords: corecursive algebra · final coalgebra · real numbers · infinite series · continued fractions · linear fractional transformations

1 Introduction

Twenty years ago, Pavlović and Pratt obtained several presentations of sets such as the real interval $[0, 1)$ as the final coalgebra of the functors $F(X) = \mathbb{N} \times X$ and $G(X) = \mathbb{N} \times X + 1$ on the category of sets, namely product with the natural numbers, where $\mathbb{N}$ is the set of natural numbers. This paper continues the exploration of continuous mathematics from the point of view of coalgebra. The difference is that we are not aiming only at final coalgebras but at the more plentiful *corecursive algebras*. We exhibit some new corecursive algebras which are connected either to continued fractions or to linear fractional transformations. We propose a general result which allows one to read off a final coalgebra from a corecursive algebra, provided some conditions are met. This general result allows us to recover the final coalgebras related to subsets of the real numbers which were found by Pavlović and Pratt, and to establish some new ones besides. For both the corecursive algebras and the final coalgebras, the structure maps are very simple.

Sources. Much of this paper is not original. The definition and basic properties of corecursive algebras are from Capretta et al. [3], connections to sums of series were perhaps first pointed out in Feys et al. [4], and the original source on final

Supported by grant #586136 from the Simons Foundation.

Published by Springer Nature Switzerland AG 2022
H. H. Hansen and F. Zanasi (Eds.): CMCS 2022, LNCS 13225, pp. 155–175, 2022.
https://doi.org/10.1007/978-3-031-10736-8_8

coalgebra presentations of the real numbers and subsets thereof is Pavlović and Pratt [10]. The main contribution of this paper is in Sects. 4 and 5, and much of Sect. 4 will also appear in the forthcoming book by Adámek et al. [1].

2 Background

This section contains the main background for the results in this paper.

2.1 The Contraction Mapping Theorem

Let (X, d) be a metric space. A *contraction maping* on X is a function $f\colon X \to X$ such that for some $\varepsilon < 1$, $d(f(x), f(y)) < \varepsilon \cdot d(x, y)$ for all x, y.

Proposition 2.1 *[Banach's Contraction Mapping Theorem]. A contraction mapping on a nonempty complete metric space X has a unique fixed point.*

2.2 Continued Fractions

Our work calls on results from the theory of continued fractions. We begin with notation pertaining to sums of finite lists of non-zero real numbers. Define

$$[b] = b, \qquad [b_0, \ldots, b_{k+1}] = b_0 + 1/[b_1, \ldots, b_k].$$

This notation makes sense for lists of real numbers, but we are only going to use it with lists of positive natural numbers.

The next result contains most of what we need.

Proposition 2.2.(1) *For all infinite sequences $a_0, a_1, \ldots, a_k, \ldots$ of natural numbers with $a_k > 0$ for $k \geq 1$, the limit*

$$\lim_{k\to\infty} [a_0, \ldots, a_k]$$

exists.

(2) *Let $a_0, a_1, \ldots, a_k, \ldots$ be a sequence of natural numbers with $a_k > 0$ for $k \geq 1$. Then*

$$\lim_{k\to\infty} [a_0, \ldots, a_k] = a_0 + (\lim_{k\to\infty} [a_1, \ldots, a_k])^{-1}.$$

(3) *Again, for an infinite sequence of natural numbers with $a_k > 0$ for $k \geq 1$, $\lim_{k\to\infty}[a_0, \ldots, a_k]$ is an irrational number.*

(4) *For every irrational number r there is a unique sequence $a_0, a_1, \ldots, a_k, \ldots$ of natural numbers with $a_k > 0$ for $k \geq 1$ such that $r = \lim_{k\to\infty}[a_0, \ldots, a_k]$.*

Proof. For the proof of (1), see Hardy and Wright [5, Thm. 165], Loya [7, Cor. 8.13], or Niven et al. [9, Thm. 7.6].

Part (2) is an easy consequence of part (1) and the definition of $[a_0, \ldots, a_k]$.

We are going to see an elementary proof of (3) in Lemma 4.5.

Here is a sketch of a coalgebraic argument for (4). Let $\mathbb{A}_{>1}$ be the set of irrational numbers > 1. Let $\mathbb{N}_{\geq 1}$ be the set of natural numbers ≥ 1. The functor $F(X) = \mathbb{N}_{\geq 1} \times X$ has as a final coalgebra the set $(\mathbb{N}_{\geq 1})^\infty$ of infinite sequences of elements of $\mathbb{N}_{\geq 1}$, with structure given by head and tail. Consider the coalgebra $e\colon \mathbb{A}_{>1} \to \mathbb{N} \times \mathbb{A}_{>1}$ given by

$$e(r) = (\lfloor r \rfloor, 1/(r - \lfloor r \rfloor)).$$

(The notation $\lfloor r \rfloor$ is for the greatest integer $\leq r$, so $\lfloor 2.1 \rfloor = \lfloor 2 \rfloor = 2$.) By finality, there is a unique map $e^\dagger\colon \mathbb{A}_{>1} \to (\mathbb{N}_{\geq 1})^\infty$. For $r \in \mathbb{A}_{>1}$, $e^\dagger(r)$ is the *canonical continued fraction representation of* x. It has the property that $\mathsf{head}(e^\dagger(r)) = \lfloor r \rfloor$, and $\mathsf{tail}(e^\dagger(r)) = e^\dagger(1/(r - \lfloor r \rfloor))$.

Write $e^\dagger(r)$ as $(a_0, a_1, \ldots, a_k, \ldots)$. By Loya [7, Thm. 8.14],

$$\lim_{k \to \infty} [a_0, a_1, \ldots, a_k] = r.$$

This is what is behind the existence of the continued fraction representation. The uniqueness of the infinite sequence $(a_0, a_1, \ldots, a_k, \ldots)$ mentioned in (4) follows from Hardy and Wright [5, Thm. 170], or Niven et al. [9, Thm. 7.10].

This completes our sketch of the proof.

Incidentally, $\lim_{k \to \infty}[a_0, a_1, \ldots, a_k]$ is denoted in several ways in the literature, such as

$$a_0 + \frac{1}{a_1 +} + \frac{1}{a_2 +} \cdots \qquad \text{or} \qquad a_0 + \cfrac{1}{a_1 + \cfrac{1}{a_2 + \cfrac{1}{\cdots}}}.$$

2.3 Corecursive Algebras

The notion of a corecursive algebra originates in Capretta et al. [3].

Definition 2.3. *Let $H\colon \mathscr{A} \to \mathscr{A}$ be an endofunctor on any category. An algebra $\alpha\colon HA \to A$ is* corecursive *if for every coalgebra $e\colon X \to HX$ there is a unique* coalgebra-to-algebra morphism $e^\dagger\colon X \to A$. *This means that $e^\dagger = \alpha \cdot He^\dagger \cdot e$:*

$$\begin{array}{ccc} X & \xrightarrow{\;e\;} & HX \\ {\scriptstyle e^\dagger}\downarrow & & \downarrow{\scriptstyle He^\dagger} \\ A & \xleftarrow[\;\alpha\;]{} & HA \end{array}$$

The map $e^\dagger$ is also called the solution of e in the algebra (A, α).

We call on a basic result on corecursive algebras related to fixed points, and a result from [1] which relates final coalgebras and corecursive algebras under hypotheses.

A *fixed point* of a functor is an algebra (A, α) whose structure α is invertible. We thus also have a coalgebra (A, α^{-1}).

Proposition 2.4 *[3]. If a corecursive H-algebra (A, α) is a fixed point, then (A, α^{-1}) is a final coalgebra. If (A, α) is a final coalgebra, then (A, α^{-1}) is a corecursive algebra.*

Proof. In the first assertion, a coalgebra-to-algebra morphism $e^\dagger: (X, e) \to (A, \alpha)$ is the same as a coalgebra-to-coalgebra morphism $e^\dagger: (X, e) \to (A, \alpha^{-1})$. In the second assertion, we use a result known as Lambek's Lemma (initial algebra structures are isomorphisms) in dual form.

Lemma 2.5 *[1]. Let (A, α) be a corecursive H-algebra, and let (B, β) be a fixed point of H which is a subalgebra via the monic $m: (B, \beta) \rightarrowtail (A, \alpha)$. Suppose that for every coalgebra $e: X \to HX$, the unique solution $e^\dagger: X \to A$ factors through m. Then (B, β^{-1}) is a final H-coalgebra.*

Proof. Fix a coalgebra $e: X \to HX$, and factor the solution $e^\dagger$ through m:

$$e^\dagger = \left(X \xrightarrow{\;\widehat{e}\;} B \overset{m}{\rightarrowtail} A\right).$$

By the solution property of $e^\dagger$, the outside of the diagram below commutes:

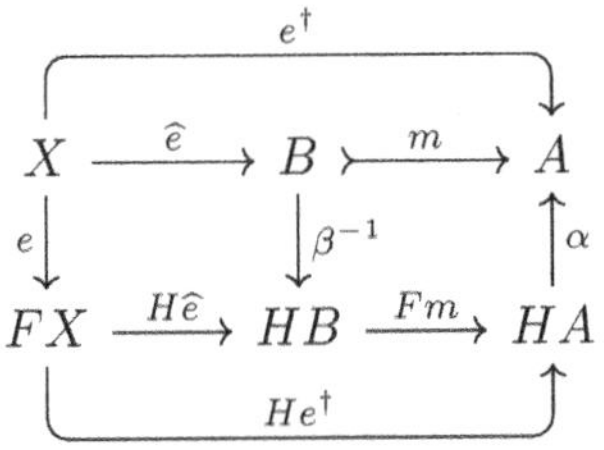

The right-hand square commutes since m is a homomorphism of algebras. The upper part is our factorization of $e^\dagger$. The lower part is the upper part with H applied. Notice that

$$\begin{aligned} m \cdot \widehat{e} &= e^\dagger && \text{definition of } m \text{ and } \widehat{e} \\ &= \alpha \cdot He^\dagger \cdot e && \text{using the outside of the figure} \\ &= \alpha \cdot Fm \cdot H\widehat{e} \cdot e && \text{using the bottom part} \\ &= m \cdot \beta \cdot H\widehat{e} \cdot e && m \text{ is an algebra homomorphism} \end{aligned}$$

Since m is monic, $\widehat{e} = \beta \cdot H\widehat{e} \cdot e$. This is just to say that $\widehat{e}$ is a coalgebra morphism.

For the uniqueness of $\widehat{e}$, let $f: (X, e) \to (B, \beta^{-1})$ be any coalgebra morphism. Then $m \cdot f$ satisfies the equation that uniquely defines $e^\dagger$ in the corecursive algebra A. By the uniqueness of $e^\dagger$, $m \cdot f = e^\dagger$. But then $m \cdot f = m \cdot \widehat{e}$. Since m is monic, $f = \widehat{e}$.

3 Summation of Geometric Series Related to Corecursive Algebra Structure

This section contains a corecursive algebra related to infinite sums of real numbers. First fix a real number $0 \leq \delta < 1$. The rest of this example depends on this parameter. Let K be the real interval $[0, 1 - \delta]$.

For the functor, we take $H\colon \mathsf{Set} \to \mathsf{Set}$ to be given by $HX = K \times X$. For a function $f\colon X \to Z$, $Hf\colon HX \to HZ$ is given by $Hf(y, x) = (y, f(x))$. We have an algebra $(\mathbb{I}, \iota)$, where $\mathbb{I}$ is the unit interval $[0, 1]$, and

$$\iota\colon H[0,1] = K \times [0,1] \to [0,1]$$

is given by $\iota(y, x) = y + \delta x$, for $y \in K$ and $x \in \mathbb{I}$. By viewing $\mathbb{I}$ as a space with the usual metric, K as a space with the discrete metric and $K \times [0, 1]$ with the product (maximum) metric, we see that ι is a *short* (also called *non-distance-increasing*) map.

Proposition 3.1. *$\iota\colon H\mathbb{I} \to \mathbb{I}$ is a corecursive algebra for H.*

Proof. Let X be a set, and consider a coalgebra $(X, e\colon X \to K \times X)$. The function set $\mathbb{I}^X = [0,1]^X$ is a complete metric space under the metric $d(f, g) = \sup_{x \in X} d_I(f(x), g(x))$. This space is non-empty due to the constant functions. We have an endofunction $\Phi\colon \mathbb{I}^X \to \mathbb{I}^X$ given by $\Phi(f) = \iota \cdot Hf \cdot e$. (To check that Φ maps into $\mathbb{I}^X$, take any $x \in X$. For this x, write $e(x)$ as (y, z), where $y \in [0, 1-\delta]$. Then $\Phi(f)(x) = y + \delta f(z) \leq (1 - \delta) + (\delta \cdot 1) = 1$.) And Φ is a contraction, since $\delta < 1$ and we are using the discrete metric on K. The fixed points of Φ are exactly the coalgebra-to-algebra morphisms $X \to \mathbb{I}$. So the Contraction Mapping Theorem (Proposition 2.1) implies our result.

Remark 3.2. We used that $\mathbb{I}$ and $H\mathbb{I}$ come with metrics such that $\mathbb{I}$ is non-empty and complete, ι is short, and Φ is a contraction. In fact, this is all we need to know: see [2] for a general result in this direction.

Also, the proof above works even when $X = \emptyset$. In that case $\mathbb{I}^X = \{!_{\mathbb{I}}\}$, where $!_{\mathbb{I}}\colon \emptyset \to \mathbb{I}$ is the empty function, Φ is the unique function $\mathbb{I}^X \to \mathbb{I}^X$, $!_{\mathbb{I}}$ is its fixed point, and $!_{\mathbb{I}} = e^\dagger$ by initiality of $\emptyset$.

Here is a way to think about what is going on. Given a coalgebra $e\colon X \to K \times X$, denote its components as $u\colon X \to K$ and $\mathsf{next}\colon X \to X$. As we know, a solution of e is a function $e^\dagger\colon X \to \mathbb{I} = [0,1]$ satisfying the equation $e^\dagger = \iota \cdot He^\dagger \cdot e$. That is, for all x,

$$e^\dagger(x) = u(x) + \delta \cdot e^\dagger(\mathsf{next}(x)). \tag{3.1}$$

We used the Contraction Mapping Theorem to prove that every system of equations e has a unique solution $e^\dagger$. But we could have simply explicitly defined the solution. For each $x \in X$, let

$$\begin{aligned} s(x) &= u(x) + \delta \cdot u(\mathsf{next}(x)) + \delta^2 \cdot u(\mathsf{next}^2(x)) + \cdots + \delta^\ell \cdot u(\mathsf{next}^\ell(x)) + \cdots \\ &= \textstyle\sum_{\ell=0}^{\infty} \delta^\ell \cdot u(\mathsf{next}^\ell(x)) \end{aligned}$$

(Note that the infinite sum is $\leq (1-\delta)\sum_{\ell=0}^{\infty}\delta^\ell = 1$.) Then s with this definition is a solution to e, because for all x,

$$\begin{aligned} s(x) &= \textstyle\sum_{\ell=0}^{\infty} \delta^\ell \cdot u(\mathsf{next}^\ell(x)) \\ &= u(x) + \textstyle\sum_{\ell=1}^{\infty} \delta^\ell \cdot u(\mathsf{next}^\ell(x)) \\ &= u(x) + \textstyle\sum_{\ell=0}^{\infty} \delta^{\ell+1} \cdot u(\mathsf{next}^{\ell+1}(x)) \\ &= u(x) + \delta \cdot \textstyle\sum_{\ell=0}^{\infty} \delta^\ell \cdot u(\mathsf{next}^\ell(\mathsf{next}(x))) \\ &= u(x) + \delta \cdot s(\mathsf{next}(x)) \end{aligned}$$

Then knowing that there is a unique solution to e, our function s is *the* solution.

Our development allows us to use the fixed point equations instead of infinite summation.

4 Corecursive Algebras Related to Subsets of the Reals, and Associated Final Coalgebra Structures

This section, the centerpiece of our paper, exhibits some corecursive algebra structures on certain subsets of the reals. Figure 1 summarizes our results. We discuss them first, mentioning points of notation and also a comparison with the first results in the area, those of Pavlović and Pratt [10].

We have two sets of results, one set using continued fractions and the other set using linear fractional transformations. The two sets correspond to the two charts in the figure. In the top chart, we first show that the set $\mathbb{R}_{\geq 0}$ of non-negative real numbers is a corecursive algebra of $FX = \mathbb{N} \times X$ with structure $\alpha(n,r) = n + \frac{1}{1+r}$. This result is new. By applying Lemma 2.5, we infer that the set $\mathbb{A}$ of positive irrationals is a final coalgebra of this same functor using the inverse of the restriction of the same map α. We exhibit this structure explicitly, using $\lfloor x \rfloor$ to denote the largest natural number $\leq x$ and $x \bmod 1$ to denote $x - \lfloor x \rfloor$. We have an isomorphism $f : \mathbb{R}_{\geq 0} \cong (0,1]$ given by $f(x) = \frac{1}{1+x}$. This restricts to an isomorphism, also denoted by $f : \mathbb{A} \cong \mathbb{B}$, where $\mathbb{B}$ is the Baire space, the set of irrationals in $[0,1]$. It transfers a final F-coalgebra structure from $\mathbb{A}$ to $\mathbb{B}$. We give the explicit formulas for the structure morphisms, thereby recovering Theorem 3.2 of [10].

Moving to the right-hand column of top chart, we change the functor from F to $G(X) = \mathbb{N} \times X + 1$. Adding the extra point allows us to extend the F-coalgebra structure on $\mathbb{R}_{\geq 0}$ to a G-coalgebra structure, and we do this by mapping the extra point (the element of the summand 1) to 0. This is what we mean by the notation $[\alpha, 0] \colon \mathbb{N} \times \mathbb{R}_{\geq 0} + 1 \to \mathbb{R}_{\geq 0}$. This map is a corecursive G-algebra whose structure is an isomorphism, hence we have a final G-coalgebra (by taking the inverse). The inverse of $[\alpha, 0]$ structure here is given explicitly in Fig. 1. Then we use the isomorphism $f : \mathbb{R}_{\geq 0} \cong (0,1]$ mentioned above to get a final coalgebra structure on the half-open interval $(0,1]$. Incidentally, this work in the right-hand column is a little easier than the work in the left-hand column because the latter calls on a classical fact about irrational numbers which we prove from first principles in Lemma 4.5.

The second chart turns to a different method of finding corecursive algebra structure, using linear fractional transformations. The main result here is a corecursive algebra structure for F on the closed unit interval $\mathbb{I} = [0,1]$. We show in Proposition 5.10, the F-algebras $\mathbb{R}_{\geq 0}$ and $\mathbb{I}$ are not isomorphic. The structure $\tau\colon \mathbb{N} \times \mathbb{I} \to \mathbb{I}$ is not injective. But its restriction to $\mathbb{N} \times [0,1)$ is a bijection, hence $[0,1)$ is a fixed point of F. We observe that the solution of every coalgebra in $(\mathbb{I}, \tau)$ takes values in $[0,1)$. That is, the solution morphism $e^{\dagger}: X \to \mathbb{I}$ factors through the inclusion $[0,1) \hookrightarrow \mathbb{I}$. So we again use Lemma 2.5 to obtain a final F-coalgebra structure on $[0,1)$ We also have an isomorphic copy of the final F-coalgebra, obtained by using the isomorphism $h : [0,1) \to \mathbb{R}_{\geq 0}$ given by $h(r) = r/(1-r)$.

4.1 Verifications

Proposition 4.1. *Concerning the functions in Fig. 1:*

(1) *The function* $\alpha\colon \mathbb{N} \times \mathbb{R}_{\geq 0} \to \mathbb{R}_{\geq 0}$ *is injective.*
(2) *The function* $\sigma\colon \mathbb{N} \times [0,1) \to [0,1)$ *is a bijection.*
(3) *Let* $f\colon \mathbb{R}_{\geq 0} \to (0,1]$ *be* $f(x) = \frac{1}{1+x}$. *Then* f *is a bijection, and the diagram on the left below commutes:*

$$
\begin{array}{ccc}
\mathbb{N} \times \mathbb{R}_{\geq 0} & \xrightarrow{\alpha} & \mathbb{R}_{\geq 0} \\
{\scriptstyle \mathbb{N}\times f}\downarrow & & \downarrow{\scriptstyle f} \\
\mathbb{N} \times (0,1] & \xrightarrow{\beta} & (0,1]
\end{array}
\qquad\qquad
\begin{array}{ccc}
\mathbb{R}_{\geq 0} & \xrightarrow{\xi} & \mathbb{N} \times \mathbb{R}_{\geq 0} \\
{\scriptstyle h^{-1}}\downarrow & & \uparrow{\scriptstyle \mathbb{N}\times h} \\
[0,1) & \xrightarrow{\zeta} & \mathbb{N} \times [0,1)
\end{array}
\tag{4.1}
$$

Also, $h : [0,1) \to \mathbb{R}_{\geq 0}$ *given by* $h(x) = \frac{x}{1-x}$ *is a bijection, and the diagram on the right above commutes.*
(4) *We have the following facts about inverses:* $[\alpha, 0]^{-1} = \chi$, $[\beta, 1]^{-1} = \varrho$, $\sigma^{-1} = \zeta$, *and* $\vartheta^{-1} = \xi$.

Proof. (1): For α, note that for all r, $0 < \frac{1}{1+r} \leq 1$. If $\alpha(n,r) = \alpha(m,s)$ is a natural number, then $r = 0 = s$, and also $\alpha(n,r) = n+1$ and $\alpha(m,r) = m+1$. In this case, $n = m$. When $\alpha(n,r) = \alpha(m,s)$ is not a natural number,

$$\frac{1}{1+r} = \alpha(n,r) \bmod 1 = \alpha(m,s) \bmod 1 = \frac{1}{1+s}.$$

And so $r = s$. In this case we also have $n = m$, since $0 < \frac{1}{1+r} \leq 1$.

In part (2) concerning $\sigma\colon \mathbb{N} \times [0,1) \to [0,1)$, notice that $\sigma = i \cdot g \cdot h$, where $h\colon \mathbb{N} \times [0,1) \to [1,\infty)$ is $h(n,r) = 1 + n + r$, $g\colon [1,\infty) \to (0,1]$ is $g(x) = 1/x$, and $i\colon (0,1] \to [0,1)$ is $i(x) = 1 - x$. All three maps are bijections.

Using Theorem 4.2 (and thus continued fractions):

	$FX = \mathbb{N} \times X$	$GX = \mathbb{N} \times X + 1$
corecursive algebra:		
carrier	$\mathbb{R}_{\geq 0}$ = reals ≥ 0	$\mathbb{R}_{\geq 0}$
structure	$\alpha(n, r) = n + \frac{1}{1+r}$	$[\alpha, 0]$
final coalgebra from Lemma 2.5:		
carrier	$\mathbb{A}$ = irrationals > 0	$\mathbb{R}_{\geq 0}$
inverse structure	α_0 = restriction of α to $\mathbb{A}$	$[\alpha, 0]$
structure	$\gamma(x) = (\lfloor x \rfloor, \frac{1}{x \bmod 1} - 1)$	$\chi(0) \in 1$ in $G(\mathbb{R}_{\geq 0})$ $\chi(x) = (x - 1, 0)$ for $x \geq 1$ in $\mathbb{N}$ else $\chi(x) = (\lfloor x \rfloor, \frac{1}{x \bmod 1} - 1)$
final coalgebra isomorphic copy:		
carrier	$\mathbb{B}$ = irrationals $\cap [0, 1]$	$(0, 1]$
inverse structure	β_0 = restriction of β to $\mathbb{B}$	$[\beta, 1]$, where $\beta(n, r) = 1/(1 + n + r)$
structure	$\delta(x) = (\lfloor \frac{1}{x} \rfloor - 1, \frac{1}{x} \bmod 1)$	$\varrho(1) \in 1$ in $G((0, 1])$ $\varrho(x) = ((1/x) - 2, 1)$ if $1/x \in \mathbb{N} \setminus \{1\}$ else $\varrho(x) = (\lfloor \frac{1}{x} \rfloor - 1, \frac{1}{x} \bmod 1)$

Using Theorem 5.8 (and thus linear fractional transformations):

	$FX = \mathbb{N} \times X$
corecursive algebra:	
carrier	$\mathbb{I} = [0, 1]$
structure	$\tau(n, r) = \frac{n+r}{1+n+r}$
final coalgebra from Lemma 2.5:	
carrier	$[0, 1)$
inverse structure	$\sigma(n, r) = \frac{n+r}{1+n+r}$
structure	$\zeta(x) = (\lfloor \frac{x}{1-x} \rfloor, \frac{x}{1-x} \bmod 1)$
final coalgebra isomorphic copy:	
carrier	$\mathbb{R}_{\geq 0}$
inverse structure	$\vartheta(n, r) = n + \frac{r}{1+r}$
structure	$\xi(x) = (\lfloor x \rfloor, \frac{x \bmod 1}{1 - x \bmod 1})$

Fig. 1. Corecursive algebras obtained, and some final coalgbras derived from them.

For (3), the inverse of f is given by $f^{-1}(x) = \frac{1-x}{x}$. For $n \in \mathbb{N}$ and $r \in \mathbb{R}_{\geq 0}$,

$$\begin{aligned} f(\alpha(n,r)) &= f(n + \tfrac{1}{1+r}) \\ &= \tfrac{1}{(1+n+\frac{1}{1+r})} \\ &= \beta(n, f(r)) \\ &= \beta((\mathbb{N} \times f)(n,r)) \end{aligned}$$

We turn to the square on the right. The inverse of h is given by $h^{-1}(x) = \frac{x}{1+x}$. To see that the square commutes, let $x \in \mathbb{R}_{\geq 0}$. Then

$$\zeta(h^{-1}(x)) = (\lfloor x \rfloor, x \bmod 1)$$

And so

$$(\mathbb{N} \times h) \cdot (\zeta \cdot h^{-1})(x) = (\lfloor x \rfloor, \frac{x \bmod 1}{1 - x \bmod 1}) = \xi(x).$$

Finally, we turn to point (4). Throughout our work, we elide the coproduct injections; we have already done so in the chart. For example, when x is a natural number ≥ 1, we wrote $\chi(x) = (x-1, 0)$, and technically we mean $\mathsf{inl}((x-1,0)) \in \mathbb{N} \times \mathbb{R}_{\geq 0} + 1$. And we write the element of 1 as $*$.

Let us first check that $[\alpha, 0]^{-1} = \chi$. First, $([\alpha, 0] \cdot \chi)(0) = 0$. For $x \geq 1$ in $\mathbb{N}$, $([\alpha, 0] \cdot \chi)(x) = \alpha(x-1, 0) = (x-1)+1 = x$. For all other x,

$$[\alpha, 0](\chi(x)) = \lfloor x \rfloor + \frac{1}{1 + (\frac{1}{x \bmod 1} - 1)} = \lfloor x \rfloor + (x \bmod 1) = x.$$

In the other direction, we have a few cases. With $* \in 1$, $[\alpha, 0](*) = 0$, and so $\chi \cdot [\alpha, 0](0) = *$. When $r = 0$, $\alpha(n,r) = n+1$, and so $\chi \cdot [\alpha, 0]((n,r)) = (n,r)$. When $r > 0$,

$$\chi([\alpha, 0](n,r)) = (\lfloor n + \tfrac{1}{1+r} \rfloor, \frac{1}{(n + \frac{1}{1+r}) \bmod 1} - 1) = (n,r).$$

We next show that $\varrho = [\beta, 1]^{-1}$. First, $([\beta, 1] \circ \varrho)(1) = 1$. For $x \in (0,1]$ such that $1/x \in \mathbb{N} \setminus \{1\}$,

$$([\beta, 1] \circ \varrho)(x) = [\beta, 1]((1/x) - 2, 1) = \frac{1}{1 + (1/x) - 2 + 1} = x.$$

Finally, for all other x,

$$([\beta, 1] \circ \varrho)(x) = \beta(\lfloor \frac{1}{x} \rfloor - 1, \frac{1}{x} \bmod 1) = \frac{1}{1 - \lfloor \frac{1}{x} \rfloor - 1 + \frac{1}{x} \bmod 1} = x.$$

In the other direction, for $* \in 1$ in $G((0,1])$, $(\varrho \circ [\beta, 1])(*) = \varrho(1) = *$. For $(n, r) \in \mathbb{N} \times (0, 1)$, note that since $r \in (0, 1)$, $\frac{1}{1/(1+n+r)} \notin \mathbb{N}$. So

$$\begin{aligned}\varrho([\beta, 1](n, r)) &= \varrho(\tfrac{1}{1+n+r}) \\ &= (\lfloor 1 + n + r \rfloor - 1, (1 + n + r) \bmod 1) \\ &= (n, r)\end{aligned}$$

Finally, for $(n, 1)$, $(\varrho \circ [\beta, 1])(n, 1) = \varrho(\frac{1}{2+n})$, so since $2 + n \in \mathbb{N} \setminus \{1\}$, $\varrho(\frac{1}{2+n}) = (2 + n - 2, 1) = (n, 1)$.

We verify that $\zeta = \sigma^{-1}$. Notice that for h as in part (3), $h(\frac{n+r}{n+1+r}) = h(h^{-1}(n + r)) = n + r$. Also, we have

$$\zeta(x) = (\lfloor h(x) \rfloor, h(x) \bmod 1).$$

Thus for $n \in \mathbb{N}$ and $r \in [0, 1)$,

$$\begin{aligned}\zeta(\sigma(n, r)) &= (\lfloor h(\tfrac{n+r}{n+1+r}) \rfloor, h(\tfrac{n+r}{n+1+r}) \bmod 1) \\ &= (\lfloor n + r \rfloor, (n + r) \bmod 1) \\ &= (n, r)\end{aligned}$$

In the other direction, for all $x \in [0, 1)$,

$$\sigma(\zeta(x)) = \sigma(\lfloor h(x) \rfloor, h(x) \bmod 1) = h(x)/(1 + h(x)) = h^{-1}(h(x)) = x.$$

We used the fact that $\lfloor h(x) \rfloor + (h(x) \bmod 1) = h(x)$.

We conclude by verifying that $\vartheta^{-1} = \xi$:

$$\begin{aligned}\vartheta(\xi(x)) &= \lfloor x \rfloor + h^{-1}(\tfrac{x \bmod 1}{1 - x \bmod 1}) \\ &= \lfloor x \rfloor + h^{-1}(h(x \bmod 1)) \\ &= \lfloor x \rfloor + (x \bmod 1) \\ &= x\end{aligned}$$

In the other direction, use $h^{-1}(r) \in [0, 1)$, so $h^{-1}(r) \bmod 1 = h^{-1}(r)$:

$$\xi(\vartheta(n, r)) = \xi(n + h^{-1}(r)) = (n, h(h^{-1}(r)) = (n, r).$$

Theorem 4.2. *The F-algebra $(\mathbb{R}_{\geq 0}, \alpha)$ is corecursive.*

Proof. It is convenient to work with a naturally isomorphic functor, $HX = \mathbb{N}_{\geq 1} \times X$, where $\mathbb{N}_{\geq 1}$ is the set of positive natural numbers. The natural isomorphism $\eta\colon F \to H$ has components $\eta_X(n, x) = (n + 1, x)$. There is a bijective correspondence between F-coalgebras $e\colon X \to FX$ and H-coalgebras

$\eta_X \cdot e\colon X \to HX$. This extends to a bijective correspondence between solutions with respect to F and those with respect to H; indeed in the diagram below the upper square commutes if and only if the outside commutes:

$$\begin{array}{ccc}
X & \xrightarrow{e^\dagger} & \mathbb{R}_{\geq 0} \\
{\scriptstyle e}\downarrow & & \uparrow{\scriptstyle \alpha} \\
FX & \xrightarrow{Fe^\dagger} & F\mathbb{R}_{\geq 0} \\
{\scriptstyle \eta_X}\downarrow & & \uparrow{\scriptstyle \eta^{-1}_{\mathbb{R}_{\geq 0}}} \\
HX & \xrightarrow{He^\dagger} & H\mathbb{R}_{\geq 0}
\end{array}$$

Furthermore, the function $j(x) = x + 1$ is an isomorphism $j\colon \mathbb{R}_{\geq 0} \to \mathbb{R}_{\geq 1}$, and it is an H-algebra isomorphism $j\colon (\mathbb{R}_{\geq 0}, \alpha \cdot \eta^{-1}_{\mathbb{R}_{\geq 0}}) \cong (\mathbb{R}_{\geq 1}, \alpha')$, where $\alpha'(n, r) = n + 1/r$. As a result, it is sufficient to prove that the H-algebra $(\mathbb{R}_{\geq 1}, \alpha')$ is corecursive.

Given a coalgebra $e\colon X \to \mathbb{N}_{\geq 1} \times X$, again denote its components as $u\colon X \to \mathbb{N}_{\geq 1}$ and $\mathsf{next}\colon X \to X$. For each $x \in X$, we have an infinite sequence a^x_i, $i \in \mathbb{N}$, where

$$a^x_i = u(\mathsf{next}^i(x)). \tag{4.2}$$

Thus, we may use Proposition 2.2(1) to define $e^\dagger$ by

$$e^\dagger(x) = \lim_{k\to\infty} [a^x_0, \ldots, a^x_k].$$

By Proposition 2.2(2),

$$e^\dagger(x) = u(x) + 1/(e^\dagger(\mathsf{next}(x))).$$

(See also Niven et al. [9, Lem. 7.8].) This states that $e^\dagger\colon X \to \mathbb{R}_{\geq 1}$ is a solution of e in the H-algebra $(\mathbb{R}_{\geq 1}, \alpha')$.

We conclude by checking the uniqueness of $e^\dagger$. Let $s\colon X \to \mathbb{R}_{\geq 1}$ be any solution of e. Then for all x, $s(x) = u(x) + 1/s(\mathsf{next}(x))$. By Loya [7, Thm. 8.14], $s(x) = \lim_{k\to\infty}[a^x_0, \ldots, a^x_k]$. This is also $e^\dagger(x)$. This completes the proof.

Corollary 4.3. *The G-algebra $(\mathbb{R}_{\geq 0}, [\alpha, 0])$ is corecursive.*

Proof. Let $e\colon Y \to \mathbb{N} \times Y + 1$ be a coalgebra. For $k \geq 0$, we say that $y \in Y$ is *k-grounded* if there are $y = y_1, y_2, \ldots, y_{k+1} \in Y$ and $n_1, n_2, \ldots, n_k \in \mathbb{N}$ such that $e(y_i) = (n_i, y_{i+1}) \in \mathbb{N} \times Y$ for $i < k$, and $e(y_{k+1}) = * \in 1$.

Let $Z \subseteq Y$ be the set of elements which are not k-grounded for any k. Then there is a unique map $f\colon Z \to \mathbb{N} \times Z$ such that the diagram below commutes, where $j\colon Z \to Y$ is the inclusion:

$$\begin{array}{ccccc}
Z & \xrightarrow{f} & \mathbb{N} \times Z & \xrightarrow{\mathsf{inl}} & \mathbb{N} \times Z + 1 \\
{\scriptstyle j}\downarrow & & & & \downarrow{\scriptstyle \mathbb{N}\times j+1} \\
Y & & \xrightarrow{\quad e \quad} & & \mathbb{N} \times Y + 1
\end{array}$$

Indeed, f is the domain-codomain restriction of e. Hence (Z, f) is a coalgebra for the functor $FX = \mathbb{N} \times X$, and so we have a unique solution $f^\dagger : Z \to \mathbb{R}_{\geq 0}$.

If $e(y)$ is not the element of the singleton set 1, let us write $e(y)$ as $(u(y), \mathsf{next}(y))$, where $u(y) \in \mathbb{N}$, and $\mathsf{next}(y) \in Y$. Notice that if y is k-grounded and $k \geq 1$, then $\mathsf{next}(y)$ is $(k-1)$-grounded. By Theorem 4.2, $(\mathbb{R}_{\geq 0}, \alpha)$ is a corecursive algebra for $\mathbb{N} \times X$. Let $f^\dagger : Z \to \mathbb{R}_{\geq 0}$ be the unique solution of f. Then the map $e^\dagger : Y \to \mathbb{R}_{\geq 0}$ below is easily seen to be the unique solution of e:

$$e^\dagger(y) = \begin{cases} f^\dagger(y) & \text{if } y \in Z, \\ 0 & \text{if } y \text{ is 0-grounded}, \\ u(y) + 1/(e^\dagger(\mathsf{next}(y)) + 1) & \text{if } y \text{ is } k\text{-grounded for } k \geq 1. \end{cases}$$

This concludes the proof.

We continue with the right-hand column of the top chart in Fig. 1.

Corollary 4.4. *The following are final G-coalgebras:*

(1) $\mathbb{R}_{\geq 0}$ *with structure* $[\alpha, 0]^{-1} = \chi$.
(2) $(0, 1]$ *with structure* $[\beta, 1]^{-1} = \varrho$.

Proof. Using Proposition 4.1 (1), the algebra structure $[\alpha, 0]$ is a bijection: α is injective, the only value which it does not take is 0, and this is rectified in $[\alpha, 0]$. Proposition 2.4 and Corollary 4.3 thus imply that $(\mathbb{R}_{\geq 0}, [\alpha, 0]^{-1})$ is a final coalgebra for G. We have seen that $[\alpha, 0]^{-1} = \chi$ in Proposition 4.1(4).

For (2), we have seen that f in (4.1) is an isomorphism of F-algebras. Since $f(0) = 1$, f is also an isomorphism of G-algebras, $f : \mathbb{R}_{\geq 0} \to (0, 1]$. We have already checked that $\varrho = [\beta, 1]^{-1}$. Thus f is an isomorphism of G-coalgebras $(\mathbb{R}_{\geq 0}, \chi) \cong ((0, 1], \varrho)$.

We turn back to the left-hand column of the top chart in Fig. 1. Our next result is also found in the standard theory of continued fractions. We present an elementary proof.

Lemma 4.5. *Let* $e\colon X \to \mathbb{N} \times X$ *be a coalgebra, and let* $e^\dagger$ *be the solution in* $(\mathbb{R}_{\geq 0}, \alpha)$. *For all* $x \in X$, *the number* $e^\dagger(x)$ *is irrational.*

Proof. The proof recalls Euclid's proof of the irrationality of $\sqrt{2}$. Suppose not. Let $Y \subseteq X$ be the set of $x \in X$ such that $e^\dagger(x)$ is rational. Note that each $e^\dagger(x)$ is non-zero. For each $x \in Y$, write $e^\dagger(x)$ as p_x/q_x, where this fraction is in lowest terms. Let $x \in Y$ minimize the sum $p_x + q_x$. Write $e(x)$ as (n, y). By the definition of $e^\dagger$,

$$e^\dagger(x) = \alpha(n, e^\dagger(y)) = n + \frac{1}{1 + e^\dagger(y)}.$$

Thus, $e^\dagger(y)$ is rational, too. That is, $y \in Y$. So we have p_y and q_y such that $e^\dagger(y) = p_y/q_y$, and $\gcd(p_y, q_y) = 1$. Our choice of x tells us that $p_x + q_x \leq p_y + q_y$. Now

$$\frac{p_x}{q_x} = n + \frac{1}{1 + \dfrac{p_y}{q_y}}.$$

Hence

$$\frac{p_x - nq_x}{q_x} = \frac{q_y}{p_y + q_y}.$$

The fraction on the right is in lowest terms, because $\gcd(q_y, p_y + q_y) = \gcd(q_y, p_y) = 1$. Therefore the denominator on the right must divide that on the left. So

$$p_y + q_y \le q_x < p_x + q_x$$

The second inequality above is strict because $p_x > 0$, since it is the numerator of a fraction in lowest terms representing a non-zero number. And the overall strict inequality above gives a contradiction.

Corollary 4.6 (**[10]**). *$FX = \mathbb{N} \times X$ has the following final coalgebras:*

(1) *The set $\mathbb{A}$ of positive irrational numbers, with structure γ.*
(2) *The Baire space $\mathbb{B}$ (the set of irrational numbers in $[0,1]$), with structure δ.*

Proof. (1) We first check that $\mathbb{A}$ is a fixed point of $\mathbb{N} \times X$. By Proposition 4.1(1), $\alpha \colon \mathbb{N} \times \mathbb{R}_{\geq 0} \to \mathbb{R}_{\geq 0}$ is injective. Hence so is its restriction $\alpha_0 : \mathbb{N} \times \mathbb{A}$. Moreover, every irrational number x is $\alpha(n, r)$ for $n = \lfloor x \rfloor$ and $r = 1/(x-n) - 1$; indeed, observe that r is a positive irrational and that $\alpha(n, r) = x$. Thus, α_0 is a bijection $\alpha_0 \colon \mathbb{N} \times \mathbb{A} \to \mathbb{A}$. Further, the inclusion $i \colon \mathbb{A} \hookrightarrow \mathbb{R}_{\geq 0}$ is an algebra homomorphism $i \colon (\mathbb{A}, \alpha) \to (\mathbb{R}_{\geq 0}, \alpha)$. By Lemma 2.5 and Theorem 4.2, $(\mathbb{A}, \alpha_0^{-1})$ is a final coalgebra.

(2) The isomorphism of F-algebras which we saw in Proposition 4.1(3) restricts to the isomorphism below:

$$\begin{array}{ccc} \mathbb{N} \times \mathbb{A} & \xrightarrow{\alpha_0} & \mathbb{A} \\ {\scriptstyle \mathbb{N} \times f} \downarrow & & \downarrow {\scriptstyle f} \\ \mathbb{N} \times \mathbb{B} & \xrightarrow{\beta_0} & \mathbb{B} \end{array} \tag{4.3}$$

We use f for the map $f \colon \mathbb{A} \to \mathbb{B}$ with the definition as before: $f(x) = \frac{1}{1+x}$. Thus, f is an isomorphism of F-coalgebras, hence $(\mathbb{B}, \beta_0)$ is a final F-coalgebra.

We saw in Proposition 4.1 that $[\alpha, 0]^{-1} = \chi$ and $[\beta, 1]^{-1} = \varrho$. These facts imply that $\alpha_0^{-1} = \gamma$, and $\beta_0^{-1} = \delta$.

5 Using Linear Fractional Transformations

We return to our results as presented in Fig. 1. We have completed the discussion of all of the results in the top chart, and we turn to the bottom chart.

In what follows, we regard the unit interval $\mathbb{I} = [0,1]$ as a metric space with the usual metric $d(x, y) = |x - y|$. For any set X, the function space $\mathbb{I}^X$ is a complete space, where the distances are defined using the pointwise supremum:

$$sd(f, g) = \sup_{x \in X} |f(x) - g(x)|. \tag{5.1}$$

5.1 The Algebra $\tau : \mathbb{N} \times \mathbb{I} \to \mathbb{I}$ Presented in Terms of Linear Fractional Transformations

Let $\mathbf{SL}(2, \mathbb{Z})$ be the group of 2×2 matrices M with integer coefficients and determinant 1. Let $s \colon \mathbb{N} \to \mathbf{SL}(2, \mathbb{Z})$ be

$$s(n) = \begin{pmatrix} 1 & n \\ 1 & n+1 \end{pmatrix}.$$

We are going to write $\mathbf{S}$ for the subsemigroup of $\mathbf{SL}(2, \mathbb{Z})$ generated by the image of s. So $\mathbf{S}$ is the set of non-empty finite products $s(n_1) \times s(n_2) \times \cdots \times s(n_k)$. We regard s as a function $s : \mathbb{N} \to \mathbf{S}$. For $M \in \mathbf{S}$ and $x \in \mathbb{I}$, let $\gamma(M, x) \in \mathbb{I}$ be

$$\gamma\left(\begin{pmatrix} a & c \\ b & d \end{pmatrix}, x\right) = \frac{ax + c}{bx + d}. \tag{5.2}$$

To check that $\gamma(M, x) \in \mathbb{I}$, we check this first for matrices $M = s(n)$ and $x \in \mathbb{I}$: if $a = b = 1$, $c \geq 1$, and $d = c + 1$ in (5.2), then $ax + c \geq 0$, $bx + d > 0$, and $ax + c < bx + d$. So $\gamma(M, x) \in [0, 1)$. Then recall that $\mathbf{SL}(2, \mathbb{Z})$ gives rise to a group action: for $M_1, M_2 \in \mathbf{S}$,

$$\gamma(M_1 \times M_2, x) = \gamma(M_1, \gamma(M_2, x)). \tag{5.3}$$

So indeed $\gamma(M, x) \in \mathbb{I}$ for $M \in \mathbf{S}$. Moreover, for $n \in \mathbb{N}$ and $x \in \mathbb{I}$, we have

$$\tau(n, x) = \frac{n + x}{n + 1 + x} = \frac{x + n}{x + (n + 1)} = \gamma\left(\begin{pmatrix} 1 & n \\ 1 & n+1 \end{pmatrix}, x\right) = \gamma(s(n), x). \tag{5.4}$$

That is, the diagram below commutes:

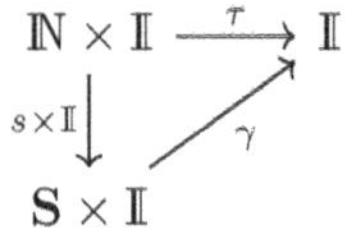

Let $e : X \to \mathbb{N} \times X$. We would like to use the general method from Proposition 3.1 in order to find a unique solution morphism $e^\dagger$ in the algebra $(\mathbb{I}, \tau)$. We consider $\phi \colon \mathbb{I}^X \to \mathbb{I}^X$ given by

$$\phi(f) = \tau \cdot (\mathbb{N} \times f) \cdot e \tag{5.5}$$

We would like it to be the case that ϕ is a contraction on the complete metric space $(\mathbb{I}^X, d)$, with d as in (5.1). As we shall see, this is not always true, and most of our work below concerns the situation when ϕ is *not* a contraction. But we do have the following fact:

$$\phi(f) = \gamma \cdot (\mathbf{S} \times f) \cdot (s \times X) \cdot e \tag{5.6}$$

The reason that (5.5) implies (5.6) comes from the diagram below. In it, the upper-right square commutes (trivially), and we have seen that the bottom part commutes.

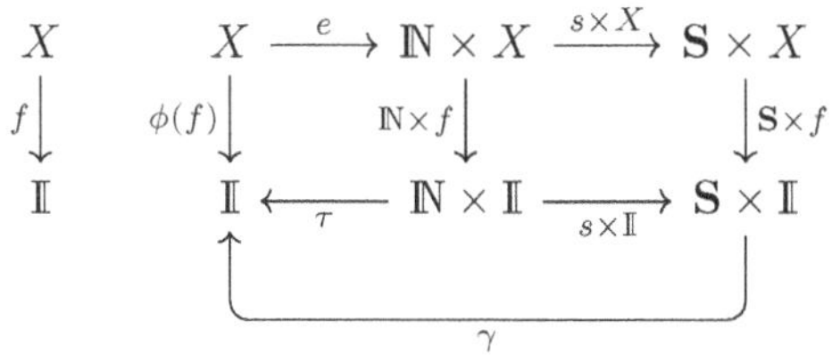

5.2 Contractivity Constants Associated to Matrices

Let $f : X \to Y$ be a map of bounded metric spaces. We define the *Lipschitz constant* of f, $\mathrm{Lip}(f)$, by

$$\sup_{x,y\in X, x\neq y} \frac{d(f(x), f(y))}{d(x,y)}.$$

It is easy to see that $\mathrm{Lip}(g \cdot f) \leq \mathrm{Lip}(g) \cdot \mathrm{Lip}(f)$. For a matrix M in $\mathbf{S}$, we associate the function $f_M : \mathbb{I} \to \mathbb{I}$ by $f_M(x) = \gamma(M, x)$. By (5.3), $f_{M\times N} = f_M \cdot f_N$. Then we define the *contractivity of* M, $\mathrm{con}(M)$, by

$$\mathrm{con}(M) = \mathrm{Lip}(f_M).$$

What interests us is $\mathrm{con}(s(n)) = \sup_{x,y\in[0,1]} |f_{s(n)}(x) - f_{s(n)}(y)|/|x-y|$.

Proposition 5.1. (1) *For all* M *and* N, *con*$(M \times N) \leq$ *con*$(M) \cdot$ *con*(N).
(2) *For all natural numbers* n, *con*$(s(n)) = \frac{1}{(n+1)^2}$.
(3) *If* $n_1, \ldots, n_k$ *is a non-empty sequence of numbers which is not all* 0, *then*

$$con(s(n_1) \times \cdots \times s(n_k)) \leq \frac{1}{4}.$$

Proof. For part (1), $\mathrm{con}(M \times N) = \mathrm{Lip}(f_{M\times N}) = \mathrm{Lip}(f_M \cdot f_N)$. And then as we have seen, this last number is $\leq \mathrm{Lip}(f_M) \cdot \mathrm{Lip}(f_N) = \mathrm{con}(M) \cdot \mathrm{con}(N)$.

For (2), we saw in (5.4) that $f_{s(n)}(x) = \gamma(s(n), x) = (n+x)/(x+(n+1))$, and then a routine calculation shows that

$$\frac{\left| \dfrac{x+n}{x+(n+1)} - \dfrac{y+n}{y+(n+1)} \right|}{|x-y|} = \frac{1}{(x+n+1)(y+n+1)}.$$

And so

$$\mathrm{Lip}(f_{s(n)}) = \sup_{x,y\in[0,1], x\neq y} \frac{1}{(x+n+1)(y+n+1)} = \frac{1}{(n+1)^2}.$$

Part (3) then follows from the previous parts; note that for $n \geq 1$, $\frac{1}{(n+1)^2} \leq \frac{1}{4}$.

Remark 5.2. The idea of deriving contractivity constants for matrices comes from Heckmann [6, Proposition 3.1]. Indeed, that result gives an upper bound on $\text{con}(M)$ for all M in $\mathbf{SL}(2,\mathbb{R})$, the set of 2×2 real matrices with non-zero determinant. We could have simply used Heckmann's result in what follows; it is much more useful than what we saw in Proposition 5.1(2). On the other hand, Heckmann changes the metric on $\mathbb{I}$. That is, in [6], the metric on $\mathbb{I}^X$ is not the metric one would at first expect. So this is why we have chosen to present Proposition 5.1(2). Incidentally, much of the work below comes from the fact that $f_{s(0)}$ is not a contraction on $\mathbb{I}^X$ (since its contractivity is 1). Unfortunately, this would not change if we adopted Heckmann's metric on this space.

5.3 A Corecursive Algebra

Consider a F-coalgebra $e : X \to \mathbb{N} \times X$. As earlier in this paper, we denote the components of e as $u\colon X \to \mathbb{N}$ and $\mathsf{next}\colon X \to X$. So for all $x \in X$, $e(x) = (u(x), \mathsf{next}(x))$. We shall show that there is a unique coalgebra-to-algebra map $e^\dagger : (X, e) \to (\mathbb{I}, \gamma)$. Let

$$Z = \{x \in X : u(x) = 0\} \qquad Y = \{x \in X : (\forall j \geq 0)(\exists k \geq j)\ \mathsf{next}^k(x) \in \overline{Z}\}$$
$$\overline{Z} = \{x \in X : u(x) > 0\} \qquad \overline{Y} = \{x \in X : (\exists j \geq 0)(\forall k \geq j)\ \mathsf{next}^k(x) \in Z\}$$

"Z" here stands for "zero". $\overline{Y}$ is the set of $x \in X$ such that for all but finitely many n, $u(\mathsf{next}^n(x)) = 0$. Note that $\overline{Z} = X \setminus Z$ and $\overline{Y} = X \setminus Y$. The ideas which led to Y and $\overline{Y}$ may be found in Lemmas 5.6 and 5.7.

Proposition 5.3. *If $Z = \emptyset$, then the map $\phi : \mathbb{I}^X \to \mathbb{I}^X$ from (5.6) is a contraction.*

Proof. Fix $f, g \in \mathbb{I}^X$ and also $x \in X$. Recall that $e(x) = (u(x), \mathsf{next}(x))$. Using (5.6), we have

$$|\phi(f)(x) - \phi(g)(x)| = |\gamma(s(u(x)), f(\mathsf{next}(x))) - \gamma(s(u(x)), g(\mathsf{next}(x)))|$$

For our fixed x, $u(x) > 0$. So by Proposition 5.1(2), $\text{con}(s(u(x))) \leq \frac{1}{4}$. The definition of $\text{con}(s(u(x)))$ tells us that

$$\begin{aligned} &|\gamma(s(u(x)), f(\mathsf{next}(x))) - \gamma(s(u(x)), g(\mathsf{next}(x)))| \\ &\leq \tfrac{1}{4}|f(\mathsf{next}(x)) - g(\mathsf{next}(x))| \\ &\leq \tfrac{1}{4}d(f,g) \end{aligned}$$

This for all $x \in X$ shows that in the space $\mathbb{I}^X$, $d(\phi(f), \phi(g)) \leq \frac{1}{4}d(f,g)$.

Proposition 5.4. *If $Z = X$, then constant function $e^\dagger(x) = 0$ is the unique solution of e. More generally, if $x \in X$ has the property that for all $k \geq 0$, $\mathsf{next}^k(x) \in Z$, then for all solutions $e^\dagger$, we have $e^\dagger(x) = 0$.*

Remark 5.5. It might be helpful to consider as a special case the system of equations with $X = \{x_n : n \in \mathbb{N}\}$:

$$x_n = x_{n+1}/(x_{n+1} + 1)$$

Proposition 5.4 implies that the only solution in $([0,1], \gamma)$ is the constant $e^\dagger(x_n) = 0$. Our argument below shows that the only solution in $\mathbb{R}_{\geq 0}$ is the constant 0. But there are other solutions if we allow negative real numbers, such as $e^\dagger(x_n) = \frac{b}{a-bn}$ when a and b are natural numbers and b is not a divisor of a.

Proof. We only prove the first statement in our result; the second is an elaboration of it. For a solution map $e^\dagger$, we write $x^\dagger$ for $e^\dagger(x)$. So for all x, $x^\dagger \geq 0$, and also

$$x^\dagger = (\mathsf{next}\ x)^\dagger/(1 + (\mathsf{next}\ x)^\dagger). \tag{5.7}$$

We cannot have $x^\dagger \geq 1$, since the values of $\frac{x}{x+1}$ for non-negative x lie in $[0, 1)$. Now an induction on $m \geq 1$ shows that we cannot have $x^\dagger \geq \frac{1}{m}$ for any $x \in X$. For suppose that $x^\dagger \geq \frac{1}{m+1}$. Due to (5.7), we get $(\mathsf{next}\ x)^\dagger \geq \frac{1}{m}$. This completes our induction. It follows that $x^\dagger = 0$ for all $x \in X$.

Lemma 5.6. *If $X = \overline{Y}$, then e has a unique solution $e^\dagger$.*

Proof. Let $X_0 = \{x \in X : (\forall k \geq 0)\ \mathsf{next}^k(x) \in Z\}$. Then we set $e^\dagger(x) = 0$ for $x \in X_0$. For all other x, there is some least number $j > 0$ such that $\mathsf{next}^j(x) \in X_0$. We take $e^\dagger(x) = \gamma(s(u(x)), e^\dagger(\mathsf{next}(x)))$. So formally we are defining $e^\dagger$ by recursion on the least j such that $\mathsf{next}^j(x) \in X_0$. It is easy to see that $e^\dagger(x)$ is a solution to e, and Proposition 5.4 implies that it is the only solution.

Lemma 5.7. *If $X = Y$, then e has a unique solution $e^\dagger$.*

Proof. We define an $(\mathbf{S}\times-)$-coalgebra structure $e^* : Y \to \mathbf{S}\times Y$ with the property that e and e^* have *the same solutions*, where a solution to e is $e^\dagger : X \to \mathbb{I}$ satisfying $e^\dagger = \tau \cdot (\mathbb{N} \times e^\dagger) \cdot e$, and a solution to e^* is $e^\dagger : X \to \mathbb{I}$ satisfying $e^\dagger = \gamma \cdot (\mathbf{S} \times e^\dagger) \cdot e^*$. We shall arrange that for all $x \in Y$,

$$\mathrm{con}(\pi(e^*(x))) \leq \frac{1}{4}, \tag{5.8}$$

where $\pi : \mathbf{S} \times Y \to \mathbf{S}$ is the projection. This last point implies that e^* has a unique solution, using Proposition 5.3 (or rather the analog of it for coalgebras of $\mathbf{S} \times -$), and the argument which we saw in the proof of Proposition 3.1.

The idea in the construction of e^* is that we should take some $x \in Z$ and replace the right-hand side of $e(x) = (0, \mathsf{next}(x))$ with pair consisting of a matrix of contractivity $\leq \frac{1}{4}$, and then an element of X other than $\mathsf{next}(x)$. For example, if $\mathsf{next}(x) \in \overline{Z}$, then we would like to have

$$e^*(x) = (s(0) \times s(u(\mathsf{next}(x))), \mathsf{next}^2(x))$$

(In this proof, the function next is from the original equation morphism e; the new e^* has its own next morphism, but we will not introduce notation for it.) But of course $\mathsf{next}(x)$ could itself belong to Z, so we need a more complicated definition; we need infinitely many elements of Z reachable from each $x \in X$. This is what is behind the definition of the set Y above. And we need to see that the replacement which we define does not change the solutions.

For $x \in X = Y$, take the least number $n(x) \geq 0$ such that $\mathsf{next}^{n(x)}(x) \in \overline{Z}$. (Such a number $n(x)$ exists since $x \in Y$. Also, $n = 0$ iff $x \in \overline{Z}$.) Write x' for $\mathsf{next}^{n(x)}(x)$ to save on notation. Let

$$e^*(x) = (M(x), \mathsf{next}(x')),$$

where $M(x)$ is the following matrix:

$$M(x) = s(0)^{n(x)} \times s(u(x')). \tag{5.9}$$

Then (5.8) for x follows from Proposition 5.1.

We must check that e and e^* have the same solutions. Let $e^\dagger$ be a solution to e. We claim that for all natural numbers m and all x such that $n(x) = m$,

$$e^\dagger(x) = \gamma(M(x), e^\dagger(\mathsf{next}(x'))). \tag{5.10}$$

We argue by induction on m. When $m = n(x) = 0$, $x \in \overline{Z}$, $M(x) = s(u(x))$, $x' = x$, and

$$\begin{aligned} e^\dagger(x) &= \tau(u(x), e^\dagger(\mathsf{next}(x))) && e^\dagger \text{ is a solution to } e \\ &= \gamma(s(u(x)), e^\dagger(\mathsf{next}(x))) && \text{by (5.4)} \\ &= \gamma(M(x), e^\dagger(\mathsf{next}(x'))) \end{aligned}$$

Now assume about m that (5.10) holds whenever $n(x) = m$. Let x be such that $n(x) = m + 1$. Write y for $\mathsf{next}(x)$. Then $x \in Z$, and $M(x) = s(0) \times M(y)$. Moreover, $n(y) = m$, so that $x' = y'$, and $\mathsf{next}^{n(y)+1}(y) = \mathsf{next}^{n(x)+1}(x)$. Let $r = e^\dagger(x') = e^\dagger(y')$. Let us start with a calculation which we will also use later in this proof:

$$\gamma(M(x), r) = \gamma(s(0) \times M(y), r) = \gamma(s(0), \gamma(M(y), r)) \tag{5.11}$$

Then we have

$$\begin{aligned} e^\dagger(x) &= \tau(0, e^\dagger(y)) && e^\dagger \text{ is a solution to } e \\ &= \gamma(s(0), \gamma(M(y), e^\dagger(y'))) && \text{by induction hypothesis} \\ &= \gamma(M(x), e^\dagger(x')) && \text{by (5.11)} \end{aligned}$$

This is (5.10) for x. This induction shows that $e^\dagger$ is a solution to e^*.

In the other direction, assume that $e^\dagger$ is a solution to e^*. That is, (5.10) holds for all $x \in X$. We claim that for all $x \in X$,

$$e^\dagger(x) = \tau(u(x), e^\dagger(\mathsf{next}(x))). \tag{5.12}$$

If $n(x) = 0$, then $M(x) = s(u(\mathsf{next}(x))$ and $x' = x$. By (5.10) and (5.4),

$$e^\dagger(x) = \gamma(s(u(x)), e^\dagger(\mathsf{next}(x))) = \tau(u(x), e^\dagger(\mathsf{next}(x))).$$

If $n(x) > 0$, then $x \in Z$, and $u(x) = 0$. Let us again write y for $\mathsf{next}(x)$, so that $M(x) = s(0) \times M(y)$, and $n(y) + 1 = n(x)$, and $x' = y'$. And we write r for $e^\dagger(x') = e^\dagger(y')$. So

$$\begin{aligned} e^\dagger(x) &= \gamma(M(x), e^\dagger(x')) && \text{by (5.10)} \\ &= \gamma(M(x), r) \\ &= \gamma(s(0), \gamma(M(y), r)) && \text{see (5.11)} \\ &= \tau(u(x), e^\dagger(y)) && \text{by (5.10) for } y \end{aligned}$$

We have (5.12). Thus e and e^* have the same solutions.

Theorem 5.8. *$(\mathbb{I}, \tau)$ is a corecursive algebra for $\mathbb{N} \times X$.*

Proof. Let $e : X \to \mathbb{N} \times X$. Both Y and $\overline{Y}$ are closed under next. So the coalgebra (X, e) splits into disjoint subcoalgebras with carriers Y and $\overline{Y}$. Each of these has a unique solution, by Lemmas 5.6 and 5.7. So X itself has a unique solution.

Corollary 5.9 ([10]). *The functor $FX = \mathbb{N} \times X$ has the following final coalgebras:*

(1) *The half-open interval $[0, 1)$ with structure ζ as given in Fig. 1.*
(2) *The set $\mathbb{R}_{\geq 0}$ with structure ξ as given in Fig. 1.*

Proof. (1) We again use Lemma 2.5, this time in conjunction with Theorem 5.8. The point is that τ in Theorem 5.8 does not take the value 1, so we have a map $\sigma : \mathbb{N} \times [0, 1) \to [0, 1)$, using the same formula as for τ. By Proposition 4.1(2), σ is a bijection. In other words, $([0, 1), \sigma)$ is a fixed point of F. Also, for each coalgebra $e \colon X \to \mathbb{N} \times X$, its solution $e^\dagger : X \to \mathbb{I}$ $(\mathbb{I}, \tau)$ does not take the value 1. (This is because $e^\dagger = \tau \cdot (\mathbb{N} \times e^\dagger) \cdot e$, and τ does not take the value 1.) Therefore $e^\dagger$ factors through the inclusion $m \colon [0, 1) \to \mathbb{I}$. This verifies that $([0, 1), \sigma^{-1})$ is a final F-coalgebra. Finally, we have seen in Proposition 4.1(2) that $\sigma^{-1} = \zeta$.

(2) The bijection h in Proposition 4.1(3) is an isomorphism of coalgebras, and so $(\mathbb{R}_{\geq 0}, \xi)$ is a final coalgebra.

5.4 Addendum

As we close the section and then the paper, we make two additional points.

First, the reader has likely noticed the missing second column in the bottom chart in Fig. 1. For the functor $GX = \mathbb{N} \times X + 1$, we can take any $s \in [0, 1]$ and get a corecursive G-algebra using the structure $[\tau, s]$. This is equivalent to saying that the F-algebra $(\mathbb{I}, \tau)$ is *completely iterative* (see [8]). However, we are not able to take this corecursive algebra $(\mathbb{I}, [\tau, s])$ and extract a final G-coalgebra using Lemma 2.5. Such a coalgebra would be (the inverse of) a subalgebra (A, α) of $(\mathbb{I}, [\tau, s])$ which includes the image of every G-coalgebra and with a bijective structure. Let $e : [0, 1) \to \mathbb{N} \times [0, 1) + 1$ be $\mathsf{inl} \cdot \zeta$. It is easy to check that the solution $e^\dagger$ is the inclusion $[0, 1) \hookrightarrow \mathbb{I}$. And so A is either $[0, 1)$ or $\mathbb{I}$ itself. Now A cannot be $\mathbb{I}$, since $[\tau, s]$ is not injective due to $\tau(1, 0) = \tau(0, 1)$. Further, A cannot be $[0, 1)$, since the algebra structure would have to be $[\sigma, s] : [0, 1)\times\mathbb{N}+1 \to [0, 1)$ with σ the restriction of τ to $[0, 1)$, and $s \in [0, 1)$, which is not injective. But as σ is surjective, s has a preimage s' under σ. Thus $[\sigma, s]$ cannot be injective.

Finally, we comment on the corecursive algebras in Theorems 4.2 and 5.8.

Proposition 5.10. *The F-algebras $(\mathbb{R}_{\geq 0}, \alpha)$ and $(\mathbb{I}, \tau)$ are not isomorphic.*

Proof. Assume towards a contradiction that $j : \mathbb{I} \to \mathbb{R}_{\geq 0}$ were an isomorphism of F-algebras. Then j is an isomorphism in Set, so it is a bijection. Let $q \in [0, 1]$ be any number such that $q \neq 1$ and such that $j(q)$ is rational. Recall from Corollary 5.9(1) that we have a final F-coalgebra $([0, 1), \zeta)$. Since $(\mathbb{I}, \tau)$ is corecursive, there is a unique $\zeta^\dagger : [0, 1) \to \mathbb{I}$ such that $\zeta^\dagger = \tau \cdot (\mathbb{N} \times \zeta^\dagger) \cdot \zeta$. The inclusion $i : [0, 1) \to \mathbb{I}$ satisfies this condition, and so $\zeta^\dagger = i$. But $(\mathbb{R}_{\geq 0}, \alpha)$ is also corecursive, and so there is a unique $\zeta^\ddagger : [0, 1) \to \mathbb{R}_{\geq 0}$ so that $\zeta^\ddagger = \alpha\cdot(\mathbb{N}\times\zeta^\ddagger)\cdot\tau$. Since j is an algebra morphism, $\zeta^\ddagger = j \cdot \zeta^\dagger = j \cdot i$. Then $\zeta^\ddagger(q) = (j \cdot i)(q) = j(q)$. So $\zeta^\ddagger(q)$ is rational. But this contradicts Lemma 4.5.

6 Conclusion

The main ideological point in this paper is that in looking at algebraic treatments of topics in continuous mathematics, one should not look only at final coalgebras but also should consider corecursive algebras. We demonstrated this by looking at infinite sums, and more extensively at corecursive algebra structures related to the real numbers. On a technical level, the main contribution of this paper is the work in Sect. 4, giving new corecursive algebra structures and new final coalgebra structures on sets of real numbers. One might have thought that since the original paper on this topic, Pavlović and Pratt [10], used continued fractions, that all future work in the area would do so. Thus the use of the linear fractional transformations in Theorem 5.8 is surprising because it avoids continued fractions.

The ideological point of the paper relates to the feeling that "... coinduction is new only by name, while it had actually been around for a long time, concealed within the infinitistic methods of *mathematical analysis*." [10, p. 106] We agree with this point. What we have not done, and thus what cries out to do next, is to show the use of the structures which we brought to light. A possible application of this work would be to give algebraic proofs of correctness of one or another algorithm for infinite-precision arithmetic using continued fractions.

Acknowledgments. We are grateful to Jiří Adámek and Stefan Milius for discussions of many matters related to this paper. We thank Helle Hansen for pointing out serious errors in a previous version, and for her encouragement. We also thank the referees for their criticism of an earlier version of the paper, and for their suggestions on how to improve it. All errors and shortcomings of this paper are our own.

References

1. Adámek, J., Milius, S., Moss, L.S.: Initial Algebras and Terminal Coalgebras: The Theory of Fixed Points of Functors. Cambridge University Press, Cambridge (in preparation)
2. Adámek, J., Milius, S., Velebil, J.: Elgot algebras. Log. Methods Comput. Sci. **2**(5:4), 1–31 (2006)
3. Capretta, V., Uustalu, T., Vene, V.: Corecursive algebras: a study of general structured corecursion. In: Oliveira, M.V.M., Woodcock, J. (eds.) SBMF 2009. LNCS, vol. 5902, pp. 84–100. Springer, Heidelberg (2009). https://doi.org/10.1007/978-3-642-10452-7_7
4. Feys, F.M.V., Hansen, H.H., Moss, L.S.: Long-term values in Markov decision processes, (co)algebraically. In: Cîrstea, C. (ed.) CMCS 2018. LNCS, vol. 11202, pp. 78–99. Springer, Cham (2018). https://doi.org/10.1007/978-3-030-00389-0_6
5. Hardy, G.H., Wright, E.M.: An Introduction to the Theory of Numbers, 4th edn. Oxford University Press, Oxford (1960)
6. Heckmann, R.: Contractivity of linear fractional transformations. Theoret. Comput. Sci. **279**(1–2), 65–82 (2002)
7. Loya, P.: Amazing and Aesthetic Aspects of Analysis. Springer, New York (2017). https://doi.org/10.1007/978-1-4939-6795-7
8. Milius, S.: Completely iterative algebras and completely iterative monads. Inform. Comput. **196**, 1–41 (2005)
9. Niven, I., Zuckerman, H.S., Montgomery, H.L.: An Introduction to the Theory of Numbers, 5th edn. Wiley, New York (1991)
10. Pavlović, D., Pratt, V.: The continuum as a final coalgebra. Theoret. Comput. Sci. **280**(1–2), 105–122 (2002)

Stick Breaking, in Coalgebra and Probability

Bart Jacobs(✉)

Institute for Computing and Information Sciences,
Radboud University, Nijmegen, The Netherlands
bart@cs.ru.nl

Abstract. Stick breaking is an elementary operation that has been formulated and used within stochastic process theory. This paper extracts the essentials of stick breaking in terms of isomorphisms between discrete probability distributions (with full support) and sequences of numbers between zero and one. This works for both finite and infinite distributions. Stick breaking is a repetitive construction with a strong coalgebraic flavour. Indeed, it is shown that stick breaking turns discrete distributions with infinite full support on the natural numbers into a final coalgebra. Once isolated as a separate construction, the usefulness of stick breaking is illustrated in the description of various probability distributions, such as binomial & multinomial and beta & Dirichlet.

1 Introduction

Consider the following mixture of paints, of four different colours: a quarter of red (R), a third of green (G), also a quarter of blue (B) and finally a sixth of yellow (Y). We write this 'convex' combination as:

$$\tfrac{1}{4}|R\rangle + \tfrac{1}{3}|G\rangle + \tfrac{1}{4}|B\rangle + \tfrac{1}{6}|Y\rangle.$$

The ket notation $|-\rangle$ is meaningless syntactic sugar, used to separate the fractions from the colours. This combination is called 'convex' since the probabilities add up to one. We call this convex combination a (discrete, finite) probability distribution over the set of colours $\{R, G, B, Y\}$. Let's write $\mathcal{D}_{fs}(\{R, G, B, Y\})$ for the set of all such distributions:

$$\mathcal{D}_{fs}(\{R,G,B,Y\}) = \Big\{ r_0|R\rangle + r_1|G\rangle + r_2|B\rangle + r_3|Y\rangle \;\Big|\; r_0, r_1, r_2, r_3 \in (0,1) \text{ with } r_0 + r_1 + r_2 + r_3 = 1 \Big\}.$$

We use the subscript *fs*, for 'full support'; this means that none of the r_i may be zero. It is needed below to prevent division by zero. We enforce fullness of support by requiring that the r_i are in the open unit interval $(0, 1) \subseteq \mathbb{R}$, without endpoints.

Published by Springer Nature Switzerland AG 2022
H. H. Hansen and F. Zanasi (Eds.): CMCS 2022, LNCS 13225, pp. 176–193, 2022.
https://doi.org/10.1007/978-3-031-10736-8_9

The above equation describes the set of distributions (on these four colours) as a *simplex*, of dimension *three*. Indeed, it is easy to see that one of the r_i is superfluous, since it is determined by the others. Explicitly, there is an isomorphism:

$$\mathcal{D}_{fs}\big(\{R, G, B, Y\}\big) \cong \Big\{(r_0, r_1, r_2) \in (0,1)^3 \,\Big|\, r_0 + r_1 + r_2 < 1\Big\}.$$

The above set on the right-hand-side is clearly a proper subset of the cube $(0,1)^3$. In essence, the *stick breaking* construction that plays a central role in this paper provides an isomorphism:

$$\mathcal{D}_{fs}\big(\{R, G, B, Y\}\big) \cong (0,1)^3. \tag{1}$$

This may not be immediate at first sight. One has to do (appropriate) rescaling.

There is an intuitive explanation of stick breaking in terms of successively breaking up a stick. We adapt this account to the above set of four colours. We start from three numbers $s_0, s_1, s_2 \in (0,1)$ and intend to turn them into a distribution on the set of colour $\{R, G, B, Y\}$.

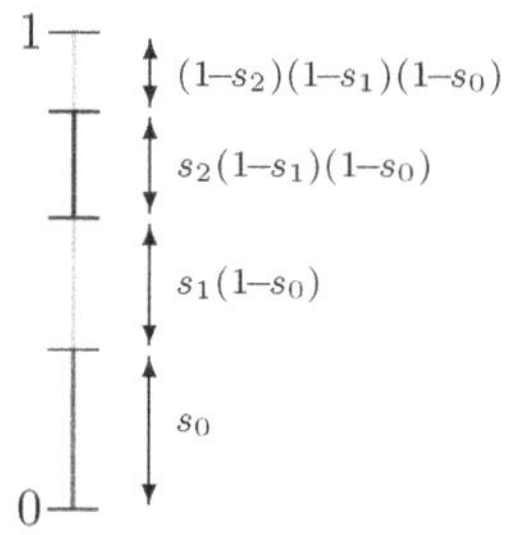

Imagine a stick of length one, as described vertically on the right. We take our first number $s_0 \in (0,1)$ and decide to paint the lower part/proportion s_0 red. We now have an unpainted part of length $1 - s_0$. We paint the s_1 proportion of it green. The newly painted part then has length $s_1(1 - s_0)$. The unpainted part is now $(1 - s_2)(1 - s_0)$. We paint the s_2-proportion of this remainder blue. The final remainder is then of length $(1 - s_2)(1 - s_2)(1 - s_0)$. We paint it yellow. Note that the resulting distribution has full support.

This construction can also be described in terms of breaking a stick, at each position where we have a change of colour in the above picture. The effect is a map $(0,1)^3 \to \mathcal{D}_{fs}\big(\{R, G, B, Y\}\big)$. We leave it at this stage to the reader to define an (inverse) map, in the opposite direction. Details will be provided in Sect. 3.

Stick breaking emerged in the description of stochastic processes, see [29] for an early source and [11] for an overview. The stick breaking isomorphism in (1) is applied to an iterated product (power) of spaces $(0,1)$ on the right-hand-side, without any dependencies. One can take for instance a (tensor) product of beta distributions on this product of $(0,1)$'s, and then transfer the result to a distribution on a space of the form $\mathcal{D}_{fs}(X)$ via stick breaking. In this way one obtains the (continuous) Dirichlet distribution (on discrete distributions) via multiple beta distributions and stick breaking. This is a known result—but not a very *well* known one—which we redescribe in Sect. 6 in the present setting.

Interestingly, the stick breaking construction can also be used for *infinite* products. It then yields an isomorphism $\mathcal{D}^{\infty}_{fs}(\mathbb{N}) \cong (0,1)^{\mathbb{N}}$, where we write $\mathcal{D}^{\infty}_{fs}$ for (discrete) distributions with infinite, full support. The above stick breaking construction is clearly repetitive, which suggests a coalgebraic structure. Indeed, as we shall see, the set of distributions $\mathcal{D}^{\infty}_{fs}(\mathbb{N})$ carries a coalgebra, which is even final.

It is especially this infinite form of stick breaking that is exploited in [29], and other sources like [6,11,24,25], to describe stochastic processes via infinite products followed by stick breaking. We give an impression of how this works, but only scratch the surface. The contribution of this paper lies in extracting the stick breaking operation from stochastic applications, in studying stick breaking on its own right, from a coalgebraic perspective, both in finite and infinite form, and then in re-applying the resulting insights in a few probabilistic illustrations.

The paper first fixes notation for discrete probability distributions, in order to introduce stick breaking in a coalgebraic setting, in Sect. 3. Then, after describing the essentials of multisets (bags) in Sect. 4, stick breaking is used to express multinomial draws from an urn in terms of successive binomial draws. This shows how drawing several balls from an urn with balls of multiple colours can be mimicked via urns with balls having only two colours (say black and white). This is *a priori* not entirely trivial.

The paper then moves on to continuous probability. It first shows how to express Dirichlet distributions as parallel beta distributions, followed by stick breaking—in analogy with the connection between multinomials and binomials via stick breaking. Insiders of the field will probably say "sure, we are aware of such connections", but to (relative) outsiders they may provide useful insight. At this stage it is assumed that the reader has a basic level of familiarity with these standard distributions. Subsequently, the use of infinite stick breaking is illustrated for the definition of stochastic processes in terms of countably many parallel beta distributions. In this setting a mean is calculated. In the end it is shown that this mean arises by finality from a very simple coalgebraic construction.

2 Discrete Probability Distributions

Let X be an arbitrary set. There are two equivalent ways of describing (discrete, finite) probability distributions on X.

- As finite convex combinations $r_1|x_1\rangle + \cdots r_n|x_n\rangle$ of elements $x_i \in X$, with probabilities $r_i \in [0,1]$ satisfying $\sum_i r_i = 1$.
- As functions $\omega\colon X \to [0,1]$ with finite support $\mathrm{supp}(\omega) := \{x \in X \mid \omega(x) \neq 0\}$ and with $\sum_x \omega(x) = 1$.

We freely switch between these two descriptions. We write $\mathcal{D}(X)$ for the set of such distributions on X, and $\mathcal{D}_{\mathit{fs}}(X) \subseteq \mathcal{D}(X)$ for the subset of distributions with *full support*, that is, with $\mathrm{supp}(\omega) = X$. Thus, writing $\mathcal{D}_{\mathit{fs}}(X)$ only makes sense when the set X is finite.

This $\mathcal{D}$ is a monad on the category **Sets**. We make occasional use of this fact, so we do not spell out the details here; we refer to external sources instead, like [13,14].

We shall write $\mathcal{D}^\infty(X)$ for arbitrary functions $\omega\colon X \to [0,1]$ with $\sum_x \omega(x) = 1$. We then put no restriction on the support of ω, but it is not hard to show that when $\sum_x \omega(x) = 1$ the support is countable, or finite.

We write $\mathcal{D}^{\infty}_{fs}(X) \subseteq \mathcal{D}^{\infty}(X)$ for the subset of distributions with *full support*. This only makes sense if the set X is countable. We use it especially for $X = \mathbb{N}$.

3 Stick Breaking

We fix a set A and consider the functor $A \times (-)\colon \textbf{Sets} \to \textbf{Sets}$. It is well known that the final coalgebra of this functor is the set $A^{\mathbb{N}}$ of infinite sequences of $(a_n)_{n\in\mathbb{N}}$ of elements $a_n \in A$.

Similarly, the functor $A + A \times (-)$ has the set $A^{\infty} := A^{+} + A^{\mathbb{N}}$ of non-empty finite and infinite sequences as final coalgebra. Details can be found in any introductory text to coalgebra, see *e.g.* [1,13,17,27].

In the sequel we take $A = (0,1)$, the open unit interval $(0,1) \subseteq \mathbb{R}$ of numbers between zero and one. We introduce stick breaking first in the infinite case. Subsequently, the finite case is handled.

3.1 Infinite Stick Breaking

By finality we introduce a function $f\colon \mathcal{D}^{\infty}_{fs}(\mathbb{N}) \to (0,1)^{\mathbb{N}}$ in the following diagram.

$$\begin{array}{ccc} (0,1) \times \mathcal{D}^{\infty}_{fs}(\mathbb{N}) & \xrightarrow{\;\mathrm{id}\times f\;} & (0,1) \times (0,1)^{\mathbb{N}} \\ {\scriptstyle shift}\uparrow & & \uparrow{\scriptstyle \cong\ \langle head,\, tail\rangle} \\ \mathcal{D}^{\infty}_{fs}(\mathbb{N}) & \xrightarrow{\quad f\quad} & (0,1)^{\mathbb{N}} \end{array} \tag{2}$$

The shift coalgebra on the left is defined as:

$$shift(\omega) := \left(\omega(0),\ \sum_{n\in\mathbb{N}} \frac{\omega(n+1)}{1-\omega(0)} \big| n \big\rangle \right). \tag{3}$$

This shift operation does three things: (1) it takes the head $\omega(0)$ of the infinite sequence $\omega = (\omega(0), \omega(1), \ldots)$; (2) it shifts the remaining tail one position forwards, so that $\omega(1)$ becomes the new head; (3) it renormalises this tail to a new distribution via division by $1 - \omega(0) = \sum_{n\geq 1} \omega(n)$.

Since each $\omega \in \mathcal{D}^{\infty}_{fs}(\mathbb{N})$ has full support, each probability $\omega(n)$ is non-zero, for $n \in \mathbb{N}$. But then none of these $\omega(n)$ can be equal to one. This ensures that the shift map is well-defined.

Proposition 1. *The function $f\colon \mathcal{D}^{\infty}_{fs}(\mathbb{N}) \to (0,1)^{\mathbb{N}}$ introduced in (2) by finality is an isomorphism. We shall write $sb = f^{-1}$ for the inverse and call it (infinite)* stick breaking.

As a result, the shift coalgebra is also final—and thus an isomorphism.

Proof. Via commutation of Diagram (2) we get:

$$f(\omega) = \left(\omega(0),\ \frac{\omega(1)}{1-\omega(0)},\ \frac{\omega(2)}{1-\omega(0)-\omega(1)}, \ldots, \frac{\omega(i)}{1-\sum_{j<i}\omega(j)}, \ldots \right).$$

For instance, the second entry is obtained as:

$$\frac{\left(\frac{\omega(2)}{1-\omega(0)}\right)}{1-\frac{\omega(1)}{1-\omega(0)}} = \frac{\omega(2)}{1-\omega(0)-\omega(1)}.$$

In the other direction one obtains stick breaking as:

$$sb\Big(r_0, r_1, \ldots\Big) := r_0\big|0\big\rangle + r_1(1-r_0)\big|1\big\rangle + \cdots + r_i \prod_{j<i}(1-r_j)\big|i\big\rangle + \cdots \tag{4}$$

If we abbreviate $\rho := sb\big(r_0, r_1, \ldots\big)$ then we get as basic property, for each $i \in \mathbb{N}$

$$1 - \sum_{j \le i} \rho(j) = \prod_{j \le i} (1-r_j). \tag{5}$$

This follows by induction on i. The statement trivially holds for $i = 0$. Next,

$$\begin{aligned} 1 - \sum_{j \le i+1} \rho(j) &= \left(1 - \sum_{j \le i} \rho(j)\right) - \rho(i+1) \\ &\overset{\text{(IH)}}{=} \prod_{j \le i} (1-r_j) - r_{i+1} \prod_{j \le i} (1-r_j) \\ &= (1-r_{i+1}) \prod_{j \le i} (1-r_j) = \prod_{j \le i+1} (1-r_j). \end{aligned}$$

We can now see that the sequence ρ forms a proper distribution:

$$\begin{aligned} \sum_{i \in \mathbb{N}} \rho(i) &= \lim_{i \to \infty} \sum_{j \le i} \rho(j) = 1 - \lim_{i \to \infty} 1 - \sum_{j \le i} \rho(j) \\ &\overset{(5)}{=} 1 - \lim_{i \to \infty} \prod_{j \le i} (1-r_j) = 1 - 0 = 1. \end{aligned}$$

This works because an infinite product of numbers $s_i \in (0,1)$ is zero.

It is not hard to see that these two functions $f \colon \mathcal{D}^{\infty}_{fs}(\mathbb{N}) \to (0,1)^{\mathbb{N}}$ and $sb \colon (0,1)^{\mathbb{N}} \to \mathcal{D}^{\infty}_{fs}(\mathbb{N})$ are each other's inverses. □

Example 2. Consider the infinite distribution:

$$\omega = \sum_{n \in \mathbb{N}} \tfrac{2}{5} \cdot \left(\tfrac{3}{5}\right)^n \big|n\big\rangle = \tfrac{2}{5}\big|0\big\rangle + \tfrac{6}{25}\big|1\big\rangle + \tfrac{18}{125}\big|2\big\rangle + \tfrac{54}{625}\big|3\big\rangle + \cdots$$

We can see that it is a distribution via the familiar formula:

$$\sum_{n \ge 0} r^n = \frac{1}{1-r} \qquad \text{for } r \in (0,1). \tag{6}$$

Then:

$$\sum_{n \ge 0} \omega(n) = \tfrac{2}{5} \cdot \sum_{n \ge 0} \left(\tfrac{3}{5}\right)^n \overset{(6)}{=} \tfrac{2}{5} \cdot \frac{1}{1 - 3/5} = \frac{2}{5-3} = 1.$$

The sequence of numbers in $(0,1)$ corresponding to ω is constant:

$$sb^{-1}(\omega) = (\tfrac{2}{5}, \tfrac{2}{5}, \tfrac{2}{5}, \ldots).$$

In general, for $r \in (0,1)$, we have $sb(r, r, r, \ldots) = \sum_{n \geq 0} r(1-r)^n |n\rangle$.

3.2 Finite Stick Breaking

Having seen the isomorphism $\mathcal{D}^{\infty}_{\mathrm{fs}}(\mathbb{N}) \cong (0,1)^{\mathbb{N}}$ in Proposition 1 one wonders if it can be restricted to distributions with finite support. We write $\underline{n} = \{0, 1, \ldots, n-1\}$, where $n \in \mathbb{N}$, for a chosen set with n elements.

We look at the set of distributions $\mathcal{D}_{\mathrm{fs}}(\underline{n})$. Given a distribution $\omega \in \mathcal{D}_{\mathrm{fs}}(\underline{n})$ we can apply a shift operation like in (3), to peel off the first element $\omega(0)$. However, what remains is a (full) distribution on $\underline{n-1}$. This gives a function $\mathcal{D}_{\mathrm{fs}}(\underline{n}) \to (0,1) \times \mathcal{D}_{\mathrm{fs}}(\underline{n-1})$, for $n > 0$. This is not a coalgebra, in the ordinary sense—but it may be understood as a coalgebra in dependent type theory.

We need a trick. We incorporate $\mathcal{D}_{\mathrm{fs}}(\underline{n})$ into a subset of $\mathcal{D}^{\infty}(\mathbb{N})$, namely the subset were probabilities may be zero, but once they are zero, they remain zero in all subsequent positions. We use the following *ad hoc* notation.

$$\begin{aligned} \mathcal{D}^{\infty}_{\mathrm{fs}<}(\mathbb{N}) &:= \big\{\omega \colon \mathbb{N} \to [0,1) \,\big|\, \textstyle\sum_n \omega(n) = 1 \text{ and } \forall n.\, \omega(n) = 0 \Rightarrow \forall m > n.\, \omega(m) = 0\big\} \\ &\subseteq \mathcal{D}^{\infty}(\mathbb{N}). \end{aligned}$$

Notice that the 'shortest' list in $\mathcal{D}^{\infty}_{\mathrm{fs}<}(\mathbb{N})$ is of the form $r|0\rangle + (1-r)|1\rangle$ for $r \in (0,1)$. For each $n > 1$ there is an inclusion $\mathcal{D}_{\mathrm{fs}}(\underline{n}) \hookrightarrow \mathcal{D}^{\infty}_{\mathrm{fs}<}(\mathbb{N})$.

We can now define a shift map of the following form, for $n > 0$.

$$\mathcal{D}^{\infty}_{\mathrm{fs}<}(\mathbb{N}) \xrightarrow{\;\;shift\;\;} (0,1) + (0,1) \times \mathcal{D}^{\infty}_{\mathrm{fs}<}(\mathbb{N})$$

This function is defined as:

$$shift(\omega) := \begin{cases} r & \text{if } \omega = r|0\rangle + (1-r)|1\rangle \\ \big(\omega(0), \sum_n \frac{\omega(n+1)}{1-\omega(0)} |n\rangle\big) & \text{otherwise} \end{cases} \tag{7}$$

In the first case we have reached a distribution of minimal size. The second case is as in (3).

As mentioned in the beginning of this section, the set $(0,1)^{\infty} = (0,1)^{+} + (0,1)^{\mathbb{N}}$ of non-empty finite and infinite sequences of numbers in the open interval $(0,1)$ forms a final coalgebra of the functor $(0,1) + (0,1) \times (-)$. By finality we thus get a map g in:

$$\begin{array}{ccc} (0,1) + (0,1) \times \mathcal{D}^{\infty}_{\mathrm{fs}<}(\mathbb{N}) & \xrightarrow{\;\mathrm{id}+(\mathrm{id}\times g)\;} & (0,1) + (0,1) \times (0,1)^{\infty} \\ \Big\uparrow{\scriptstyle shift} & & \cong \Big\uparrow{\scriptstyle next} \\ \mathcal{D}^{\infty}_{\mathrm{fs}<}(\mathbb{N}) & \xrightarrow{\qquad g \qquad} & (0,1)^{\infty} = (0,1)^{+} + (0,1)^{\mathbb{N}} \end{array} \tag{8}$$

It is not hard to see that the map g sends a distribution $\frac{1}{16}|0\rangle+\frac{1}{4}|1\rangle+\frac{3}{16}|2\rangle+\frac{1}{2}|3\rangle$ to the sequence $\langle\frac{1}{16},\frac{4}{15},\frac{3}{11}\rangle \in (0,1)^{\infty}$.

We now get the finite analogue of Proposition 1. The proof is essentially as in the infinite case, and is left to the reader.

Proposition 3. *For each $n > 1$ the function g defined in (8) restricts to a function $\mathcal{D}_{fs}(\underline{n}) \to (0,1)^{+}$. In fact, it forms an isomorphism $g\colon \mathcal{D}_{fs}(\underline{n}) \xrightarrow{\cong} (0,1)^{n-1}$. Its inverse $sb\colon (0,1)^{n-1} \xrightarrow{\cong} \mathcal{D}_{fs}(\underline{n})$ will be called* stick breaking. *It is given by:*

$$
\begin{aligned}
& sb(r_0,\dots,r_{n-2}) \\
&= r_0\big|0\big\rangle + r_1(1-r_0)\big|1\big\rangle + r_2(1-r_1)(1-r_0)\big|2\big\rangle + \cdots + \\
&\qquad r_{n-2}(1-r_{n-3})\cdots(1-r_0)\big|n-2\big\rangle + (1-r_{n-2})\cdots(1-r_0)\big|n-1\big\rangle.
\end{aligned}
$$

□

Example 4. For instance,

$$
\begin{aligned}
sb(\tfrac{1}{4},\tfrac{1}{3},\tfrac{3}{4}) &= \tfrac{1}{4}|0\rangle + \tfrac{1}{4}|1\rangle + \tfrac{3}{8}|2\rangle + \tfrac{1}{8}|3\rangle \\
sb(\tfrac{7}{8},\tfrac{2}{3},\tfrac{3}{4}) &= \tfrac{7}{8}|0\rangle + \tfrac{1}{12}|1\rangle + \tfrac{1}{32}|2\rangle + \tfrac{1}{96}|3\rangle.
\end{aligned}
$$

Stickbreaking does *not* preserve convex combinations. For instance:

$$
\begin{aligned}
\tfrac{1}{4}\cdot sb(\tfrac{1}{4},\tfrac{1}{3},\tfrac{3}{4}) + \tfrac{3}{4}\cdot sb(\tfrac{7}{8},\tfrac{2}{3},\tfrac{3}{4}) &= \tfrac{23}{32}|0\rangle + \tfrac{1}{8}|1\rangle + \tfrac{15}{128}|2\rangle + \tfrac{5}{128}|3\rangle \\
&\neq \tfrac{23}{32}|0\rangle + \tfrac{21}{128}|1\rangle + \tfrac{45}{512}|2\rangle + \tfrac{15}{512}|3\rangle \\
&= sb\big(\tfrac{1}{4}\cdot\langle\tfrac{1}{4},\tfrac{1}{3},\tfrac{3}{4}\rangle + \tfrac{3}{4}\cdot\langle\tfrac{7}{8},\tfrac{2}{3},\tfrac{3}{4}\rangle\big).
\end{aligned}
$$

4 Multisets

A *multiset* (or *bag*) is a 'subset' except that elements may occur multiple times. We write a multiset on a set X also in ket form, as a finite formal sum $n_1|x_i\rangle + \cdots + n_k|x_k\rangle$ of elements $x_i \in X$ and natural numbers $n_i \in \mathbb{N}$. Such a multiset can equivalently be described as a function $\varphi\colon X \to \mathbb{N}$ with finite support. We write $\mathcal{M}(X)$ for the set of finite multisets on X, and $\mathcal{M}_{fs}(X) \subseteq \mathcal{M}(X)$ for the subset of multisets with full support, that is, with $\varphi(x) \neq 0$ for each $x \in X$; again, this only makes sense when the set X is finite. The multiset operation $\mathcal{M}$ is a monad on **Sets**, like $\mathcal{D}$.

We associate several numbers with a multiset $\varphi \in \mathcal{M}(X)$.

- The *size* $\|\varphi\| := \sum_{x\in X}\varphi(x)$ is the total number of elements in the multiset, including multiplicities.
- The *factorial* $\varphi\wr := \prod_{x\in X}\varphi(x)!$ is the product of (ordinary) factorials of the multiplicities.
- The *multinomial coefficient* is $(\varphi) := \frac{\|\varphi\|!}{\varphi\wr}$.

We write $\mathcal{M}[K](X) \subseteq \mathcal{M}(X)$ for the subset of multisets with size $K \in \mathbb{N}$.

4.1 Binomial and Multinomial Distributions

Drawing coloured balls from an urn is one of the most basic probabilistic models, see *e.g.* [18,21,23,26] and many other references. Here we look at draws with replacement, known as binomial draws (when there are two colours) or multinomial draws (when there are multiple colours).

For a fixed number $K \in \mathbb{N}$ we describe the familiar *binomial* distributions via a function:

$$[0,1] \xrightarrow{\;bn[K]\;} \mathcal{D}\big(\{0,1,\ldots,K\}\big)$$

It captures the probability of drawing $i \in \{0,1,\ldots,K\}$ black balls from an urn with (only) black and white balls, out of K independent draws, each with black ball probability $r \in [0,1]$. Thus:

$$bn[K](r) := \sum_{0 \le i \le K} \binom{K}{i} \cdot r^i \cdot (1-r)^{K-i} \, \big| i \big\rangle.$$

There is a *multinomial* version which assigns a probability to a multiset of size K, as a draw (with replacement) of K-many balls from an urn with balls whose colours are described by a set X. The distribution of colours over the balls in the urn is captured abstractly via a distribution $\omega \in \mathcal{D}(X)$. Multinomial distributions will thus be described as a function:

$$\mathcal{D}(X) \xrightarrow{\;mn[K]\;} \mathcal{D}\big(\mathcal{M}[K](X)\big)$$

The definition is:

$$mn[K](\omega) := \sum_{\varphi \in \mathcal{M}[K](X)} (\varphi) \cdot \prod_{x \in X} \omega(x)^{\varphi(x)} \, \big| \varphi \big\rangle.$$

The binomial version is a special case, when X is a two-element set $\underline{2}$, via the isomorphisms $\mathcal{D}(\underline{2}) \cong [0,1]$ and $\mathcal{M}[K](\underline{2}) \cong \{0,1,\ldots,K\}$. For more information, see *e.g.* [15,16].

We present one result about multinomials. It is useful to recall as preparation for a similar but more complicated result later on.

Lemma 5. *Let $\omega \in \mathcal{D}(X)$ be an 'urn' and K the size of draws. The mean of the multinomial* $mn[K](\omega)$ *is $K \cdot \omega$. Explicitly,*

$$mean\Big(mn[K](\omega)\Big) := \sum_{\varphi \in \mathcal{M}[K](X)} mn[K](\omega)(\varphi) \cdot \varphi = K \cdot \omega.$$

Strictly speaking, $K \cdot \omega$ is not a multiset, since we allow only natural numbers as multiplicities. But for a result like this one may wish to allow non-negative reals too. Once we do so, we can use inclusions $\mathcal{D}(Y) \hookrightarrow \mathcal{M}(Y)$ and see this result as an application of the multiplication map μ of the multiset monad $\mathcal{M}$, in:

$$\begin{array}{ccccc} \mathcal{D}(X) & \xrightarrow{mn[K]} & \mathcal{D}(\mathcal{M}[K](X)) & \hookrightarrow & \mathcal{M}(\mathcal{M}(X)) \\ & & & & \downarrow{\scriptstyle \mu} \\ & \searrow_{K\cdot(-)} & & \longrightarrow & \mathcal{M}(X) \end{array} \tag{9}$$

Proof. Fix an arbitrary element $y \in X$.

$$\begin{array}{rcl} \Big(mean\big(mn[K](\omega)\big)\Big)(y) \\ & = & \sum_{\varphi\in\mathcal{M}[K](X)} mn[K](\omega)(\varphi)\cdot\varphi(y) \\ & = & \sum_{\varphi\in\mathcal{M}[K](X),\,\varphi(y)\neq 0} \varphi(y)\cdot\frac{K!}{\prod_x \varphi(x)!}\cdot\prod_x \omega(x)^{\varphi(x)} \\ & = & \sum_{\varphi\in\mathcal{M}[K](X),\,\varphi(y)\neq 0} \frac{K\cdot(K-1)!}{(\varphi(y)-1)!\cdot\prod_{x\neq y}\varphi(x)!}\cdot\omega(y)\cdot\omega(y)^{\varphi(y)-1}\cdot\prod_{x\neq y}\omega(x)^{\varphi(x)} \\ & = & K\cdot\omega(y)\cdot\sum_{\varphi\in\mathcal{M}[K-1](X)}\frac{(K-1)!}{\prod_x\varphi(x)!}\cdot\prod_x\omega(x)^{\varphi(x)} \\ & = & K\cdot\omega(y)\cdot\sum_{\varphi\in\mathcal{M}[K-1](X)} mn[K-1](\omega)(\varphi) \\ & = & K\cdot\omega(y). \end{array}$$

□

5 Multinomials as Iterated Binomials

A simple question is: can we mimic a draw of multiple coloured balls from an urn in terms of draws of only two colours? More precisely, can we express a multinomial draw in terms of several binomial draws? We then encounter the problem that binomial draws use probabilities between zero and one and multinomials draws use distributions, as convex combinations. We show that stick breaking sb provides the connection. We first give a concrete formulation and then express it more abstractly.

Lemma 6. *Fix $n \geq 1$ and $K \geq 0$. For probabilities $\vec{r} = r_0, \ldots, r_{n-2} \in (0,1)^{n-1}$ and a multiset $\varphi = \sum_{i<n} k_i|i\rangle \in \mathcal{M}[K](\underline{n})$,*

$$\begin{array}{rcl} mn[K]\big(sb(\vec{r})\big)(\varphi) & = & bn[K](r_0)(k_0)\cdot bn[K-k_0](r_1)(k_1) \\ & & \cdot\ldots\cdot bn[K-\sum_{i<n-2}k_i](r_{n-2})(k_{n-2}). \end{array}$$

Notice that the last multiplicity $k_{n-1} = \varphi(n-1)$ is not used. It is superfluous if we know that the multiset has size K, since then $k_{n-1} = K - \sum_{i<n-1} k_i$.

Proof. One can use induction on n. When $n = 1$ the above equation formulates a binary multinomial as binomial, via the isomorphisms $\mathcal{D}(\underline{2}) \cong [0,1]$ and $\mathcal{M}[K](\underline{2}) \cong \{0, 1, \ldots, K\}$. Concretely:

$$mn[K]\big(r|0\rangle + (1-r)|1\rangle\big)\big(k_0|0\rangle + k_1|1\rangle\big) = bn[K](r)(k_0).$$

Next, let $\varphi = \sum_{i \leq n} k_i|i\rangle \in \mathcal{M}[K](\underline{n+1})$ and $\vec{r} = r_0, \ldots, r_{n-1} \in (0,1)^n$ be given. We use a shifted multiset $\varphi' = \sum_{i<n-1} k_{i+1}|i\rangle$ of size $K - k_0$. Then:

$$\begin{aligned}
& bn[K](r_0)(k_0) \cdot bn[K-k_0](r_1)(k_1) \cdot \ldots \cdot bn[K-\textstyle\sum_{i<n-1} k_i](r_{n-1})(k_{n-1}) \\
&\overset{(\mathrm{IH})}{=} bn[K](r_0)(k_0) \cdot mn[K-k_0]\big(sb(r_1, \ldots, r_{n-1})\big)(\varphi') \\
&= \binom{K}{k_0} \cdot r_0^{k_0} \cdot (1-r_0)^{K-k_0} \cdot (\varphi') \cdot \prod_{i>0} sb(r_1, \ldots, r_{n-1})(i)^{k_i} \\
&= \frac{K!}{k_0! \cdot (K-k_0)!} \cdot \frac{(K-k_0)!}{k_1! \cdots k_{n-1}!} \cdot r_0^{k_0} \cdot \prod_{i>0} \Big(sb(r_1, \ldots, r_{n-1})(i) \cdot (1-r_0)\Big)^{k_i} \\
&= (\varphi) \cdot \prod_{i \geq 0} sb(r_0, \ldots, r_{n-1})(i)^{k_i} \\
&= mn[K]\big(sb(\vec{r})\big)(\varphi).
\end{aligned}$$

□

We reorganise this result a bit. For $K, n \in \mathbb{N}$ with $n > 0$ we define a set of sequences of natural numbers.

$$\mathcal{S}[K](n) := \{(k_0, \ldots, k_{n-2}) \in \mathbb{N}^{n-1} \mid \forall i.\, k_i \leq K - \textstyle\sum_{j<i} k_j\}.$$

Next we define the *sequential binomial* map $sbn[K] \colon (0,1)^{n-1} \to \mathcal{D}\big(\mathcal{S}[K](n)\big)$ by:

$$\begin{aligned}
sbn[K](\vec{r})(\vec{k}) \;=\; & bn[K](r_0)(k_0) \cdot bn[K-k_0](r_1)(k_1) \\
& \cdot \ldots \cdot bn[K-\textstyle\sum_{i<n-2} k_i](r_{n-2})(k_{n-2}).
\end{aligned}$$

Theorem 7. *In the situation described above, multinomial distributions can be described as sequential binomial distributions via stick breaking, as in the following commuting diagram.*

$$\begin{array}{ccc}
\mathcal{D}\big(\mathcal{S}[K](n)\big) & \xrightarrow{\;\cong\;} & \mathcal{D}\big(\mathcal{M}[K](\underline{n})\big) \\
{\scriptstyle sbn[K]}\uparrow & & \uparrow{\scriptstyle mn[K]} \\
(0,1)^{n-1} & \xrightarrow[\cong]{\;sb\;} & \mathcal{D}(\underline{n})
\end{array}$$

Proof. This is just a fancy reformulation of Lemma 6. It uses the obvious isomorphism $\mathcal{S}[K](n) \xrightarrow{\cong} \mathcal{M}[K](\underline{n})$, given by $(k_0, \ldots, k_{n-2}) \mapsto \sum_{i<n-1} k_i|i\rangle + (K - \sum_i k_i)|n-1\rangle$, at the top, together with the functoriality of $\mathcal{D}$. □

6 Dirichlet via Parallel Beta's

This section describes an application of stick breaking in continuous probability theory. It reformulates the famous Dirichlet distribution in terms of parallel beta distributions, with stick breaking forming the connection. This is similar to the result in the previous section, since beta distributions can be understood as binary versions of Dirichlet distributions—just like binomials being binary versions of multinomials. For background information on the beta and Dirichlet distributions we refer to standard textbooks, like [2,3,8,19,30].

We shall describe these continuous distributions via the Giry monad $\mathcal{G}$, which generalises the discrete probability monad $\mathcal{D}$, see [12,14,22] for details. We shall use continuous probability distributions on subsets $S \subseteq \mathbb{R}^n$, given by a probability density function (pdf) $f\colon S \to \mathbb{R}_{\geq 0}$, satisfying $\int f(x)\,\mathrm{d}x = 1$. The distribution itself is given by a mapping from the Borel σ-algebra Σ_S of measurable subsets of S, to $[0,1]$. Thus, it is the mapping on measurable subsets $M \subseteq S$,

$$M \longmapsto \int_{x \in M} f(x)\,\mathrm{d}x.$$

We write $\mathcal{G}(S)$ for the set of such distributions. For $\phi \in \mathcal{G}(S)$ and $\chi \in \mathcal{G}(T)$ there is a parallel product $\phi \otimes \chi \in \mathcal{G}(S \times T)$ determined by $(\phi \otimes \chi)(M \times N) = \phi(M) \cdot \chi(N)$, for measurable subsets $M \subseteq S$, $N \subseteq T$.

We illustrate this for the beta distributions on $(0,1)$, which we describe as parameterised by numbers $a, b \in \mathbb{N}_{>0}$. This can be generalised to more general numbers, but we don't need that here. The pdf $\mathit{pbf}_{\mathit{Beta}}(a,b)\colon (0,1) \to \mathbb{R}_{\geq 0}$ is given by:

$$\mathit{pbf}_{\mathit{Beta}}(a,b)(r) := \frac{r^{a-1} \cdot (1-r)^{b-1}}{B(a,b)} \quad \text{where} \quad B(a,b) = \frac{(a-1)! \cdot (b-1)!}{(a+b-1)!}. \tag{10}$$

The Dirichlet distribution takes the form of a map:

$$\mathcal{M}_{\mathrm{fs}}(\underline{n}) \xrightarrow{\quad \mathit{Dir} \quad} \mathcal{G}\big(\mathcal{D}_{\mathrm{fs}}(\underline{n})\big). \tag{11}$$

For a multiset $\psi \in \mathcal{M}_{\mathrm{fs}}(\underline{n})$ we describes its pdf $\mathcal{D}_{\mathrm{fs}}(\underline{n}) \to \mathbb{R}_{\geq 0}$ as:

$$\mathit{pbf}_{\mathit{Dir}(\psi)}(\omega) := \frac{(\|\psi\|-1)!}{(\psi-\mathbf{1})\wr\!\wr} \cdot \prod_{i \in \underline{n}} \omega(i)^{\psi(i)-1} \qquad \text{where} \qquad \mathbf{1} = \sum_{i \in \underline{n}} 1|i\rangle.$$

This looks very much like the multinomial distribution $\mathit{mn}[K]$. Indeed, there is a close connection: if we view the multinomial as a map $\mathit{mn}[K]\colon \mathcal{D}_{\mathrm{fs}}(\underline{n}) \to \mathcal{D}(\mathcal{M}[K](\underline{n})) \cong \mathcal{G}(\mathcal{M}[K](\underline{n}))$ then Dirichlet is its *dagger* [4,5,10] in the opposite direction (11), using a uniform prior. Details will be elaborated elsewhere. A further basic fact is that the Kleisli composition 'multinomial after Dirichlet' yields Pólya distributions [21].

Our focus lies on the theorem below that expresses the Dirichlet distribution as parallel product $\otimes$ of beta's, connected via stick breaking. This is a known 'folklore' result, for which it is hard to find a precise reference and/or formulation, but see [11, §3.1] for a brief description. As an aside, there is also a way to express Dirichlet via gamma distributions that is more familiar, see *e.g.* [30, 7.7.1] or [7, Prop. 4.1].

Here we can precisely formulate Dirichlet via beta's because we have explicitly identified the (finite) stick breaking isomorphism $sb\colon (0,1)^{n-1} \xrightarrow{\cong} \mathcal{D}_{fs}(\underline{n})$. The formulation below uses functoriality of Giry $\mathcal{G}$.

Theorem 8. *For* $n > 0$ *and* $\psi \in \mathcal{M}(\underline{n})$,

$$\begin{aligned} Dir(\psi) = \mathcal{G}(sb)\Big(&Beta\big(\psi(0), \textstyle\sum_{i>0}\psi(i)\big) \otimes Beta\big(\psi(1), \textstyle\sum_{i>1}\psi(i)\big) \otimes \cdots \\ &\cdots \otimes Beta\big(\psi(n-3), \psi(n-2)+\psi(n-1)\big) \otimes Beta\big(\psi(n-2), \psi(n-1)\big)\Big). \end{aligned}$$

Proof. One proceeds like in the proof of Lemma 6, in combination with integration by substitution. We give an exemplaric proof, for $n = 3$, illustrating how this works.

The Dirichlet distribution involves, in this case, an integral over $\mathcal{D}_{fs}(\underline{3})$. This means that we integrate over $(0,1)$, say with a variable s_0, and then over $(0,1-s_0)$, say with s_1, and then use $s_2 = 1 - s_0 - s_1$. We thus restrict the inverse of the stick breaking isomorphism $sb\colon (0,1)^2 \xrightarrow{\cong} \mathcal{D}_{fs}(\underline{3})$ to an isomorphism:

$$D_2 := \{(s_0, s_1) \mid s_0 \in (0,1), s_1 \in (0, 1-s_0)\} \xrightarrow[\cong]{h} (0,1)^2$$

There is an isomorphism $\mathcal{D}_{fs}(\underline{3}) \cong D_2$ via dropping the last number. This function $h = (h_0, h_1)$ is thus given by:

$$h(s_0, s_1) = (s_0, \tfrac{s_1}{1-s_0}).$$

In order to do (multidimensional) integration by substitution we need the determinant of the matrix of partial derivatives of h. This is:

$$\begin{vmatrix} \frac{\partial h_0}{\partial s_0}(\vec{s}) & \frac{\partial h_0}{\partial s_1}(\vec{s}) \\ \frac{\partial h_1}{\partial s_0}(\vec{s}) & \frac{\partial h_1}{\partial s_1}(\vec{s}) \end{vmatrix} = \begin{vmatrix} 1 & 0 \\ \frac{s_1}{1-s_0} & \frac{1}{1-s_0} \end{vmatrix} = \frac{1}{1-s_0}. \qquad (*)$$

We are now ready to prove the equation in the theorem, for $n = 3$. We fix $\psi \in \mathcal{M}_{fs}(\underline{3})$. Let $M \subseteq \mathcal{D}_{fs}(\underline{3})$ be an arbitrary measurable subset; we identify it with $\overline{M} \subseteq D_2$ when needed, via the isomorphism $\mathcal{D}_{fs}(\underline{3}) \cong D_2$ described above.

$$\begin{aligned}
&\mathcal{G}(sb)\Big(\mathit{Beta}(\psi(0),\psi(1)+\psi(2)) \otimes \mathit{Beta}(\psi(1),\psi(2))\Big)(M) \\
&= \int_{(r_0,r_1)\in sb^{-1}(M)} \mathit{pbf}_{\mathit{Beta}}(\psi(0),\psi(1)+\psi(2))(r_0)\cdot \mathit{pbf}_{\mathit{Beta}}(\psi(1),\psi(2))(r_1)\,\mathrm{d}r_0,r_1 \\
&\overset{(10)}{=} \int_{(r_0,r_1)\in h(\overline{M})} \frac{r_0^{\psi(0)-1}\cdot(1-r_0)^{\psi(1)+\psi(2)-1}}{B(\psi(0),\psi(1)+\psi(2))}\cdot\frac{r_1^{\psi(1)-1}\cdot(1-r_1)^{\psi(2)-1}}{B(\psi(1),\psi(2))}\,\mathrm{d}r_0,r_1 \\
&= \int_{(s_0,s_1)\in\overline{M}} \frac{s_0^{\psi(0)-1}\cdot(1-s_0)^{\psi(1)+\psi(2)-1}}{B(\psi(0),\psi(1)+\psi(2))} \\
&\qquad\cdot\frac{\big(\frac{s_1}{1-s_0}\big)^{\psi(1)-1}\cdot\big(1-\frac{s_1}{1-s_0}\big)^{\psi(2)-1}}{B(\psi(1),\psi(2))}\cdot\frac{1}{1-s_0}\,\mathrm{d}s_0,s_1 \quad \text{via substitution, using } (*) \\
&\overset{(10)}{=} \int_{(s_0,s_1)\in\overline{M}} \frac{s_0^{\psi(0)-1}}{\frac{(\psi(0)-1)!\cdot(\psi(1)+\psi(2)-1)!}{(\psi(0)+\psi(1)+\psi(2)-1)!}}\cdot\frac{s_1^{\psi(1)-1}\cdot\big(1-s_0-s_1\big)^{\psi(2)-1}}{\frac{(\psi(1)-1)!\cdot(\psi(2)-1)!}{(\psi(1)+\psi(2)-1)!}}\,\mathrm{d}s_0,s_1 \\
&= \int_{\omega\in M} \frac{(\|\psi\|-1)!}{(\psi(0)-1)!\cdot(\psi(1)-1)!\cdot(\psi(2)-1)!}\cdot\prod_{i\in\underline{3}}\omega(i)^{\psi(i)-1}\,\mathrm{d}\omega \\
&= \int_{\omega\in M} \frac{(\|\psi\|-1)!}{(\psi-\mathbf{1})!\!\!!}\cdot\prod_{i\in\underline{3}}\omega(i)^{\psi(i)-1}\,\mathrm{d}\omega \\
&= \mathit{Dir}(\psi)(M).
\end{aligned}$$

The proof in general, for arbitrary $n > 0$, works in the same way, but involves much more book keeping. □

7 Infinite Stick Breaking and Beta Distributions

In the literature on stochastic processes infinite stick breaking $sb\colon (0,1)^{\mathbb{N}} \xrightarrow{\cong} \mathcal{D}^{\infty}_{\mathrm{fs}}(\mathbb{N})$ from Proposition 1 is used as construction to produce (continuous) distributions on (discrete, infinite) distributions in $\mathcal{D}^{\infty}_{\mathrm{fs}}(\mathbb{N})$. For numbers $a_n, b_n \in \mathbb{N}_{>0}$ one can define:

$$sbB(a,b) := \mathcal{G}(sb)\left(\bigotimes_{n\in\mathbb{N}} \mathit{Beta}(a_n,b_n)\right) \in \mathcal{G}\big(\mathcal{D}^{\infty}_{\mathrm{fs}}(\mathbb{N})\big). \tag{12}$$

The abbreviation *sbB* stands for 'stick break Beta'; it is described as 'stick-breaking prior' in [11, §1.1]. When we pull out the parameters we get a function:

$$\big(\mathbb{N}_{>0}\big)^{\mathbb{N}} \times \big(\mathbb{N}_{>0}\big)^{\mathbb{N}} \xrightarrow{\quad sbB \quad} \mathcal{G}\big(\mathcal{D}^{\infty}_{\mathrm{fs}}(\mathbb{N})\big) \tag{13}$$

Examples of such stochastic processes $sbB(a,b)$ are Dirichlet-Poisson [6,9,20] and Pitman-Yor [24,25]. For instance, in the Dirichlet-Poisson case the sequence a is constantly one, and the sequence b is also constant, determined by a parameter. For Pitman-Yor only a is constant. These stochastic processes are used for infinite mixture models, as "stick breaking priors", see [11] for an overview.

As an aside, the probabilities in distributions in $\mathcal{D}_{fs}^{\infty}(\mathbb{N})$ are sometimes used in descending order, see *e.g.* [20, Appendix], so that what is commonly called Dirichlet-Poisson is a quotient of our general formulation (12). However, here we abstract away from such matters and will simply work with the above formulation.

We concentrate on one small thing, namely computing the mean of a stick break beta process (12). This allows us to conclude this article with a coalgebraic observation. Thus, in the style of Diagram (9) our goal is to describe the composite:

$$\begin{array}{ccccc} (\mathbb{N}_{>0})^{\mathbb{N}} \times (\mathbb{N}_{>0})^{\mathbb{N}} & \xrightarrow{sbB} & \mathcal{G}(\mathcal{D}_{fs}^{\infty}(\mathbb{N})) & \hookrightarrow & \mathcal{G}(\mathcal{G}(\mathbb{N})) \\ & & & & \downarrow \mu \\ & & & & \mathcal{G}(\mathbb{N}) \end{array} \tag{14}$$

where μ is the multiplication of the Giry monad. Interestingly, the outcome is a discrete distribution on $\mathbb{N}$.

We first observe that the mean can also be computed as Kleisli extension, which we write as $\gg\!=$. Indeed:

$$\begin{aligned} mean(sbB(a,b)) &= \mu\left(\mathcal{G}(sb)\left(\bigotimes_{n\in\mathbb{N}} Beta(a_n,b_n)\right)\right) \\ &= sb \gg\!= \left(\bigotimes_{n\in\mathbb{N}} Beta(a_n,b_n)\right). \end{aligned}$$

We first calculate the latter expression in the finite case. For instance, at position $0 \in \underline{3}$ of the distribution in $\mathcal{D}_{fs}(\underline{3})$ one has:

$$\begin{aligned} &\Big(sb \gg\!= \big(Beta(a_0,b_0) \otimes Beta(a_1,b_1)\big)\Big)(0) \\ &= \int_0^1\int_0^1 sb(r_0,r_1)(0)\cdot pbf_{Beta}(a_0,b_0)(r_0)\cdot pbf_{Beta}(a_1,b_1)(r_1)\, dr_1\, dr_0 \\ &= \int_0^1 r_0 \cdot \frac{r_0^{a_0-1}\cdot(1-r_0)^{b_0-1}}{B(a_0,b_0)}\cdot\left(\int_0^1 pbf_{Beta}(a_1,b_1)(r_1)\, dr_1\right) dr_0 \\ &= \frac{B(a_0+1,b_0)}{B(a_0,b_0)} \overset{(10)}{=} \frac{a_0!\cdot(b_0-1)!}{(a_0+b_0)!}\cdot\frac{(a_0+b_0-1)!}{(a_0-1)!\cdot(b_0-1)!} = \frac{a_0}{a_0+b_0}. \end{aligned}$$

Similarly, at position 1,

$$\begin{aligned} &\Big(sb \gg\!= \big(Beta(a_0,b_0) \otimes Beta(a_1,b_1)\big)\Big)(1) \\ &= \int_0^1\int_0^1 sb(r_0,r_1)(1)\cdot pbf_{Beta}(a_0,b_0)(r_0)\cdot pbf_{Beta}(a_1,b_1)(r_1)\, dr_1\, dr_0 \\ &= \int_0^1 (1-r_0)\cdot\frac{r_0^{a_0-1}\cdot(1-r_0)^{b_0-1}}{B(a_0,b_0)}\cdot\int_0^1 r_1\cdot\frac{r_1^{a_1-1}\cdot(1-r_1)^{b_1-1}}{B(a_1,b_1)}\, dr_1\, dr_0 \\ &= \frac{B(a_0,b_0+1)}{B(a_0,b_0)}\cdot\frac{B(a_1+1,b_1)}{B(a_1,b_1)} = \frac{b_0}{a_0+b_0}\cdot\frac{a_1}{a_1+b_1}. \end{aligned}$$

Thus, Kleisli extension $\gg\!=$ gives the following distribution on $\underline{3}$.

$$\begin{aligned} sb \gg\!= \Big(\mathit{Beta}(a_0, b_0) \otimes \mathit{Beta}(a_1, b_1)\Big) \\ = \frac{a_0}{a_0+b_0}\big|0\big\rangle + \frac{a_1 b_0}{(a_0+b_0)(a_1+b_1)}\big|1\big\rangle + \frac{b_0 b_1}{(a_0+b_0)(a_1+b_1)}\big|2\big\rangle. \end{aligned}$$

This reveals the pattern. It can be extended to infinity.

Lemma 9. *For sequences* $a, b \in \big(\mathbb{N}_{>0}\big)^{\mathbb{N}}$ *the mean of stick-break-beta yields the following distribution in* $\mathcal{D}^{\infty}_{\mathit{fs}}(\mathbb{N})$.

$$\mathit{mean}\big(\mathit{sbB}(a, b)\big) = sb \gg\!= \left(\bigotimes_{n\in\mathbb{N}} \mathit{Beta}(a_n, b_n)\right) = \sum_{n\in\mathbb{N}} \frac{a_n \prod_{i<n} b_i}{\prod_{i\leq n}(a_i+b_i)}\big|n\big\rangle.$$

□

For instance, for Poisson-Dirichlet we have $a_n = 1$ and $b_n = t$, where $t \in \mathbb{N}_{>0}$ is a parameter. The resulting mean is the infinite discrete distribution:

$$\sum_{n\in\mathbb{N}} \frac{t^{n-1}}{(1+t)^n}\big|n\big\rangle = \frac{1}{1+t}\sum_{n\in\mathbb{N}} \left(\frac{t}{1+t}\right)^n \big|n\big\rangle$$

We conclude by returning to a coalgebraic narrative. It turns out that the non-entirely trivial distribution in Lemma 9 can be obtained by finality from a completely trivial and standard coalgebra, involving the derivative a' of a sequence/stream a, see [28] for many more examples.

Proposition 10. *Consider the finality diagram:*

$$\begin{array}{ccc} (0,1) \times \Big(\big(\mathbb{N}_{>0}\big)^{\mathbb{N}} \times \big(\mathbb{N}_{>0}\big)^{\mathbb{N}}\Big) & \overset{\mathrm{id}\times h}{\dashrightarrow} & (0,1) \times \mathcal{D}^{\infty}_{\mathit{fs}}(\mathbb{N}) \\ {\scriptstyle c}\uparrow & & \uparrow{\scriptstyle \cong\ \mathit{shift}} \\ \big(\mathbb{N}_{>0}\big)^{\mathbb{N}} \times \big(\mathbb{N}_{>0}\big)^{\mathbb{N}} & \overset{h}{\dashrightarrow} & \mathcal{D}^{\infty}_{\mathit{fs}}(\mathbb{N}) \end{array}$$

The coalgebra c on the left is defined as:

$$c(a, b) := \left(\frac{a_0}{a_0 + b_0}, a', b'\right) \qquad \text{where} \qquad \begin{cases} a'_n = a_{n+1} \\ b'_n = b_{n+1}. \end{cases}$$

The function $h\colon \big(\mathbb{N}_{>0}\big)^{\mathbb{N}} \times \big(\mathbb{N}_{>0}\big)^{\mathbb{N}} \to \mathcal{D}^{\infty}_{\mathit{fs}}(\mathbb{N})$ *obtained by finality is then the mean of stick-break-Beta, as described in Lemma 9.*

Proof. We recall that the *shift* coalgebra (3), in the rectangle on the right, is final, by Proposition 1. Let's write:

$$h(a, b) = \mathit{mean}\big(\mathit{sbB}(a, b)\big) = \sum_{n\in\mathbb{N}} \frac{a_n \prod_{i<n} b_i}{\prod_{i\leq n}(a_i+b_i)}\big|n\big\rangle.$$

It suffices to show that this h makes the above rectangle commute. We look at the first and second projections separately.

$$\big(\pi_1 \circ \mathit{shift} \circ h\big)(a,b) \overset{(3)}{=} h(a,b)(0) = \frac{a_0}{a_0 + b_0} = \big(\pi_1 \circ c\big)(a,b).$$

And:

$$\begin{aligned}
\big(\pi_2 \circ \mathit{shift} \circ h\big)(a,b) \overset{(3)}{=} \sum_{n\in\mathbb{N}} \frac{h(a,b)(n+1)}{1 - h(a,b)(0)}\big|n\big\rangle &= \sum_{n\in\mathbb{N}} \frac{\frac{a_{n+1}\prod_{i<n+1} b_i}{\prod_{i\leq n+1}(a_i+b_i)}}{1 - \frac{a_0}{a_0+b_0}}\big|n\big\rangle \\
&= \sum_{n\in\mathbb{N}} \frac{\frac{a_{n+1}\prod_{i<n+1} b_i}{\prod_{i\leq n+1}(a_i+b_i)}}{\frac{b_0}{a_0+b_0}}\big|n\big\rangle \\
&= \sum_{n\in\mathbb{N}} \frac{a_{n+1}\prod_{0<i<n+1} b_i}{\prod_{0<i\leq n+1}(a_i+b_i)}\big|n\big\rangle \\
&= \sum_{n\in\mathbb{N}} \frac{a'_n\prod_{i<n} b'_i}{\prod_{i\leq n}(a'_i+b'_i)}\big|n\big\rangle \\
&= h(a',b') \\
&= \big(h \circ \pi_2 \circ c\big)(a,b). \qquad \square
\end{aligned}$$

8 Concluding Remarks

This paper extracts stick breaking from stochastic process theory and investigates it in a coalgebraic setting. This works smoothly for infinite stick breaking, yielding a new description $\mathcal{D}^{\infty}_{\mathit{fs}}(\mathbb{N})$ of the final coalgebra of the functor $(0,1)\times(-)$. In the finite case, the coalgebraic treatment of stick breaking is a bit artificial. Nevertheless, the following two stick breaking isomorphisms are both fundamental and useful.

$$(0,1)^{n-1} \xrightarrow{\;\cong\;} \mathcal{D}_{\mathit{fs}}(\underline{n}) \qquad\text{and}\qquad (0,1)^{\mathbb{N}} \xrightarrow{\;\cong\;} \mathcal{D}^{\infty}_{\mathit{fs}}(\mathbb{N}).$$

This usefulness has been illustrated by relating multinomials to iterated binomials and by relating Dirichlet to parallel Beta's. Also, one, coalgebraic, aspect of the use of infinite stick breaking in stochastic processes has been elaborated, namely the computation of the mean, via finality. This area of stochastic processes may benefit also in other ways from coalgebraic techniques.

Acknowledgements. Thanks are due to the anonymous reviewers for their useful feedback.

References

1. Adámek, J.: Introduction to coalgebra. Theory Appl. Categ. **14**(8), 157–199 (2005)
2. Bernardo, J., Smith, A.: Bayesian Theory. Wiley, Hoboken (2000). https://onlinelibrary.wiley.com/doi/book/10.1002/9780470316870. https://doi.org/10.1002/9780470316870
3. Billingsley, P.: Probability and Measure. Wiley-Interscience, New York (1995)
4. Cho, K., Jacobs, B.: Disintegration and Bayesian inversion via string diagrams. Math. Struct. Comput. Sci. **29**(7), 938–971 (2019). https://doi.org/10.1017/s0960129518000488
5. Clerc, F., Danos, V., Dahlqvist, F., Garnier, I.: Pointless learning. In: Esparza, J., Murawski, A.S. (eds.) FoSSaCS 2017. LNCS, vol. 10203, pp. 355–369. Springer, Heidelberg (2017). https://doi.org/10.1007/978-3-662-54458-7_21
6. Crane, H.: The ubiquitous Ewens sampling formula. Stat. Sci. **31**(1), 1–19 (2016). https://doi.org/10.1214/15-STS529
7. Danos, V., Garnier, I.: Dirichlet is natural. In: Ghica, D. (ed.) Mathematical Foundations of Programming Semantics. Electronic Notes in Theoretical Computer Science, vol. 319, pp. 137–164. Elsevier, Amsterdam (2015)
8. Feng, S.: The Poisson-Dirichlet Distribution and Related Topics. Probability and its Applications, Springer, Heidelberg (2010). https://doi.org/10.1007/978-3-642-11194-5
9. Ferguson, T.: A Bayesian analysis of some nonparametric problems. Ann. Stat. **1**(2), 209–230 (1973). https://doi.org/10.1214/aos/1176342360
10. Fritz, T.: A synthetic approach to Markov kernels, conditional independence, and theorems on sufficient statistics. Adv. Math. **370**, 107239 (2020). https://doi.org/10.1016/J.AIM.2020.107239
11. Ishwaran, H., James, L.: Gibbs sampling methods for stick-breaking priors. J. Am. Stat. Assoc. **96**(453), 161–173 (2001). https://doi.org/10.1198/016214501750332758
12. Jacobs, B.: Measurable spaces and their effect logic. In: Logic in Computer Science. IEEE Computer Science Press (2013)
13. Jacobs, B.: Introduction to Coalgebra. Towards Mathematics of States and Observations. Tracts in Theoretical Computer Science, vol. 59. Cambridge University Press, Cambridge (2016)
14. Jacobs, B.: From probability monads to commutative effectuses. J. Log. Algebraic Methods Program. **94**, 200–237 (2018). https://doi.org/10.1016/j.jlamp.2016.11.006
15. Jacobs, B.: From multisets over distributions to distributions over multisets. In: Logic in Computer Science. IEEE Computer Science Press (2021). https://doi.org/10.1109/lics52264.2021.9470678
16. Jacobs, B.: Urns & tubes. Compositionality (2022, to appear). https://arxiv.org/abs/2110.02155
17. Jacobs, B., Rutten, J.: A tutorial on (co)algebras and (co)induction. In: Sangiorgi, D., Rutten, J. (eds.) Advanced Topics in Bisimulation and Coinduction. Tracts in Theoretical Computer Science, vol. 52, pp. 38–99. Cambridge University Press, Cambridge (2011)
18. Johnson, N., Kotz, S.: Urn Models and Their Application: An Approach to Modern Discrete Probability Theory. Wiley, Hoboken (1977)
19. Kallenberg, O.: Foundations of Modern Probability. Probability Theory and Stochastic Modelling, vol. 99. Springer, Cham (2021). https://doi.org/10.1007/978-3-030-61871-1

20. Kingman, J.: Random partitions in population genetics. Proc. Royal Soc. Series A **361**, 1–20 (1978). https://doi.org/10.1098/rspa.1978.0089
21. Mahmoud, H.: Pólya Urn Models. Chapman and Hall, London (2008)
22. Panangaden, P.: Labelled Markov Processes. Imperial College Press, London (2009)
23. Pishro-Nik, H.: Introduction to probability, statistics, and random processes. Kappa Research LLC (2014). https://www.probabilitycourse.com
24. Pitman, J.: Random discrete distributions invariant under size-biased permutation. Adv. Appl. Probab. **28**(2), 525–539 (1995). https://doi.org/10.2307/1428070
25. Pitman, J., Yor, M.: The two-parameter Poisson-Dirichlet distribution derived from a stable subordinator. Ann. Probab. **25**(2), 855–900 (1997). https://doi.org/10.1214/aop/1024404422
26. Ross, S.: A First Course in Probability, 10th edn. Pearson Education, London (2018)
27. Rutten, J.: Universal coalgebra: a theory of systems. Theor. Comput. Sci. **249**, 3–80 (2000). https://doi.org/10.1016/S0304-3975(00)00056-6
28. Rutten, J.: A coinductive calculus of streams. Math. Struct. Comput. Sci. **15**(1), 93–147 (2005). https://doi.org/10.1017/S0960129504004517
29. Sethuraman, J.: A constructive definition of Dirichlet priors. Stat. Sin. **4**, 639–650 (1994). https://doi.org/10.21236/ada238689
30. Wilks, S.: Mathematical Statistics. Wiley, Hoboken (1962)

Author Index

The manufacturer's authorised representative in the EU is Springer Nature Customer Service Centre GmbH, Europaplatz 3, 69115 Heidelberg, Germany. If you have any concerns regarding our products, please contact ProductSafety@springernature.com

Printed and bound by CPI Group (UK) Ltd, Croydon, CR0 4YY
30/09/2026
02225195-0015